The Nile Waters

Joseph Awange

The Nile Waters

Weighed from Space

Springer

Joseph Awange
School of Earth and Planetary Sciences
(Spatial Sciences Discipline)
Curtin University
Perth, WA, Australia

ISBN 978-3-030-64754-4 ISBN 978-3-030-64756-8 (eBook)
https://doi.org/10.1007/978-3-030-64756-8

This Springer imprint is published by the registered company Springer Nature Switzerland AG
The registered company address is: Gewerbestrasse 11, 6330 Cham, Switzerland

To the memories of my late parents (James Odhiambo Awange and Margaret Atieno Awange) and my late grandmothers (Amelea Gwada and Rosalina Omulo Odera). The love you showered on me is still fresh like the Nile waters. You are greatly missed.

Joseph L. Awange
Perth, Australia, October 2020

Foreword

As I read "The Nile Waters, Weighed from Space", I heard the ghosts of the Colorado River and of the Rio Grande River. These once mighty rivers are but a mirage of their former selves. Their flows are drained by thirsty peoples who are dependent on the fresh waters to drink, to irrigate crops, and to create hydroelectric power. Like the Colorado and Rio Grande deltas, I imagined a Nile River delta void of its fluvial discharge. What is the fate of the Nile River? What will become of its waters as dams are built upstream and reservoirs filled? What will become of the sediments as the flows are regulated and the floodplain soils are no longer delivered to downstream farmers? I tell my students that vision is easy but implementation is hard. It is easy for me to visualize a pristine Nile River where the ecology is set free. But what about the hard challenges of ensuring that people who live all throughout the Nile Basin also have water?

Implementing solutions to these challenges requires water knowledge. This is the fundamental point made by Prof. Awange in his book. He is showing us the present state of water knowledge in the Nile River basin. Unfortunately, much of Africa's waters are not well measured in-situ. In the Congo Basin, there once were hundreds of ground-based stations measuring rainfall, temperatures, and river water flows.

Now, there are less than a dozen such stations. The Nile has suffered a similar fate. Fortunately, spaceborne measurements of the hydrologic cycle are now routine. In Prof. Awange's book, you will read about satellite altimeters that ping river surfaces to reliably measure the waters upping and downing. He tells us about spaceborne radar devices that measure rainfall throughout the seasons, and of satellite imagers that inform about the vegetation greenery. "Weighing" the Nile from space is very apropos. You will read about a satellite mission called GRACE which is, in fact, measuring the mass of the earth beneath each orbit. Differences between these orbitally measured masses result from changes in the hydrosphere. As the waters move from the rains over the Ethiopian highlands, flow to and from the Nile Basin groundwaters, and eventually make their way down the Blue Nile to join the White Nile, the mass beneath the GRACE satellites changes. Prof. Awange shows us these GRACE-measured changes in water masses.

The present state of water knowledge requires the incorporation of these measurements into models that show how and when water moves from one storage to the next. Here again Prof. Awange presents the state-of-the-art modeling that incorporates specialized statistical methods known as data assimilation. The satellite measurements and the robust statistical models show these hydrologic interconnections throughout the Nile River basin. All of this then leads to a multi-decadal view of 30 to 40 years of spaceborne water measurements being assimilated into hydrologic models. As you read, you will find out about reservoirs filling, about rains that are timed with multi-year oscillations in ocean temperatures, about droughts that have starved crops, and about peoples and governments who are grappling with all of it.

The Nile River basin, like the Congo River basin that I study, is at crossroads. A time in history where our environment needs us. Africa and its water resources can benefit by leapfrogging the installation and operation of expensive, tedious multitudes of in-situ hydrologic stations. Instead, Africa's countries can embrace the satellite hydrology era. This won't eliminate the need for ground-based measurements given that those are required to verify what we see from space. Moreover, Africa's researchers will use these space technologies, will create their own statistical algorithms, and will build models of hydrologic knowledge. This is a technology and thus a knowledge that is the foundation to answering questions regarding the fate of the Nile River. Arguably, the Nile River is the most unique river in the world. It spans extremes of rainfall from being measured by meters to being measured by centimeters, from the humid tropics to the driest of deserts. Yet, thirsty people live throughout this basin, and therefore the demands on its water resources are uneven. Knowing the water amounts throughout the entire Nile Basin is a critical step for governments and international treaties to avoid the "Tragedy of the Commons".

Africa can embrace this future through the leadership of Prof. Awange and others like him who have devoted their careers to Africa's waters. His students and research associates are compounding this water knowledge. As you read "The Nile Waters, Weighed from Space" please know that you are learning from *the* world's leading authority on doing just that, weighing the Nile from space!

USA
October 2011/2020

Doug Alsdorf, Ph.D.
Professor of Geophysics at the Ohio State University

Preface

Exploiting *advanced statistical techniques* and the *latest state-of-the-art multi-mission satellites, surface models, and reanalysis products*, this book provides the first comprehensive weighing of the changes in the Nile River Basin (NRB; ~ 3,400,000 km^2)'s stored waters' compartments; surface, soil moisture and groundwater, and their association to climate variability/change and anthropogenic impacts on the one hand. On the other hand, it argues on the need for *equitable use of the NRB's waters* by all 11 countries within its basin, and *doing away with obsolete Nile treaties* that were signed by Britain, Egypt, and Sudan, which prohibit the use of the Nile by eight upstream countries. With Ethiopia's construction of Africa's largest dam (GERD; Grand Ethiopian Renaissance Dam) along the Blue Nile, which is expected to take several years to fill, the Nile is back on the news. Combined with Uganda's Nalubaale, Kiira, and Bujagali dams on the White Nile, these human-induced impacts (i.e., damming), coupled with those of climate variability/change are expected to exacerbate tension with the low stream countries (Egypt and Sudan) fearing the cut in the Nile's total volume. Furthermore, the Nile river, arguably the world's longest river (6800 km), impacts on the livelihood of over 300 million people of 11 countries within its basin. This population is expected to double in the next 25 years [7] thereby putting extreme pressure on its water resources. An in-depth analysis of changes in the Nile's stored waters, therefore, is essential to inform its management and sustainable equitable use. Owing to its sheer size, however, obtaining in-situ data from "boots on the ground" is practically impossible, paving way to the space-based weighing of the Nile River Basin using a suite of high spatio-temporal remotely sensed and reanalysis products, as well as those of hydrological models.

Due to the extended area covered by the NRB, it experiences various geographic and climatic conditions; i.e., rainfall rate over the southern part of the basin is around 1500 mm/year, much bigger than that over the northern parts with about 20–50 mm/year [2–4]. Correspondingly, evaporation rates and temperatures vary across the basin. The Nile water level fluctuation depends not only on the regional climate variations but also on large-scale ocean-atmosphere coupling factors such as El Niño Southern Oscillation (ENSO), which affect the basin differently [6]. In addition to climatic variabilities, the Nile's water is also affected by ever-increasing human impacts, e.g., excessive dam constructions as mentioned above and irrigation, especially over the countries in the southern parts of the basin, causing significant challenges to the northern countries [1, 5, 8]. A combination of all these factors shows the vulnerability of the region in facing water scarcity challenges, especially in the absence of accurate water monitoring system and correspondingly poor future planning.

In this book, which provides a pioneering weighing of the Nile River Basin from space, its current state, overview of the obsolete Nile treaties, challenges and potentials vital for its sustainable use are presented. The book will be useful to those in water resources management and policy formulations, hydrologists, environmentalists, engineers, and researchers.

Kyoto, Japan Joseph Awange
Recife, Brazil
Karlsruhe, Germany
Kisumu, Kenya
Perth, Australia

References

1. Ahmed AA, Ismail UAE (2008) Sediment in the Nile river system, Consultancy Study requested by UNESCO
2. Awange JL, Ferreira VG, Forootan E, Khandu, Andam-Akorful SA, Agutu NO, He XF (2016) Uncertainties in remotely sensed precipitation data over Africa. Int J Climatol 36:303–323, http://dx.doi.org/10.1002/joc.4346
3. Awange JL, Hu KX, Khaki M (2019) The newly merged satellite remotely sensed, gauge and reanalysis-based Multi-Source Weighted-Ensemble Precipitation: Evaluation over Australia and Africa (1981–2016), Sci Total Environ 670:448–465, https://doi.org/10.1016/j.scitotenv.2019.03.148
4. Camberlin P (2009) Nile basin climates. Dumont, Henri J. The Nile: Origin, Environments, Limnology and Human Use, Springer, pp 307–333, Monographiae Biologicae
5. Conway D (2002) Extreme rainfall events and lake level changes in east africa: Recent events and historical precedents. In EO Odada, DO Olago (eds.) The East African Great Lakes: Limnology, palaeolimnology and biodiversity. Advances in Global Change Research V. 12. Kluwer, Dordrecht. pp 63–92
6. Khaki M, Awange J, Forootan E, Kuhn M, (2018b) Understanding the association between climate variability and the Nile's water level fluctuations and water storage changes during 1992–2016, Sci Total Environ 645:1509–1521, https://doi.org/10.1016/j.scitotenv.2018.07.212
7. Nunzio J (2013) Conflict on the Nile: the future of transboundary water disputes over the world's longest river. Future Directions International, Dalkeith
8. Taye MT, Ntegeka V, Ogiramoi NP, Willems P (2011) Assessment of climate change impact on hydrological extremes in two source regions of the Nile River Basin. Hydrol Earth Syst Sci 15:209–222, http://dx.doi.org/10.5194/hess-15-209-2011

Contents

Part I
Water Resources, Challenges and Space Sensing

Chapter 1
Global Freshwater Resources

Several authors, politicians, leaders of international organizations and journalists have cautioned the world community that the increasing scarcity of freshwater resources might lead to national and international conflicts. When relating this to climate change forecasts – most of which indicate that climate change will have a significant impact on the availability of freshwater resources, on water quality, and on the demand for water – this is alarming news for humankind as it threatens human security.

– Molen and Hildering [1]

Accounting for water for environmental requirements shows that abstraction of water for domestic, food and industrial uses already have a major impact on ecosystems in many parts of the world, even those not considered "water scarce" [2]. In the absence of coordinated planning and international cooperation at an unprecedented scale, therefore, the next half century will be plagued by a host of severe water-related problems, threatening the well being of many terrestrial ecosystems and drastically impairing human health, particularly in the ***poorest regions*** *of the world. [3]*

1.1 Diminishing Freshwater Resources

1.1.1 Status

Freshwater, influenced globally by climate variability/change [4–9], human use [10–12] and knowledge deficiency resulting from inadequate hydrometeorological observation stations [13–16], is one of the basic necessities without which human beings cannot survive since it is key to the sustainability of all kinds of life forms. Water has multiple uses namely; nutritional, domestic, recreational, navigational, waste

J. Awange, *The Nile Waters*,
https://doi.org/10.1007/978-3-030-64756-8_1

disposal and ecological as it is a habitat for living and non-living organisms (biodiversity), etc. And, because it is indispensable to different sectors including manufacturing, agriculture, fisheries, wildlife survival, tourism and hydroelectric power generation, just to list but a few, it is a vital factor of economic production. For many countries, most freshwater endowments encompass surface waters, groundwater, wetlands and glaciers. Surface water bodies include lakes, rivers, swamps, springs, dams and water pans dispersed within different basins. In general, people living in the vast arid and semi-arid parts of the world rely heavily on groundwater resources. Gleeson et al. [17] puts the number of people living in regions where groundwater is threatened to be 1.7 billion. Furthermore, groundwater is also an important supplementary source of water for many urban households in most developing countries.

At a global scale, although much of the Earth is covered by water, most of it is unsuitable for human consumption, since 96% of it is found in the saline oceans. According to the U.N., only 2.5% of the roughly 1.4 billion cubic kilometers of water on Earth is freshwater, and approximately 68.9% of the freshwater is trapped in glacial ice or permanent snow in mountainous regions—the Arctic and Antarctica. Roughly 30.8% is groundwater, much of which is inaccessible to humans, and the remainder 0.3% comprise surface waters in lakes and rivers [18]. Of the 0.3% available for human and animal consumption, much is inaccessible due to unreachable underground locations and depths [19]. Jury and Vaux Jr. [3] caution that focusing on the global freshwater storage resource alone is misleading because much of the water is inaccessible. They suggest that humanity's freshwater resource consisting of rainfall used to grow crops, accessible groundwater, and surface water be considered.

1.1.2 Water Scarcity

Although no common definition of water scarcity exists, Rijsberman [2] defines a water insecure person as an individual who does not have access to safe and affordable water to satisfy his or her needs for drinking, washing, or livelihood. A water scarce area is then said to be an area where a large number of people are water insecure for a significant period of time, e.g., Rijsberman [2]. Rijsberman [2] points out that an area qualifies as a water-scarce area depending on, e.g., (i) the definition of people's needs and whether the definition takes into account the needs of the environment, the water for nature, (ii) the fraction of the resource that is made available (or could be made available) to satisfy these needs, and (iii), the temporal and spatial scales used to define scarcity.

Even though Rijsberman [2] argues that at a global scale, and from a supply and demand perspective, it is still debatable whether water scarcity is fact or fiction, it is incontestable that fresh water is increasingly becoming a scarce resource and shortages could drive conflict as well as negatively hit food and energy production [3, 20, 21] (see Fig. 1.1). That water shortage is emerging as one of the leading challenges of the 21st century has been documented, e.g., in [2, 22–24]. To underscore the seriousness of water shortage, the World Bank in its 1992 World Development Report

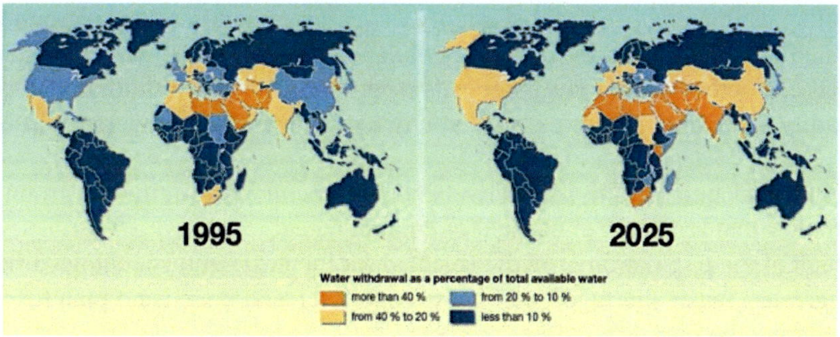

Fig. 1.1 Increased global water stress. The Nile River Basin can be categorized as a medium to high water withdrawal area. *Source* [41]

pointed out that 22 countries faced severe water shortage while further 18 were in danger of facing shortages if fluctuation in rainfall persisted [25]. Recent studies, e.g., [26–31] point to fluctuation in rainfall in parts of Greater Horn of Africa where most low income earners depend on rain fed agriculture for food production, thus suggesting that the water shortage problem is not fading away any time soon. A more gloomy picture is the estimate that by 2050, about two billion people will be short of water, a potential cause of conflict [32]. That a large population of the world will face water scarcity is supported by several studies, e.g., [33–37], with the most likely to be affected dwelling in Africa, Asia and the Middle East, see, e.g., [24, 38–40]. Already, model projections, see e.g., [7] suggest a 40% of the population languishing under water scarcity, with 57% likely to be water stressed by 2015 and 69% by 2075 [23, 33].

By reviewing several publications on water in relation to poverty and environmental degradation nexus, Duraiappah [25] presents activities that lead to water shortages. ***First***, there is the issue of the commercial interest of the rich and wealthy driven primarily by power, greed and wealth that benefits from market and institutional failures (e.g., absence or misuse of water property rights). ***Second***, Duraiappah [25] attribute over usage of water supply by the small holders (the poor) to water subsidies that provide incentives. Of the two groups, the poor are more likely to be affected by water shortages as compared to the rich, a situation that could contribute to environmental degradation on the one hand, which would lead to poverty on the other hand, i.e., endogenous poverty causing environmental degradation, see, e.g., Duraiappah [25].

This picture leads Jury and Vaux Jr. [3] to warn that "without immediate action and global cooperation, water supply and water pollution crisis of unimaginable dimensions will confront humanity, limiting food production, drinking water access, and the survival of innumerable species on the planet". They list the following four factors to support their hypothesis that the world is headed towards a future where billions of people will be forced to live in places where their food and water requirements will not be met [3]:

1. Unlike estimates of the global supply of scarce minerals or underground fuels, which are surrounded by uncertainty, planetary supplies of water are relatively well characterized. No large deposits of groundwater await human detection in readily accessible locations, so that any new resources discovered will be very expensive to develop.
2. Many vital human activities have become dependent on utilizing groundwater supplies that are being exhausted or contaminated.
3. Much of the population growth projected for the next century will occur in areas of greatest water shortages, e.g., Africa and Asia, and there is no plan for accommodating the increases.
4. Global economic forces are luring water and land from food production into more lucrative activities such as biofuel, see, e.g., [42–44], while at the same time encouraging pollution that impairs drinking water quality for a large and ever-growing segment of the population.

Evidently, the management of water resources' conflicts, focusing on negotiation, mediation and decision-making processes, in order to prevent, manage and resolve water conflicts is emerging as a contemporary and topical research issue.

Physical water scarcity is evident in densely populated arid areas in many parts of Central and West Asia, and North Africa with projected availabilities of less than 1000 m^3/capita/year [2]. This has a wide range of negative impacts and ramifications. For instance, it results in higher incidences of waterborne, water-related or sanitation-related diseases such as malaria, diarrhoea and skin infections. In addition, as has been pointed out before, increasing cases of water conflicts, especially between pastoralist and farming communities along lower and upper river basins, as witnessed in water stressed countries like Kenya [45] (with a renewable freshwater per capita endowment estimated at about 548 m^3/capita/year [46]), are also likely to heighten food insecurity. Alcamo [33] point to the fact that 12% of the global basins are already vulnerable to water withdrawals. On the gender scale, since women are the primary collectors, users and managers of water for domestic use in most developing countries, water scarcity disproportionately affects them because it is they who have to trek long distances, often all day, in search of water. Against the above background, is the growing realization today that availability of sufficient, accessible and quality potable water is not only a matter of great socio-economic and political importance, but also one of fundamental human rights, see e.g., [47–50].

1.1.3 Impacts of Climate Variability/Change on Freshwater

Intergovernmental panel of climate change [51] projects a temperature increase of 1.5 °C by 2030, an increase which will definitely have an enormous impact on the world's freshwater. That temperature has been on the rise due to human's carbon footprint had been noted in the previous IPCC (2007 and 2013) reports [52]. Infact, [53] points to the fact that the global and ocean temperatures rose by 0.85 °C [0.65–

1.06 °C] over the period 1880–2012, 0.89 °C [0.69–1.08 °C] from 1901–2012, and 0.72 °C [0.49–0.89 °C] between 1951 and 2012. Change in climate resulting in the temperature rise above as well as climate variability exemplified through indicators such as global teleconnections, i.e., ENSO [54, 55] impact on the availability of freshwater. For instance, climate variability influence extreme whether/climate conditions resulting, e.g., in increased frequency and severity of droughts [31, 56, 57]. Droughts majorly affect the surface water although a long spell could also impact on groundwater [58, 59]. Moreover, the influence of climate variability/change on the hydrological cycle impacts on precipitation, which is the main recharge of freshwater.

1.1.4 Water-Poverty-Environment Nexus

Water plays a significant role both in poverty alleviation and environmental related issues. With regard to poverty, food policy contributes toward the overall goal of eradicating extreme poverty and hunger, and is dependent on water as one of the drivers for realizing such a goal, see e.g., Hanjra and Qureshi [60]. It provides a key component of agricultural requirement for food production as well as for domestic use. If people are able to provide food, then absolute poverty measure of a society could arguably be low. Indeed, that water and food security issues are closely linked has been pointed out, e.g., by [10, 60]. Rijsberman [2] is even more direct by stating that water will be the major constraint for agriculture in the coming decades, more so in the continents that experience high percentage of poverty such as Asia and Africa.

The problem, however, is that more focus is placed on the water crisis, which is likely to be fuelled by increased population leading to about 1.2 billion lacking water. To the contrary, less documentation exist for the large part of the population that live in rural areas below the poverty line. Duraiappah [25], through literature review, identified

- water shortage, and,
- water pollution

as the two major issues within the water sector that plays an important role in the poverty-environmental degradation nexus. Understanding the link between poverty and environment in relation to water, therefore, calls for a more closer look at the whole issue of water scarcity in relation to the poor rather than a holistic approach where such scarcity is viewed in terms of the total population. This in essence calls for poverty eradication measures to incorporate issues that will address water scarcity and insecurity. This is partly because if the poor are unable to access clean and safe water, food productivity will be hampered leading to hunger and malnutrition hence less productivity. Also, lack of access to safe drinking water by the poor will only aggravate the risk of water borne diseases such as cholera. When the low income group are faced by such water-borne diseases, their productivity deteriorates and they risk losing their jobs, and hence sources of income. The expected outcome of loss in income is that the low income earners will experience economic and social hardship, which over time results in poverty, thus exemplifying how environmental degradation causes poverty, e.g., [22, 25].

1.2 Water Resource Monitoring

1.2.1 Need for Monitoring

The importance of water as a resource, therefore, calls for sound environmental conservation measures that enhance its protection and management. It is in relation to this that the World Bank, as an emerging priority of its lending framework, decided to broaden the development focus in its 1993 "Water resource management policy paper" to include the *protection* and *management* of water resources in an environmentally sustainable, socially acceptable, and economically efficient manner [46]. The protection and management of water resources calls for an elaborate and well established management and monitoring program, e.g., [61, 62].

Information about water resources and the environment is inherently geographic. Maps, whether on paper or in digital Geographical Information System (GIS, see e.g., [61, 62]) formats, continue to be the medium for the expression of engineering plans and designs. This is because we are basically concerned about the spatial distribution and character of the land and its waters. Johnson [63] argues that weather patterns, rainfall and other precipitation, and resultant water runoff are primary driving forces for land development, water supplies, and environmental impacts and pollution. Our water resources systems comprise dams and reservoirs, irrigated lands and canals, water supply collection and distribution systems, sewers and storm water systems, and flood plains. These systems are designed in response to a complex mix of topography and drainage patterns, population and land use, sources of water, and related environmental factors [63, 64].

In general, the planning and engineering design processes used in the development and management of water resources involve different levels of data abstraction. Data are collected and used to characterize the environment at some level of detail, or scale. In seeking to make decisions about plans and designs, data must be collected to describe the resource, and procedures or models must be developed to predict the resultant changes. These data and models help us understand the real world, and this understanding guides our decision making [63].

According to Taylor and Alley [65], essential components of a water level monitoring program include; *selection of observation wells*, *determination of the frequency of water level measurements*, *implementation of quality assurance*, and *establishment of effective practices for data reporting*. In selecting the observation wells, the authors state that the decisions made about the number and locations of observation wells are crucial to any water-level data collection program [65], see also [59]. In regard to locations, Global Navigation Satellite Systems (GNSS) satellites [66, 67] could contribute in generating a fast and accurate survey of well location-based data. These data could then be integrated with other information such as water level in a GIS system to enhance the accessibility of water level data, where the GIS plays the role of depicting the locations of the observed wells relative to pertinent geographic, geologic, or hydrologic features, e.g., [65].

Taylor and Alley [65] present areas where the monitored ground water levels could be used. Some of these include: determination of the hydraulic properties of aquifers (aquifer tests); mapping of the altitude of the water table or potentiometric surface; monitoring of the changes in groundwater recharge and storage, e.g., [59]; monitoring of the effects of climatic variability; monitoring of the regional effects of groundwater development; statistical analysis of the water level trends; monitoring of the changes in groundwater flow directions; monitoring of the groundwater and surface water interaction; and numerical (computer) modeling of groundwater flow or contaminant transport.

Information on the spatial and temporal behaviour of terrestrial water storage, therefore, is crucial for the management of local, regional and global water resources. This information will [68]:

- Enhance sustainable utilization of water resources by, e.g., farmers, urban consumers, miners, etc.
- Guide water resource managers and policy makers in the formulation of policies governing its sustainable use, conservation and management. In particular, State water managers are more informed in regulating the utilization of water, e.g., for industrial and irrigation purposes.
- Benefit local environmental monitoring, management policies and practices that ensures a balance between sustainable utilization and environmental conservation and protection. Changes in water availability impacts upon the environment in several ways, e.g., any significant imbalance in its level affects the ecological system by influencing salinity, land subsidence, and the vulnerability of wetlands ecosystem among others.
- Benefit various government agencies at various levels (national, provincial, and local) by providing data that enhance and compliment their works. Such agencies include departments of *water*, *agriculture*, *weather forecasting* and *climate* studies, and so forth.

The conservation and management of water is of paramount importance in areas with arid or semi-arid climates, which include many parts of Australia, especially in times of severe drought, as experienced in Murray Darling Basin [68]. In 2006, Australia faced its worst drought in a century as was seen from daily reports that were emerging in both the local and international media. A more grim picture of the future of the water situation for Australia was to follow from the IPCC [69] report, which stated that Australia's water crisis will worsen in the coming years due to drought! There clearly exists an urgent need to have efficient monitoring technique(s) that will enhance the analysis of water scarcity at river basin or more localized scales. Indeed, this argument is supported by Rijsberman [2] who states that the global analysis of water scarcity is of very limited use in assessing whether individual or communities are water secure. To this effect, Rijsberman [2] states:

> The river basin is more and more adopted as the appropriate scale to understand the key processes with increasing water scarcity as human use goes up to the point where basins "close".

One such technique that has emerged supreme in monitoring changes in stored water at river basin scales, is the Gravity Recovery and Climate Experiment (GRACE) satellites [70] discussed in details in Sect. 3.3.3; see details in [58, 61, 62, 66, 67].

Timely and precise information on the changes in stored water at smaller (localized) scales of economical values, e.g., urban consumption, agriculture, industries, and mining to within 10–14 days (so far achievable by GRACE satellite) will enhance sustainable conservation and management of this precious *dwindling* resource.

The availability of techniques that delivers information on the changes in stored water at a more local scale, is the first step towards realizing an efficient water society.

Water resource managers are able to make decisions based on timely and accurate knowledge; thereby saving considerable resources that are often spent as a penalty of inefficient decisions based on lack of information. In the south-western wheat belt of Australia, for example, accurate knowledge of changes in stored water will be beneficial to the sustainable utilization of water, while at the same time realizing the economic contribution of wheat farming to the overall Gross Domestic Product (GDP). A blind focus on the GDP's growth without paying attention to the state of salient contributors such as water stored in aquifers is detrimental, since a fall in the amount of the available water in such areas would definitely mean reduced yields.

Since the entire system of stored water is coupled within the hydrological cycle (Fig. 3.3), hydrologists will be in a position to better understand their local hydrological cycle, thanks to information at localized levels. Hydrologists will also be able to use such information to refine and calibrate local-scale models, e.g., rainfall runoff models [71], for further improvement in their hydrological cycles. This will also contribute to our understanding of the impacts of climate change on regional and global hydrological cycles.

Environmental studies have a chance of greatly benefiting from information about changes in stored water. It is widely acknowledged that stored water (surface and groundwater) plays a key role in sustaining natural biodiversity and the functioning of the environment as a whole. Knowledge of the changes in water level is therefore essential for the very survival of the entire ecosystem, which could be adversely affected by extreme changes in stored water. In wetlands, for example, some vegetation and ecosystems have been known to respond to water level fluctuations [72].

Accurate monitoring of changes in stored water at smaller wetland scales will thus help in the preservation and conservation of such wetland ecosystems. Changes in water level also brings with it environmental phenomena such as salinity, compacting of aquifers due to the removal of water causing land subsidence, and changes in the properties of the top 5 cm of soil. Information on changes in stored water thus contributes enormously to the environmental conservation and protection. Remote sensing and GIS [61, 62] can be used to monitor the water quality. This is possible by employing multispectral, multi-temporal image data and analyzing parameters such as the distribution of suspended sediment, turbidity and chlorophyll. These indicators can be determined through regression analysis.

1.2.2 Monitoring of Stored Water at Basin Scales

In Awange and Kiema [61, 62, 66, 67], the *geoid* is introduced as a fundamental physical surface to which all observations are referred to if they depend on *gravity*, and whose shape is influenced by inhomogeneous mass distribution within the interior of the Earth [73, p. 29]. In Chap. 3, the concept of *gravity field variations* is related to hydrological processes. Measurements of the time-varying gravity field by LEO (low earth orbiting) satellites, e.g., GRACE discussed in Sect. 3.3.3 are the key to the contribution of space monitoring of changes in water levels at basin scales, see e.g., [74]. Such techniques now enable the monitoring of groundwater recharge, see e.g., [58, 75, 76], which is the most important element in groundwater resources

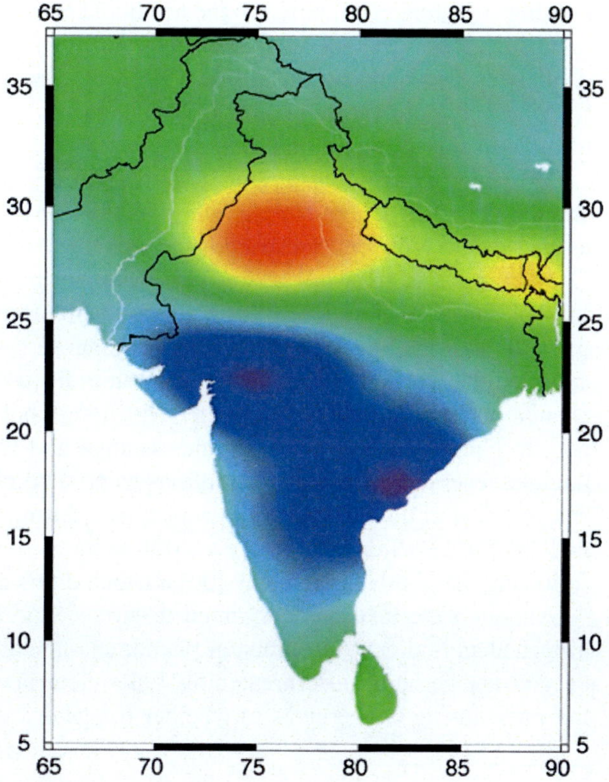

Fig. 1.2 Groundwater changes in India (2002–2008). Groundwater recharge is indicated by blue while depletion is indicated in red. *Source* NASA (I. Velicogna/UC Irvine)

management and could also be applicable to monitoring salinity management measures at the catchment level. For example, in 2009, GRACE satellites showed that north-west of India's aquifers had fallen at a rate of 0.3048 m yr^{-1} (a loss of about 109 km^3 per year) between 2002 and 2008, see Fig. 1.2.[1]

1.3 Importance of Monitoring the Nile River Basin

The ***Nile River*** (Fig. 2.1, approximately 6800 km long), located on the eastern part of Africa is a major source of fresh water for 11 countries within the Nile River Basin (NRB), where its main sources are the White and Blue Niles. Due to the extended area covered by the NRB, it experiences various geographic and climatic conditions; i.e., rainfall rate over the southern part of the basin is around 1500 mm/year, much bigger than that over northern parts with about 20–50 mm/year [77–79]. Correspondingly, evaporation rate and temperature vary across the NRB. The Nile water level fluctuation depends not only on the regional climate variations but also on large-scale ocean-atmosphere coupling factors such as El Niño Southern Oscillation (ENSO), which affect the basin differently [80]. In addition to climatic variabilities, the Nile's water is also affected by ever-increasing human impacts, e.g., excessive dam construction and irrigation, especially over the countries in southern parts of the basin, causing significant challenges to the northern countries, see, e.g., [81–83]. A combination of all these factors shows the vulnerability of the region in facing water scarcity challenges, especially in the absence of accurate water monitoring system and correspondingly poor future planning. This therefore calls for an elaborate monitoring affordable only through the use of satellites.

Moreover, ***Lake Victoria*** (Fig. 4.1), the world's second largest freshwater lake, and the largest in the developing world, is the source of the White Nile (also known as the Victoria Nile) and a resource shared by the three East African countries: Kenya, Uganda and Tanzania. It is a source of water for irrigation, transport, domestic and livestock uses, and supports the livelihood of more than 42 million people who live around it [40, 84, 85, 104]. Its fish products, (i.e., Tilapia and Nile Perch) are exported the world over [84]. Its role as an indicator of environmental and climate change on long-term scales together with its global significance are documented, e.g., in Nicholson and Yin [86] and Awange and Ong'ang'a [84]. Since the 60s, the lake level has exhibited fluctuations as pointed out by Nicholson [87, 88]. The sharpest rise in the lake water level occurred during the El'Nino rains of early 60s and 1997/1998. Some reports, e.g., [89] suggest that the lake level rose by 2.5 m following the 1960s floods. Kite [90] attributed this rise to over-lake precipitation.

Although the lake has continued to attract worldwide attention due to its significance and other environmental phenomenon such as water hyacinth, in the last decade, and perhaps most threatening, Lake Victoria water level receded at an alarming rate causing concerns as to whether the lake was actually drying up as it hap-

[1]The Economist, September 12th 2009, pp. 27–29: Briefing India's water crisis.

pened in the pre-historic time, see e.g., [10]. According to Kull [91], the lake levels dropped to more than 1.1 m below the 10-year average. Water levels have remained above average for more than 40-years, but current water levels are below normal and the lowest level since September 1961. The socio-economic impacts of this drastic fall in Lake Victoria water level have been reported in Awange et al. [10, 92]. At the time of writing this book (2020), the lake level has risen to a point where people are wondering whether it is retracing its shorelines of the 60s.[2] Continuous monitoring of the lake, therefore, is paramount to the sustainable use of the Nile River Basin as a whole. A special coverage of monitoring Lake Victoria from space is presented in Awange [93].

1.4 Objectives and Aims of the Book

As part of the solution to water scarcity problem, quick fix approach has revolved around supply management approach where infrastructures, e.g., dams have been constructed or expanded to increase the available water supply. Rijsberman [2] argue, however, that although the quick fix approach has largely succeeded in producing cheap food, water supply, and sanitation to a large number of people, many people still do not have access to safe and affordable drinking water despite huge investments. Rijsberman [2] states:

> …close to half the world population lacks access to sanitation, many rural poor do not have access to water for productive purposes, groundwater levels in key aquifers are falling rapidly, many rivers are no longer reaching the sea, etc.

As a shift from this school of thought, Rijsberman [2] points to the emergence of integrated water resource movement that has brought about organizations such as the World Water Council and the Global Water Partnership that is pushing for a demand management approach seeking;

1. to involve users more in the management of water, often through the establishment of forms of water user associations;
2. to price water and/or make it a trade-able commodity; and,
3. to establish river basin authorities that integrate the usually fragmented government responsibilities for water into a single authority responsible for a hydrographically defined area, the river basin.

It is in support to item (3) above that satellites have been recognized as having the potential to provide space-based weighing of changes in terrestrial water storage [103]. In essence, they are tools that assist water managers in conserving and controlling the utilization of dwindling water resources in a sustainable way. Water is arguably one of the most precious resource in the world, therefore, it is logical to try to monitor its distribution as efficiently as possible, and space techniques offer such

[2] http://news.trust.org/item/20200819141141-rb3c8/.

opportunity [10, 94–97]. This is because one of the environmentally important signals detected by satellites such as GRACE is the temporal gravity field, i.e., variation induced by changes in the distribution of water on and below the Earth's surface, i.e., hydrology, e.g., [97]. Satellite altimetry provides the possibility of monitoring sea or lake surface heights as was demonstrated for Lake Naivasha [98]. Other studies undertaken with respect to use of GRACE to monitor hydrology include, e.g., [58, 68, 75, 99–101].

As already mentioned above, the hydrology of the Nile is mainly characterized and influenced by high variations in climate and altitude/topography, which have a great bearing on flow magnitudes and patterns in the different parts of the basin [102]. These differences are more pronounced in the two tributaries of the Nile; the White and Blue Niles, with the White Nile exhibiting relatively steady flows and the Blue Nile exhibiting highly seasonal flows [102]. The aim of this book, therefore, is to provide a space (satellite) weighing of entire Nile River Basin (NRB)'s total water storage (TWS; surface, groundwater, soil moisture and vegetation) changes. Owing to its large basin ($>300{,}000\,km^2$), "boots on the ground" monitoring of NRB in terms of the fluctuation of its level, impacts of climate change and anthropogenic factors on its TWS, the behaviour of vegetation within its basin, and the extremes, e.g., droughts, is practically impossible. The most viable approach is to turn to space techniques [103] to weigh the NRB's TWS changes, what is demonstrated in the present book.

References

1. Molen I, Hildering A (2005) Water: cause for conflict or co-operation? ISYP J Sci World Aff 1(2):133–143
2. Rijsberman FR (2006) Water scarcity: fact or fiction. Agric Water Manag (Elsevier) 80:5–22
3. Jury WA, Vaux HJ Jr (2007) The emerging global water crisis: managing scarcity and conflict between water users. Adv Agron (Elsevier Inc) 95:77p. https://doi.org/10.1016/50065-2113(07)95001-4
4. Anyah RO, Forootan E, Awange JL, Khaki M (2018) Understanding linkages between global climate indices and terrestrial water storage changes over Africa using GRACE products. Sci Total Environ 635:1405–1416. https://doi.org/10.1016/j.scitotenv.2018.04.159
5. Vörösmarty CJ, Green P, Salisbury J, Lammers RB (2000) Global water resources: vulnerability from climate change and population growth. Science 289(5477):284–288. https://doi.org/10.1126/science.289.5477.284
6. Phillips T, Nerem RS, Fox-Kemper B, Famiglietti JS, Rajagopalan B (2012) The influence of ENSO on global terrestrial water storage using GRACE. Geophys Res Lett 39(L16705):2012. https://doi.org/10.1029/2012GL052495
7. Schewe J, Heinke J, Gerten D, Haddeland I, Arnell NW, Clark DB, Dankers R, Eisner S, Fekete BM, Colón-Gonzlez FJ, Gosling SN, Kim H, Liu X, Masaki Y, Portmann FT, Satoh Y, Stacke T, Tang Q, Wada Y, Wisser D, Albrecht T, Frieler K, Piontek F, Warszawski L, Kabat P (2013) Multimodel assessment of water scarcity under climate change. PNAS 111(9):3245–3250. https://doi.org/10.1073/pnas.1222460110
8. Hurkmans R, Troch PA, Uijlenhoet R, Torfs P, Durcik M (2009) Effects of climate variability on water storage in the Colorado River Basin. J Hydrometeorol 10:1257–1270. https://doi.org/10.1175/2009JHM1133.1

9. Ndehedehe CE, Awange J, Kuhn M, Agutu N, Fukuda Y (2017d) Climate teleconnections influence on West Africa's terrestrial water storage. Hydrol Process 31(18):3206–3224. https://doi.org/10.1002/hyp.11237
10. Awange JL, Sharifi M, Ogonda G, Wickert J, Grafarend EW, Omulo M (2007) The Falling Lake Victoria water level: GRACE, TRIMM and CHAMP satellite analysis. Water Resour Manag. https://doi.org/10.1007/s11269-007-9191-y
11. Van Loon AF, Stahl K, Di Baldassarre G, Clark J, Rangecroft S, Wanders N, Gleeson T, Van Dijk AIJM, Tallaksen LM, Hannaford J, Uijlenhoet R, Teuling AJ, Hannah DM, Sheffield J, Svoboda M, Verbeiren B, Wagener T, Van Lanen HAJ (2016) Drought in a human-modified world: reframing drought definitions, understanding, and analysis approaches. Hydrol Earth Syst Sci 20(9):3631–3650. https://doi.org/10.5194/hess-20-3631-2016
12. Konar M, Evans TP, Levy M, Scott CA, Troy TJ, Vörösmarty CJ, Sivapalan M (2016) Water resources sustainability in a globalizing world: who uses the water? Hydrol Process 30(18):3330–3336. https://doi.org/10.1002/hyp.10843
13. Hall JW, Grey D, Garrick D, Fung F, Brown C, Dadson SJ, Sadoff CW (2014) Coping with the curse of freshwater variability. Science 346(6208):429–430. https://doi.org/10.1126/science.1257890
14. Alsdorf DE, Lettenmaier DP (2003) Tracking fresh water from space. Science 301(5639):1491–1494. https://doi.org/10.1126/science.1089802
15. Alsdorf D, Lettenmaier D, Vrsmarty C (2003) The need for global, satellitebased observations of terrestrial surface waters. Eos Trans Am Geo-Phys Union 84(29):269–276. https://doi.org/10.1029/2003EO290001
16. Vörösmarty C, Askew A, Grabs W, Barry RG, Birkett C, Dll P, Goodison B, Hall A, Jenne R, Kitaev L, Landwehr J, Keeler M, Leavesley G, Schaake J, Strzepek K, Sundarvel SS, Takeuchi K, Webster F (2001) Global water data: a newly endangered species. Eos Trans Am Geophys Union 82(5):54–58. https://doi.org/10.1029/01EO00031
17. Gleeson T, Wada Y, van Beek MFPBLPH (2012) Water balance of global aquifers revealed by groundwater footprint. Nature 488:197–200. https://doi.org/10.1038/nature11295
18. UNEP (2002) A world of salt: total global saltwater and freshwater estimates. http://www.unep.org/dewa/assessments/ecosystems/water/vitalwater/freshwater.htm. Accessed 25 Aug 2010
19. Hofman AR (2004) The connection: water and energy security. http://www.iags.org/n0813043.htm. Accessed 25 Aug 2010
20. Gleick PH (1993) Water and conflict: fresh water resources and international security. Int Secur 18(1):79–112. https://doi.org/10.2307/2539033
21. Gleick PH (2000) The world's water: Biennial report on freshwater resources 2000–2001. Island Press, Washington
22. Agola NO, Awange JL (2014) Globalized poverty and environment 21st century challenges and innovative solutions. Springer, Berlin
23. Kuylenstierna JL, Bjrklund G, Najlis P (1997) Sustainable water future with global implications: everyone's responsibility. Nat Resour Forum 21(3):181–190. https://doi.org/10.1111/j.1477-8947.1997.tb00691.x
24. Vörösmarty CJ, Douglas EM, Green PA, Revenga C (2005) Geospatial indicators of emerging water stress: an application to Africa. Ambio
25. Duraiappah A (1998) Poverty and environmental degradation: a review and analysis of the nexus. J Hydrol 113(1–4):297–306
26. Awange JL, Khandu, Forootan E, Schumacher M, Heck B (2016) Exploring hydro-meteorological drought patterns over the Greater Horn of Africa (1979–2014) using remote sensing and reanalysis products. Adv Water Resour. https://doi.org/10.1016/j.advwatres.2016.04.005
27. Awange JL, Ferreira VG, Forootan E, Khandu, Andam-Akorful SA, Agutu NO, He XF (2016) Uncertainties in remotely sensed precipitation data over Africa. Int J Climatol 36(1):303–323. https://doi.org/10.1002/joc.4346
28. Omondi P, Awange J, Ogallo LA, Ininda J, Forootan E (2013) The influence of low frequency sea surface temperature modes on delineated decadal rainfall zones in Eastern Africa region. Adv Water Resour. https://doi.org/10.1016/j.advwatres.2013.01.001

29. Omondi P, Awange J, Ogallo LA, Okoola RA, Forootan E (2012) Decadal rainfall variability modes in observed rainfall records over East Africa and their relations to historical sea surface temperature changes. J Hydrol 464–465:140–156. https://doi.org/10.1016/j.jhydrol.2012.07.003
30. Agutu NO, Awange JL, Zerihun A, Ndehedehe CE, Kuhn M, Fukuda Y (2017) Assessing multi-satellite remote sensing, reanalysis, and land surface models' products in characterizing agricultural drought in East Africa. Remote Sens Environ 194:287–302. https://doi.org/10.1016/j.rse.2017.03.041
31. Mpelasoka F, Awange JL, Zerihun A (2018) Influence of coupled ocean-atmosphere phenomena on the Greater Horn of Africa droughts and their implications. Sci Total Environ 610–611:691–702. https://doi.org/10.1016/j.scitotenv.2017.08.109
32. IRIN Humanitarian News and Analysis, UN Office for Coordination of Humanitarian Affairs (2006). Global: the global water crisis: managing a dwindling resource. http://www.irinews.org. Accessed 25 Sep 2011
33. Alcamo J, Doll P, Kaspar F, Siebert S (1997) Global change and global scenarios of water use and availability: an application of WaterGAP 1.0. University of Kassel, CESR, Kassel, Germany
34. Alcamo J, Henrichs T, Rosch T (2000) World water in 2025: global modeling and scenario analysis. In: Rijsberman FR (ed) World water scenarios analyses. World Water Council, Marseille
35. Vorosmarty CJ, Green P, Salisbury J, Lammers RB (2000) Global water resources: vulnerability from climate change and population growth. Science 289:284–288
36. Wallace JS (2000) Increasing agricultural water efficiency to meet future food production. Agric Ecosyst Environ 82:105–119
37. Wallace JS, Gregory PJ (2002) Water resources and their use in food production. Aquat Sci 64:363–375
38. Agola NO, Awange JL (2014) Globalized poverty and environment. Springer, Berlin
39. Yang H, Reichert P, Abbaspour K, Zehnder AJB (2003) A water resources threshold and its implications for food security. Environ Sci Technol 37:3048–3054
40. Awange JL, Hu KX, Khaki M (2019) The newly merged satellite remotely sensed, gauge and reanalysis-based multi-source weighted-ensemble precipitation: evaluation over Australia and Africa (1981–2016). Sci Total Environ 670:448–465. https://doi.org/10.1016/j.scitotenv.2019.03.148
41. Rekacewicz P (2006) Increased global water stress. Vital water graphics 2. Le monde diplomatique. http://www.grida.no/publications/vg/water2/page/3289.aspx. Accessed 15 Apr 2012
42. Neff RA, Parker CL, Kirschenmann FL, Tinch J, Lawrence RS (2011) Peak oil, food systems, and public health. Am J Public Health 101(9):1587–1597
43. Payne WA (2010) Farming systems and food security in sub-Saharan Africa. In: Lal R, Stewart BA (eds) Food security and soil quality. Taylor and Francis, New York
44. Othieno H, Awange JL (2016) Energy resources in Africa. Distribution, opportunities and challenges. Springer International Publishing AG, Berlin. https://doi.org/10.1007/978-3-319-25187-5
45. Carolina BF (2002) Competition over water resources: analysis and mapping of water-related conflicts in the catchment of Lake Naivasha (Kenya). MSc Thesis, ITC
46. World Bank (2003) Water resource and environment. Davis R, Hirji R (eds) Technical note G.2. Lake Management
47. Birnie P, Boyle A (1993) International law & the environment. Camb Law J (Cambridge University Press) 52(3):540p
48. Gleick PH (1999) The human right to water. Water Policy 1:487–503
49. Schelton D (1991) Human rights, environmental rights, and the right to environment. Stanf J Int Law (Heinonline) 28
50. Zehnder AJB, Yang H, Schertenleib R (2003) Water issues: the need for action at different levels. Aquat Sci: Res Boundaries. 65(1):1–20. https://doi.org/10.1007/s000270300000

51. IPCC (2018) Global warming of 1.5°C. An IPCC special report on the impacts of global warming of 1.5°C above pre-industrial levels and related global greenhouse gas emission pathways, in the context of strengthening the global response to the threat of climate change, sustainable development, and efforts to eradicate poverty. In: Masson-Delmotte V, Zhai P, Pörtner H-O, Roberts D, Skea J, Shukla PR, Pirani A, Moufouma-Okia W, Péan C, Pidcock R, Connors S, Matthews JBR, Chen Y, Zhou X, Gomis MI, Lonnoy E, Maycock T, Tignor M, Waterfield T (eds)
52. IPCC (2014) Climate change 2014: synthesis report. Contribution of working groups I, II and III to the fifth assessment report of the intergovernmental panel on climate change. In: Core Writing Team, Pachauri RK, Meyer LA (eds). IPCC, Geneva, Switzerland, 151p
53. Hartmann DL, Klein Tank AMG, Rusticucci M, Alexander LV, Brönnimann S, Charabi Y, Dentener FJ, Dlugokencky EJ, Easterling DR, Kaplan A, Soden BJ, Thorne PW, Wild M, Zhai PM (2013) Observations: atmosphere and surface. In: Stocker TF, Qin D, Plattner G-K, Tignor M, Allen SK, Boschung J, Nauels A, Xia Y, Bex V, Midgley PM (Hrsg.) Climate change 2013: the physical science basis. Contribution of working group I to the fifth assessment report of the intergovernmental panel on climate change. Cambridge University Press, Cambridge
54. Saji NH, Goswami BN, Vinayachandran PN, Yamagata T (1999) A dipole mode in the tropical Indian ocean. Nature 401:360–363
55. Trenberth KE (1997) The definition of El Niño. Bull Am Meteorol Soc 78:2771–2777
56. Awange JL, Mpelasoka F, Goncalves R (2016) When every drop counts: analysis of droughts in Brazil for the 1901–2013 period. Sci Total Environ 566–567:1472–1488. https://doi.org/10.1016/j.scitotenv.2016.06.031
57. Mpelasoka F, Awange JL, Goncalves RM (2018) Accounting for dynamics of mean precipitation in drought projections: a case study of Brazil for the 2050 and 2070 periods. Sci Total Environ 622–623:1519–1531. https://doi.org/10.1016/j.scitotenv.2017.10.032
58. Agutu NO, Awange JL, Ndehedehe C, Kirimi F, Kuhn M (2019) GRACE-derived groundwater changes over Greater Horn of Africa: temporal variability and the potential for irrigated agriculture. Sci Total Environ. https://doi.org/10.1016/j.scitotenv.2019.07.273
59. Hu K, Awange JL, Kuhn M, Saleem A (2019) Spatio-temporal groundwater variations associated with climatic and anthropogenic impacts in South-West Western Australia. Sci Total Environ. https://doi.org/10.1016/j.scitotenv.2019.133599
60. Hanjra MA, Qureshi ME (2010) Global water crisis and future food security in an era of climate change. Food Policy 35(5):365–377
61. Awange JL, Kiema JBK (2013) Environmental geoinformatics. Monitoring and management. Springer, Berlin
62. Awange JL, Kiema JBK (2018) Environmental geoinformatics. Extreme hydro-climatic and food security challenges: exploiting the big data, 2nd edn. Springer International Publishers, Berlin
63. Johnson LE (2009) Geographic information systems in water resources engineering. CRC Press Taylor & Francis Group, Boca Raton. ISBN 978-1-4200-6913-6
64. Agutu NO, Awange JL, Ndehedehe C, Mwaniki MW (2020) Consistency of agricultural drought characterization over upper Greater Horn of Africa (1982–2013): topographical, gauge density, and model forcing influence. Sci Total Environ. 709. https://doi.org/10.1016/j.scitotenv.2019.135149
65. Taylor CJ, Alley WM (2001) Ground-water-level monitoring and the importance of long-term water-level data. U.S. Geological Survey Circular 1217, Denver, Colorado
66. Awange JL (2012) Environmental monitoring using GNSS. Global navigation satellite system. Springer, Berlin
67. Awange JL (2018) GNSS environmental sensing. Revolutionizing environmental monitoring, Springer, Berlin
68. Rieser D, Kuhn M, Paill R, Anjasmara IM, Awange J (2010) Relation between GRACE-derived surface mass variations and precipitation over Australia. Aust J Earth Sci 57(7):887–900. https://doi.org/10.1080/08120099.2010.512645

69. IPCC (Intergovernmental Panel on Climate Change) (2007) Contribution of working group I to the fourth assessment report
70. Tapley BD, Bettadpur S, Ries JC, Thompson PF, Watkins MM (2004) GRACE measurements of mass variability in the Earth system. Science 305:503–505. https://doi.org/10.1126/science.1099192
71. Ellett KM, Walker JP, Rodell M, Chen JL, Western AW (2005) GRACE gravity fields as a new measure for assessing large-scale hydrological models. In: Zerger A, Argent RM (eds) MODSIM 2005 international congress on modelling and simulation, December 2005, pp 2911–2917. Modelling and Simulation Society of Australia and New Zealand. ISBN: 0-9758400-2-9
72. Casanova MT (1994) Vegetative and reproductive responses of charophytes to water-level fluctuations in permanent and temporary wetlands in Australia. Aust J Mar Freshw Res 45:1409–1419
73. Leick A (2004) GPS satellite surveying, 3rd edn. Wiley, New York
74. ILEC (International Lake Environment Committee) (2005) Managing lakes and their basins for sustainable use. A report for the lake basin managers and stakeholders. International Lakes Environmental Committee Foundation, Kusatsu, Japan
75. Awange LJ, Gebremichael M, Forootan E, Wakbulcho G, Anyah R, Ferreira CG, Alemayehu T (2014) Characterization of Ethiopian mega hydrogeological regimes using GRACE, TRMM and GLDAS datasets. Adv Water Resour 74:64–78. https://doi.org/10.1016/j.advwatres.2014.07.012
76. Hu K, Awange JL, Khandu Forootan E, Goncalves RM, Fleming K (2017) Hydrogeological characterisation of groundwater over Brazil using remotely sensed and model products. Sci Total Environ 599–600:372–386. https://doi.org/10.1016/j.scitotenv.2017.04.188
77. Awange JL, Saleem A, Sukhadiya RM, Ouma YO, Kexiang H (2019a) Physical dynamics of Lake Victoria over the past 34 years (1984–2018): is the lake dying? Sci Total Environ 658:199–218. https://doi.org/10.1016/j.scitotenv.2018.12.051
78. Awange JL, Ferreira VG, Forootan E, Khandu, Andam-Akorful SA, Agutu NO, He XF (2016) Uncertainties in remotely sensed precipitation data over Africa. Int J Climatol 36:303–323. https://doi.org/10.1002/joc.4346
79. Camberlin P (2009) Nile basin climates. In: Dumont, Henri J (ed) The Nile: origin, environments, limnology and human use. Monographiae Biologicae, pp 307–333. Springer, Berlin
80. Khaki M, Awange J, Forootan E, Kuhn M (2018) Understanding the association between climate variability and the Nile's water level fluctuations and water storage changes during 1992–2016. Sci Total Environ 645:1509–1521. https://doi.org/10.1016/j.scitotenv.2018.07.212
81. Ahmed AA, Ismail UAE (2008) Sediment in the Nile river system. Consultancy Study Requested by UNESCO
82. Conway D (2002) Extreme rainfall events and lake level changes in East Africa: recent events and historical precedents. In: Odada EO, Olago DO (eds) The East African Great Lakes: limnology, palaeolimnology and biodiversity. Advances in global change research, vol 12, pp 63–92. Kluwer, Dordrecht
83. Taye MT, Ntegeka V, Ogiramoi NP, Willems P (2011) Assessment of climate change impact on hydrological extremes in two source regions of the Nile river basin. Hydrol Earth Syst Sci 15:209–222. http://dx.doi.org/10.5194/hess-15-209-2011
84. Awange JL, Ong'ang'a O (2006) Lake Victoria-ecology. Resource of the lake basin and environment. Springer, Berlin
85. Morgana B, Awange JL, Saleem A, Kexiang H (2020) Understanding vegetation variability and their "hotspots" within Lake Victoria basin for the 2003–2018 period. Appl Geogr
86. Nicholson SE, Yin X, Mamoudou BA (2000) On the feasibility of using a lake water balance model to infer rainfall: an example from Lake Victoria. Hydrol Sci J 45(1):75–95. https://doi.org/10.1080/02626660009492307
87. Nicholson SE (1998) Historical fluctuations of Lake Victoria and other lakes in the Northern Rift Valley of East Africa. In: Lehman JT (ed) Environmental change and response in East African lakes. Kluwer, Dordrecht, pp 7–35

88. Nicholson SE (1999) Historical and modern fluctuations of lakes Tanganyika and Rukwa and their relationship to rainfall variability. Clim Change 41:53–71. https://doi.org/10.1023/A:1005424619718
89. Aseto O, Ong'ang'a O, Awange JL (2003) Poverty. A challenge for the Lake Victoria basin. OSIENALA Series 5. Printed by Africa Herald Publishing House, Kendu Bay
90. Kite GW (1982) Analysis of Lake Victoria levels. Hydrol Sci J 27(2):99–110. https://doi.org/10.1080/02626668209491093
91. Kull D (2006) Connections between recent water level drops in Lake Victoria, dam operations and drought. Available at: http://www.irn.org/programs/nile/pdf/060208vic.pdf. Accessed 25 Sep 2011
92. Awange JL, Aluoch J, Ogallo L, Omulo M, Omondi P (2007) Frequency and severity of drought in the Lake Victoria region (Kenya) and its effects on food security. Clim Res 33:135–142. https://doi.org/10.3354/cr033135
93. Awange JL (2021) Lake Victoria monitored from space. Springer Nature, Berlin
94. Awange JL, Forootan E, Kuhn M, Kusche J, Heck B (2014) Water storage changes and climate variability within the Nile basin between 2002 and 2011. Adv Water Resour 73:1–25. https://doi.org/10.1016/j.advwatres.2014.06.010
95. Awange JL, Anyah RN, Agola N, Forootan E, Omondi P (2013) Potential impacts of climate and environmental change on the stored water of Lake Victoria Basin and economic implications. Water Resour Res 49:8160–8173
96. Awange JL, Fleming KM, Kuhn M, Featherstone WE, Heck B, Anjasmara I (2011) On the suitability of the 4° × 4° GRACE mascon solutions for remote sensing Australian hydrology. Remote Sens Environ 115:864–875. https://doi.org/10.1016/j.rse.2010.11.014
97. Awange JL, Sharifi MA, Baur O, Keller W, Featherstone WE, Kuhn M (2009) GRACE hydrological monitoring of Australia. Current limitations and future prospects. J Spat Sci 54(1):23–36. https://doi.org/10.1080/14498596.2009.9635164
98. Awange J, Forootan E, Kusche J, Kiema JKB, Omondi P, Heck B, Fleming K, Ohanya S, Gonçalves RM (2013) Understanding the decline of water storage across the Ramser-lake Naivasha using satellite-based methods. Adv Water Resour 60:7–23. https://doi.org/10.1016/j.advwatres.2013.07.002
99. Andersen OB, Seneviratne SI, Hinderer J, Viterbo P (2005) GRACE-derived terrestrial water storage depletion associated with the 2003 European heat wave. Geophys Res Lett 32(L18405):2–5. https://doi.org/10.1029/2005GL023574
100. Forootan E, Awange J, Kusche J, Heck B, Eicker A (2012) Independent patterns of water mass anomalies over Australia from satellite data and models. Remote Sens Environ 124:427–443. https://doi.org/10.1016/j.rse.2012.05.023
101. Khandu E, Forootan M, Schumacher J Awange, Miler Schmied H (2016) Exploring the influence of precipitation extremes and human water use on total water storage (TWS) changes in the Ganges-Brahmaputra-Meghna river basin. Water Resour Res 52(3):2240–2258
102. Nile Basin Water Resources Atlas (2017). http://atlas.nilebasin.org/
103. Alsdorf DE, Rodrguez E, Lettenmaier DP (2007) Measuring surface water from space. Rev Geophys 45(2):RG2002. https://doi.org/10.1029/2006RG000197
104. Kayombo S, Jorgensen SE (2006) Lake Victoria: experience and lessons learned brief. International Lake Environment Committee

Chapter 2
Nile River Basin and Its Challenges

> *Of course, getting all the countries to agree on what they believe is fair and equal use of the Nile's resources is no easy task. Lower countries (Egypt and Sudan) rely on some old treaty that they signed with Britain decades ago to impose unrealistic water use conditions to the upper countries. For this reason, some countries, e.g., Ethiopia and Uganda, have decided to disregard the treaty and are busy constructing large hydropower dams (i.e., GERD and Bujagali, respectively) within the Nile.*
>
> *—Joseph Awange (https://www.livescience.com/57023-nile-river-facts.html)*

2.1 Summary

Water resources within the Nile River Basin (NRB; Fig. 2.1) are under intense pressure from both humans and the changing climate. With the increasing population, and more demand for both water and electricity, the various countries within the NRB are likely to exploit the waters within their sub-basins, thus, negatively impacting on the entire Nile's stored water.

As the NRB extends over a wide latitudinal range (from $\sim$5° S to $\sim$31°N) covering the tropical and desert climatic zones, it cannot be expected to experience homogeneous climatic and rainfall patterns over its extent. The northern part of the basin, including Egypt and Sudan, is mostly dry with 92% of Egypt and 89% of Sudan located in semi-arid regions with virtually no rainfall during summer. In contrast the southern portion, which encompasses the Ethiopian Plateau, is located in humid areas and experiences heavy rainfall during summer. During the season between October and May, both regions are relatively dry due to the presence of the northeast trade winds [1]. In addition, variations in geology and soils of the basin strongly influence water flow and groundwater availability [3, 11]. Significant rainwater deficits and the variable duration of the rainy season over yearly to decadal time scales results in

J. Awange, *The Nile Waters*,
https://doi.org/10.1007/978-3-030-64756-8_2

hydrological deficits that are not necessarily reflected in a direct response of the base flow, e.g., [18]. Comprehensive information on various aspects of the Nile River, i.e. the Nile journey; the Aswan High Dam Reservoir; morphology and sediment quality of the Nile; threats to biodiversity; fish and fisheries; rain-fed agriculture, rainfall data, and fluctuations in rainfall; the impact of climate change; and hydropolitics and legal aspects can be found in [20] while [15] provides a geomorphological view of the Nile waters (Fig. 2.1).

2.2 The Nile: Arguably the World's Longest River

2.2.1 The Origins of River Nile

Most history books credit the "discovery" of the source of the Nile to the 19th century explorer John Speke (Fig. 2.4a) who captivated the Western world with this news in 1862.[1] This claim, however, no longer holds. In fact, a little-known German explorer, Bruckhart Waldekker, proved Speke wrong.[2] It is now accepted that the Nile originates from two distinct geographical zones—the basin of the "White Nile" and the "Blue Nile". The White Nile originates from the Great Lakes Region (Fig. 2.4) and is fed by the Bahr-el-Jebel water system to the North and East of the Nile Congo River divides. The Blue Nile originates in the Highlands of Ethiopia and Eritrea, so do the other major tributaries of the Nile, Atbara and the Sobat. It is opined, e.g., by [25] that the Ethiopian Highlands contributes about 85% of the total volume. Lake Victoria, though, is a major source of River Nile contributing less than 15% of Nile water. About 84% of the water in the Lake comes from the seven rivers in Kenya, even though Kenya owns only 6% of the Lake. Chapter 4 looks at Lake Victoria, with a detailed treatment presented in Awange [8]. The Nile river is arguably the longest river in the world [19], e.g., Fig. 2.2a, with its basin being the fourth largest, i.e., Fig. 2.2b.

2.2.2 The Nile River Basin

The Nile River Basin (NRB, Fig. 2.1) is one of the Earth's most impressive examples of the influence of (i) climate on the flow conditions of a water system. Tropical wet and dry climate prevail in the equatorial lakes region and part of Ethiopia, sub-tropical dry arid (desert) dominate Sudan and Egypt, sub-tropical dry semi-arid prevail in the southern part of Sudan as well as the tropical wet and tropical monsoonal around the Lake Victoria and some parts of the Ethiopian highlands [19]. Figure 2.3a adopted

---

[1] Britannica Concise.

[2] John Mbaria, "Revoke Obsolete River Nile Treaty" in the Daily Nation, Thursday, March 29, 2002.

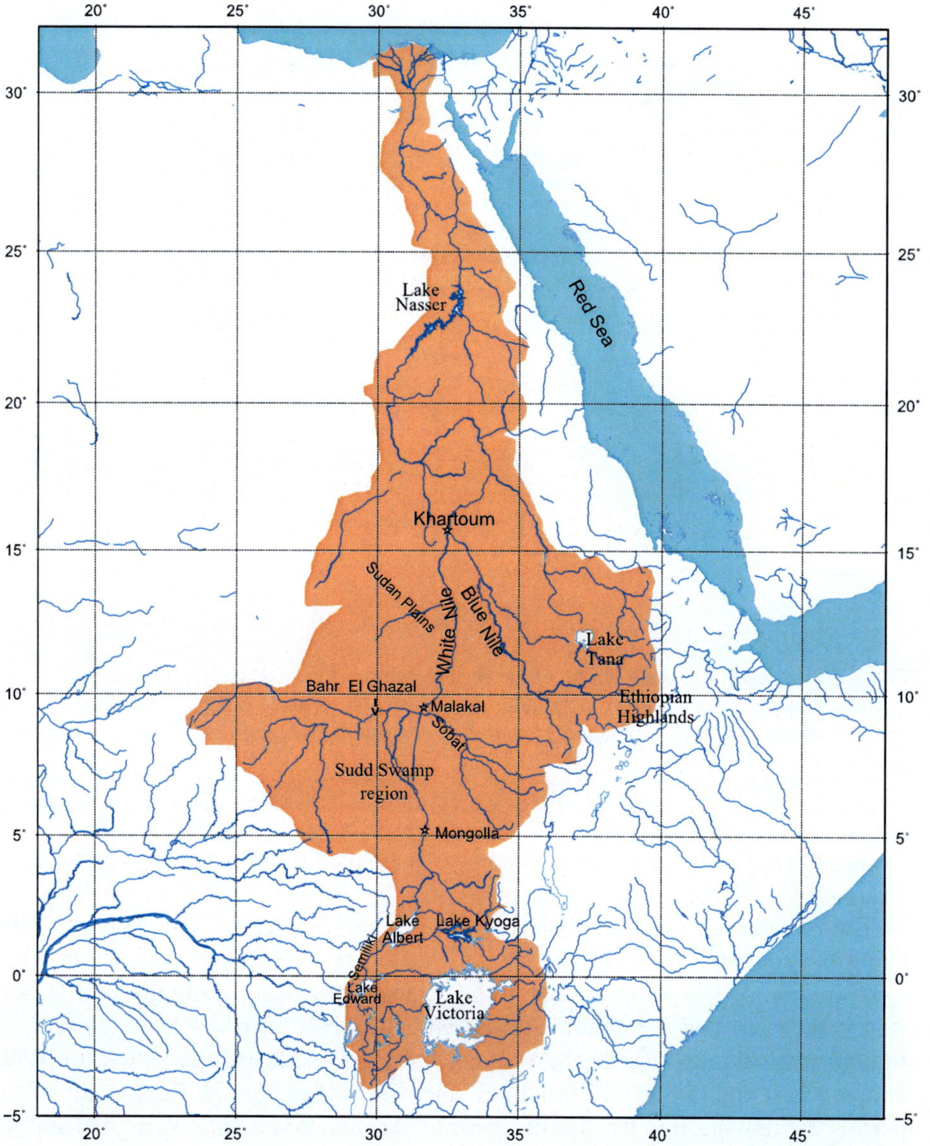

Fig. 2.1 The Nile River Basin (brown shaded region) with the major features and place names as discussed in the text. *Source* [7]

from [19] provides classifications within the NRB, and (ii), topography (Fig. 2.3c) that includes mountain ranges of the Upper Kagera, White Nile, Blue Nile and Tekeze–Atbara rivers. The upper parts of the basin have a ridged topography and steep slopes on the one hand while on the other hand, there are large plateau regions along the middle reaches of the basin and wide flood plains from the lower reaches

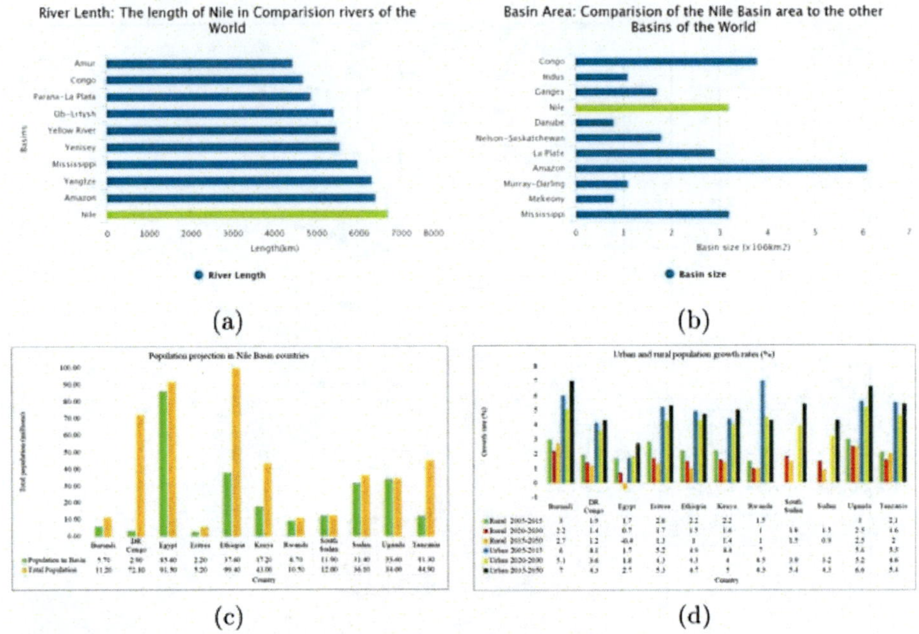

Fig. 2.2 Nile River Basin's **a** length of the river in comparison to other long global rivers, **b** area in comparison to other basins, **c** population in comparison to total countries' population, and **d**, projected rural and urban population growth. *Source* (Modified) Nile Basin Water Resource Atlas (http://atlas.nilebasin.org/)

to the delta [19]. The geology and land use/land cover of NRB are also shown in the figure [19].

The Nile has two major tributaries, the White Nile and the Blue Nile, the latter being the source of most of the river's water. The White Nile rises in the Great Lakes region of Eastern Africa (Fig. 2.4a), and flows northwards through Uganda (Fig. 2.4b, c, d) and the South Sudan. The Blue Nile starts at Lake Tana in the Ethiopian highlands, flowing into Sudan from the southeast and meets the White Nile at Khartoum in Sudan. From there, the Nile passes through Egypt and ends its journey by flowing into the Mediterranean Sea. *The White Nile (East African lake region)* includes the countries of Burundi, Rwanda, Uganda, Kenya and Tanzania, and is the home to Lake Victoria, the world's second largest freshwater lake, and the source of the White Nile (258,000 km^2, Fig. 4.1) [2, 6, 8]. From Lake Victoria (Fig. 2.4a), the water masses discharge to Lake Kyoga (Fig. 2.4b, c, d), which also receives water from its surrounding catchment (~75,000 km^2) before flowing into Lake Albert. In addition to the waters received from Lake Kyoga, Lake Albert is supplied by its upstream Semiliki Basin and the Lake Edward sub-Basin (see Fig. 2.1). Together, Lakes Edward, Albert, and George form the western edge of the Nile Basin comprising an area of 48,000 km^2 of which 7,800 km^2 is open water [26]. In the Sudd Swamp region, the water supply to the Nile benefits from two other basins, the Bahr-

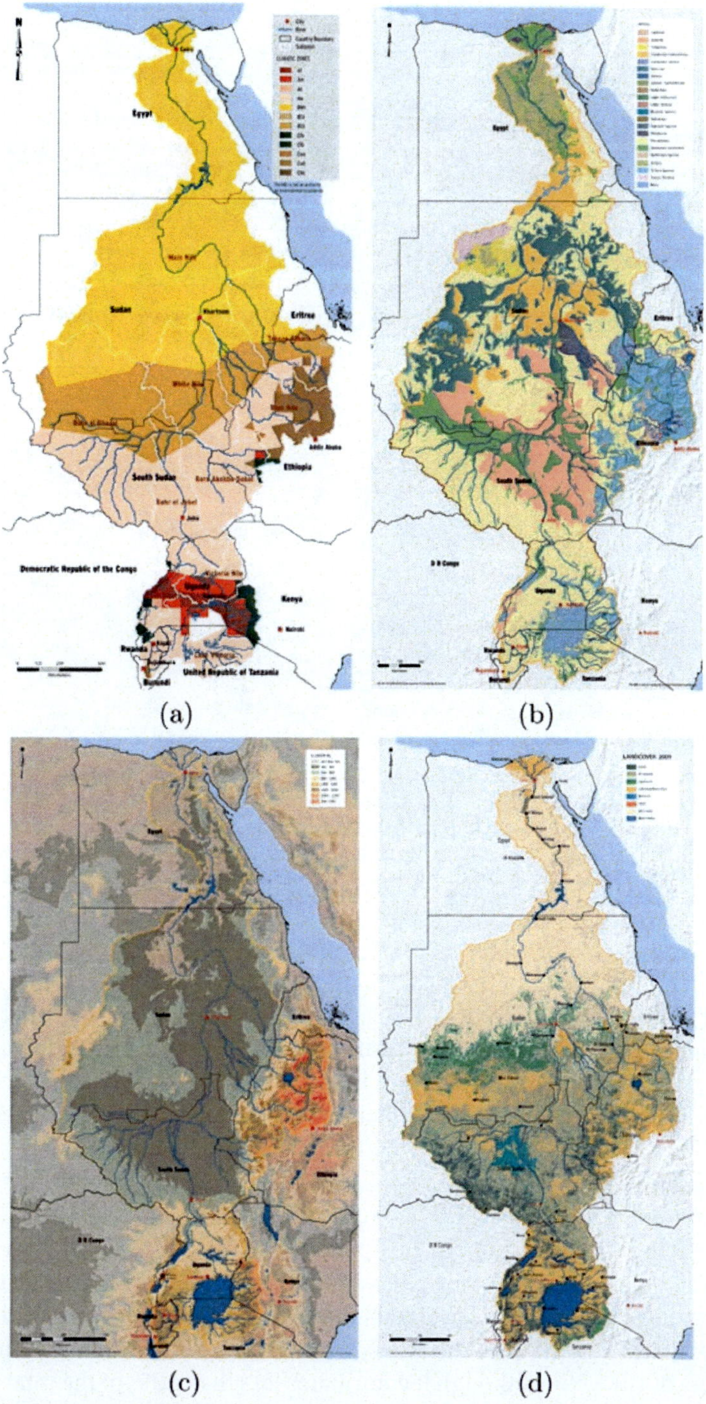

Fig. 2.3 Nile River Basin's **a** climate, **b** geology, **c** Topography, and **d**, Land use land cover. *Source* Nile Basin Water Resource Atlas (http://atlas.nilebasin.org/)

Fig. 2.4 **a** Pillar signifying the source of the White Nile erected by John H. Speke in 1858 in Lake Victoria, **b** exit of the White Nile through Nalubale dam (Uganda), **c** flow of the White Nile downstream, and **d** off goes the White Nile from Jinja towards Lake Kyoga (Uganda)

el-Ghazal (526,000 km^2) to the west and the Sobat (150,000 km^2) to the east, before exiting at Malakal.

The Bahr-el-Ghazal (BEG) sub-basin consists of a number of rivers that originate from the Congo-Nile River divide and is made up of a large area of very low slope such that nearly all the basin runoff and precipitation is evaporated (see Fig. 2.12) or leaked into swamps (i.e., about 96% is lost), with only about 0.5 km^3 leaving the basin annually [9].

The Ethiopian highlands (EH) comprises twelve significant sub-basins aggregated into four primary basins; Lake Tana region (20,000 km^2), the upper Blue Nile region (150,000 km^2), the lower Blue Nile region (60,000 km^2), and the Dinder-Rahad region (70,000 km^2). All together, they cover almost 300,000 km^2 and contribute approximately 65% of the Nile's total water [26]. Lake Tana forms the main headwaters of Blue Nile although it contributes less than 10% of the total Blue Nile's flow [26].

Lake Nasser region: The Egyptian desert starts from Khartoum, where the Nile flows northward towards Egypt through Lake Nasser (formed by the Aswan dam) and then to the Mediterranean Sea. Yates and Strzepek [26] reported a net loss of water between the joining of the Atbara river with the Nile north of Khartoum, and Lake Nasser due to evaporation and seepage.

The four main areas of interest in this book, whose changes in total water storage (TWS; surface, groundwater, soil moisture and vegetation) will be remotely sensed by Gravity Recovery and Climate Experiment (GRACE) satellites discussed in Chap. 3 due to their larger spatial coverage (i.e., over 200,000 km^2) are (see Fig. 2.1): (i) Lake Victoria Basin (LVB), which is the headwaters of the White Nile; (ii) the Bahr-el-Ghazal region (BEG), the main western tributary of the Nile and the largest sub-basin. Its supply to the Sudd wetlands is however reported by [16] to be negligible. Nonetheless, it is included in order to assess the impacts of climate variability on the NRB's total water storage. The Sudd wetlands (swamp) is the region between Mongalla and Malakal, which consists of swamps and lagoons, and is believed to be where most of the White Nile's water is lost due to evaporation [26], e.g., Fig. 2.12; (iii) the Ethiopian highlands (EH), and the headwaters of the Blue Nile; and (iv), the Egyptian desert region consisting of Lake Nasser, where significant amounts of water are lost due to evaporation, see details of regions in Conway and Hulme [9]. The characteristics of these regions are summarized in (Table 2.1).

2.3 Potentials and Challenges Facing the Nile's Waters

The Nile River Basin (NRB) is one of the largest in the world, with an area of about 3,400,000 km^2 (almost one-tenth of Africa) and traversing some 6,500 km from south to north as it winds its way across the boundaries of eleven (11) countries (Fig. 2.5a); Tanzania, Uganda, Kenya, Rwanda, Burundi, Democratic Republic of Congo (DRC), Eritrea, Ethiopia, Sudan, South Sudan, and Egypt, e.g., [21]. As it flows through these countries, it supports a livelihoods of over 300 million people with a population density in 2017 as shown in Fig. 2.5b. In Fig. 2.2c, the population living within the Nile River Basin (>300 million) is compared to the total country population (>500 million) while in Fig. 2.2d, the population growth in the basin is presented for both rural and urban regions for the periods 2005–2015, 2020–2035 and 2035–2050. The population projections from the figure indicate continued growth in the basin, which will certainly put pressure on its water resources. According to Fig. 2.2d, highest rural population growth rates in the Nile Basin are observed for the period 2005–2015 at the range of 1.5–3.0%. In the future (2020–2030 and 2035–2050) the rural population growth rate is projected to decline and the rate is expected to be negative for Egypt by 2035–2050 while in contrast, urban population growth rate is expected to increase significantly in all Nile Basin countries [19].

The Nile River Basin has diverse biodiversity (Fig. 2.6) and attractive tourist sites (Fig. 2.7). On the White Nile's portion of the Nile is the spectacular Murchison Falls' site where the river squeezes through an 8 m opening (Fig. 2.8). This site is popular

Table 2.1 Characteristics of the Nile sub-basins

	Area	Catchment	Remark
Lake Victoria	250,000 km^2	Lakes Kyoga (75,000 km^2)	Head Waters of the While Nile
		Albert, Elbert, George (48,000 km^2)	
		Semiliki Basin	
Barrel-Ghazel (BEG)	526,000 km^2	Congo-Nile River divide	About 96% of basin runoff and precipitation is lost to evaporation and leakage to swamp
Ethiopian Highland (EH)	300,000 km^2	Lake Tana Region (20,000 km^2)	Head watersa of the Blue Nile contributing about 65% of the Nile waters
		Upper Blue Nile region (150,000 km^2)	
		Lower Blue Nile region (60,000 km^2)	
		Dinda Rahad region (60,000 km^2)	
Egypt Desert Region		Lake Nasser (formed by Aswan dam)	Water loss due to evaporation

with tourists from all over the world. Given that the Nile services 11 countries with its precious resources, it is expected that disputes will often arise. For this reason, in 1999, The Nile Basin Initiative (NBI), an intergovernmental partnership among all the 11 countries was formed to play a moderating role between the countries in order to realise the sharing of its resources.

Although the challenges associated with the sharing of the Nile's water resources necessitated the formation of the NBI above, increasing threats have, however, emerged from both anthropogenic and natural factors, e.g., [14]. Anthropogenic influences have been fueled by the increasing human population that has put pressure on domestic water needs [17], the supply of hydroelectric power (see, e.g., [2, 28] and Fig. 2.9), all coupled with the need to sustain economic growth. Damming the Nile (Both the White and Blue Niles) has not only contributed to conflicts between the lower and upper countries depending on it, but has also resulted in geomorphological changes of the river itself (e.g., Fig. 2.10). For example, it can be noticed that the damming of the White Nile through the construction of Bujagali dam (Fig. 2.9a) has changed the river's geomorphology (e.g., Fig. 2.11a–d has been replaced by lake Bujagali in Fig. 2.11e–h). In Fig. 2.11f, the construction process can be evidenced by the deposition of rocks while Fig. 2.11g and h show the completed Bujagali dam. On the Blue Nile side, its damming through the construction of the largest hydro-

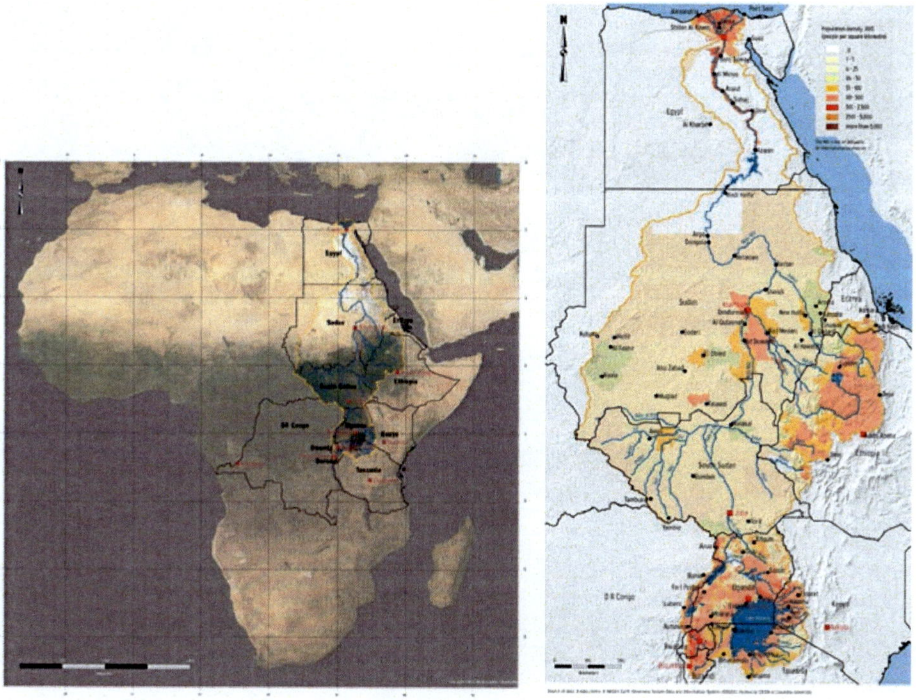

(a) (b)

Fig. 2.5 Nile River Basin's **a** countries, and **b** population density (people per square kilometer) of the 11 countries by 2017. *Source* Nile Basin Water Resource Atlas (http://atlas.nilebasin.org/)

electric dam in Africa, the Grand Ethiopian Renaissance Dam (GERD; located just over 300 mi northwest of Ethiopia's capital, Addis Ababa, Fig. 2.9b), which started in 2011 and is currently (2020) filling up, is already generating controversy between Ethiopia, Egypt and Sudan.[3]

However, not only are the demands on water increasing, but the available water supplies appear to be decreasing, e.g., through evaporation (Fig. 2.12), with environmental degradation of the upper Blue Nile catchment having increased throughout the 1980s [25] (Fig. 2.13 indicate the degradation along the White Nile). Whittington and McClelland [25] found that about 86% of the annual Nile river flow into Egypt originates from Ethiopia, and warn of significant implications for Egypt and Sudan should Ethiopia undertake any potential extractions; this issue emphasizes the need for cooperation between the three Blue Nile riparian states.

Natural factors include the changing climate, which has been the subject of numerous studies, e.g., [26, and the references therein]. Therefore, a combination of human population growth, unsustainable water usage and development, and desertification

[3]https://www.aljazeera.com/news/2020/07/egypt-seeks-urgent-clarification-gerd-rising-water-levels-200716060332700.html.

Fig. 2.6 White Nile's biodiversity

are just some of the factors that threaten the Nile's ability to supply crucially needed water to the people of the NRB.

The present-day state of the stored water variations and their relations to climate variability (e.g., El Niño and Southern Oscillation (ENSO) and the Indian Ocean Dipole (IOD)) in the Nile Basin are, however, also not well understood. Most studies dealing with the NRB to date, however, have dealt mainly with modeling the impacts of climate change (e.g., [5, 10, 13]), with very little being reported on how to monitor the spatial and temporal variations in the stored water (surface, groundwater and soil moisture) of the basin in a holistic manner, and linking them to climate variability. The reason for such few studies, e.g., [11, 23] has been partly attributed to its large size, as well as the lack of appropriate monitoring techniques that could cover such a vast spatial extent. For instance, the hydrological water balance involves the flow of surface water, the movement of deeper groundwater, and the coupling of the land, ocean and atmosphere through evaporation and precipitation. Monitoring these components requires an accuracy and completeness of geographical data coverage that challenges conventional measurement capabilities.

Fig. 2.7 Various tourists attraction sites along the banks of the White Nile

Fig. 2.8 Various scenes of the Murchison Fall along the White Nile where the river squeezes itself through a narrow stretch of 8 m

2.4 The Obsolete River Nile Treaties

For a complete and comprehensive understanding of the NRB, it is important to address the question of the Nile Treaties that were signed by Egypt and other interested parties to safeguard the interest of Egypt. These treaties' controversies have arisen in recent years, with other riparian states demanding their abrogation. This section presents an overview of the Nile Treaties and their implications for the economic development of the riparian states upstream. A comprehensive assessment of the law governing the use and management of the Nile and the development of a

(a)

(b)

Fig. 2.9 Daming the Nile river; **a** Bujagali dam along the White Nile, and **b**, Grand Ethiopian Renaissance Dam (GERD) dam along the Blue Nile (GERD; *Source* Reuters/Aljasera)

legal and institutional framework for cooperation over shared freshwater resources is presented in [24] while [22] introduces an analytic framework constructed upon the iterated Prisoners' Dilemma game to model and analyze transboundary water interactions along the Nile River.

2.4.1 Defects of Past Treaties: Legal Implications

Most of these consist only of an article in the treaties and agreements about colonial boundaries and economic territories. In chronological order [12]:

Fig. 2.10 Various scenes of pre-Bujagali dam construction. The undisturbed geomorphology of the river can clearly be seen. These sites used to be tourists attraction centers where divers charged some fee and risked their lives to swim along the dangerous river current as can be seen in the top right image. The geomorphological change of these sites become clearer in Fig. 2.11

Fig. 2.11 In contrast to Fig. 2.10, **a–d** shows the pre-Bujagali dam construction, while **e–h** show the post-Bujagali dam construction where the river geomorphology has been replaced by an artificial lake formation that has changed its ecological system

Fig. 2.12 White Nile's water exposed to evaporation

1. The Anglo Italian protocol signed on 15th April 1891.
2. The treaty between Britain and Ethiopia of 15th May 1902.
3. The agreement between Britain and the government of the independent state of the Congo signed on 9th of May 1906.
4. The 1901 agreement between Britain and Italy over the use of the River Gash.
5. The Tripartite (Britain–France–Italy) Treaty of December 13, 1906.
6. The 1925 exchange of notes between Britain and Italy concerning Lake Tsana.
7. The agreement between Egypt and Anglo Egyptian Sudan dated 7th May 1929.
8. The 1959 Nile Waters Agreement (between Egypt and Sudan).

Let us closely examine these treaties.

The Anglo-Italian Protocol of April 15, 1891: Only Article III of this treaty refers to the Nile water. The remaining articles define the colonial territorial claims of Great Britain and Italy in East Africa. Article III states that the Italian government engages not to construct on the Atbara River, in view of irrigation, any work, which might sensibly modify its flow into the Nile. This article of the treaty is curious in light of the fact that neither this river flowed in the territory claimed by Italy nor was Italy colonizing a country near the Atbara River, in order to have a claim over the river. However, the reference to the Atbara River on the part of Britain made some sense since the Sudan and Egypt, through which the Atbara flows, were within the British colonial territory. The reason for Italy to sign such an agreement foregoing its

Fig. 2.13 Degraded vegetation along the White Nile's river bank

irrigation development without receiving any benefit in return is unclear. Moreover, for Great Britain to be interested in including this reference in a treaty with a country at a distance of some thousands of kilometers from the Atbara River makes the essence of the agreement more irrelevant.

It appears that the intent of the treaty was not the use of the Nile water, but to establish a colonial boundary. Given this context, the treaty cannot be seen as an agreement over property rights to the river. Even if, say, the treaty is assumed to define use rights to the river, what was meant by the term "sensibly modify the flow into the Nile"? The volume of Atbara water used upstream to be considered, as a sensible modification by downstream users of Britain colonies is undefined.

The language used is too vague to provide the parties with clear property rights and guarantees for water use. It is not unreasonable, therefore, for the remaining riparian states to see no reason for this treaty to provide a historical base for binding present and future cooperation on Nile water use.

The Treaty between Great Britain and Ethiopia of May 1902: The aim of this treaty was to establish the border between Ethiopia and the Sudan. One of its articles, Number III, relates to the use of Nile water. The English version, as reviewed by Britain and later by the Sudan, reads as follows: "*His Majesty the Emperor Menilik II, King of Kings of Ethiopia, engages himself towards the Government of His Britannic Majesty not to construct or allow to be constructed any work across the Blue Nile, Lake Tana, or the Sobat, which would arrest the flow of their waters except in agreement with His Britannic Majesty's Government and the Government of Sudan*" The Amharic version, however, gave a different meaning and understanding to Ethiopia and "was never ratified by this country". The treaty was understood by Ethiopia as follows: "*...Securing and maintaining the prior agreement of Britain before construction of any work on the Nile tributaries; not to stop (arrest) the flow of the Nile rivers did not mean not to use; and, that the treaty was made between Britain (colonizer of the Sudan) but not with the Sudan as it was under the colonial power of Britannia. As Britain is no longer ruling the Sudan, this agreement does not hold at present.*"

The 1902 agreement has been the most controversial treaty in the history of Nile agreements as both parties claimed that their own understanding of the treaty was correct. And, not only has the claim remained controversial but it has also been the cause of disputes, which threaten the socio-political and economic dynamics of the basin environment and efforts toward future cooperation. This is because, first, referring to this agreement, Sudan has argued that Ethiopia should not use the Nile water without the permission of the Sudan. Second, the Sudanese claim has been supported by Egypt with the possibility of military retaliation if Ethiopia used the Nile water. Third, there is the possibility that this threat has played some role in Ethiopia's extremely poor record in food production, though poor agricultural policies have no doubt played the major role [4].

This controversy remained a threat to present and future cooperation over the Nile waters. By forbidding the Nile water from being arrested by Ethiopia, the treaty inadvertently advocated the principle of sustainable development. Whether or not its intent was out of concern over the impact of development of Nile waters on the environment is subject of much debate. Stopping a flow of a river creates an artificial Lake that destroys ecological systems (see e.g., Fig. 2.11) and often results in the relocation of people. However, it is unrealistic to believe that this treaty was predictive enough to anticipate the impact of water impoundment on biophysical and social systems in the basin, looking at its objectives and the time it was constituted.

The Agreement between Britain and the Government of the Independent State of the Congo on 9th May 1906: This was an agreement on the colonial boundary of the Congo between Britain and Belgium. The Congo being called an 'independent state' when the treaty was signed by the Government of Belgium on behalf of this country was hypocritical. Article III of the agreement was about the

Nile waters and it stated that: "*The Government of the independent state of the Congo undertakes not to construct, or allow to be constructed, any work over or near the Semliki or Isango River, which would diminish the volume of water entering Lake Albert except in agreement with the Sudanese Government*".

Belgium signed this unfair agreement on behalf of the Congo, despite the agreement entirely favoring the downstream users of the Nile waters and restricting the people of the Congo foregoing its Nile water use. The agreement did not require downstream users to consult the upstream countries for anything that they might do to the Nile waters. In this treaty, it is difficult to find any incentive for riparian states that might enhance their future cooperation since it involved neither the principle of equitable water use nor the approach of integrated water development.

The 1901 Agreement between Britain and Italy over the Use of the River Gash: The agreement states that "*The Government of Erythraea, while recognizing all its rights on the waters of the Gash and having regard to the requirements of the Colony, sees no difficulty in declaring that, in so far as the regime of the waters of that river are concerned, it will regulate its conduct in accordance with the principles of good neighbourship*". Evidence is scarce, however, on whether or not this treaty bound the parties to the agreement. Nevertheless, of all the treaties and agreements made during the colonial period, it could be said that this agreement was the most equitable. Because of difficulty in ensuring equitable water use, the agreement was defined and reinforced later by the "the Anglo-Egyptian Exchange of Notes" with subsequent detailed arrangements of 1925.

The exchange of notes included technical provisions suitable for practical implementation as follows: "*Quantified allocation, to each party, of water from the river Gash, Flow regime terms and conditions for water allocation, and the Amount of annual payment by the Sudan to Eritrea as a proportion of Sudanese revenues from irrigated cultivation at Kassala*". This treaty is held by some of the riparian states as not being binding because the colonial signatory governments are no longer present in the Nile basin. Nevertheless, it can be used as a basis for the effort being undertaken to establish cooperation among the riparian countries.

The Tripartite (Britain–France–Italy) Treaty of December 13, 1906: Article 4 (a) of this treaty dealt with the use of the Nile water in Ethiopia's sub-basin. It states: "To act together …to safeguard; the interests of Great Britain and Egypt in the Nile Basin, more especially as regards the *regulation of the waters of that river and its tributaries (due to consideration being paid to local interests) without prejudice to Italian interests*". This treaty denied "**the absolute sovereignty**" of Ethiopia over its water resource. It resulted in Ethiopia immediately notifying its rejection of the agreement by indicating that no country had the right to stop it from using its own water resources. Neither Ethiopia's military power nor its international political and economic influence was strong enough to protect Ethiopia's sovereign rights over its water resource. Ethiopia's rejection of this agreement was a revision, if not retraction, of the May 15, 1902 treaty signed between Ethiopia and Britain.

The 1925 Exchange of Notes between Britain and Italy Concerning Lake Tana: Britain and Italy had signed an agreement in 1919 over Lake Tana, of Ethiopia, which read in part as follows: "In view of the predominating interests of Great Britain in respect of the control of the waters of Lake Tana, Italy offers Great Britain her support, in order that she may obtain from Ethiopia the concession to *carry out works of barrage in the Lake itself* ...".

In 1925 it was expanded to say that:

> ...Italy recognizes the prior hydraulic rights of Egypt and the Sudan ...not to construct on the head waters of the Blue Nile and the White Nile (the Sobat) and their tributaries and influence any work, which might sensibly modify their flow into the main river.

Ethiopia opposed this agreement and notified the parties of its objections as follows: To the Italian government, Ethiopia said:

> The fact that you have come to an agreement, and the fact that you have thought it necessary to give us a joint notification of that agreement, make it clear that your intention is to exert pressure, and this in our view, at once raises a previous question. This question, which calls for preliminary examination, must therefore be laid before the League of Nations.

And, to the Britannia government Ethiopia said:

> The British Government has already entered into negotiations with the Ethiopian Government in regard to its proposal, and we had imagined that, whether that proposal was carried into effect or not, the negotiations would have been concluded with us; we would never have suspected that the British Government would come to an agreement with another Government regarding our Lake.

When an explanation was required from the British and the Italian governments by the League of Nations, they denied challenging Ethiopia's sovereignty over Lake Tana. This notwithstanding, however, there was no explicit mechanism enforcing the agreement. A reliable and self-enforcing mechanism that can protect the property rights of each stakeholder is essential if the principle of economically and ecologically sustainable international water development is to be applied".

2.4.2 The 1929 Treaty: Britain both Referee and Player

The Agreement between Egypt and Anglo-Egyptian Sudan of 7th May 1929:

This agreement included the following terms:

(a) Egypt and Sudan utilize 4 billion cubic meters of the Nile flow per year, respectively;
(b) The flow of the Nile during January 20 to July 15 (dry season) would be reserved for Egypt;
(c) Egypt reserves the right to monitor the Nile flow in the upstream countries;
(e) Egypt assumed the right to undertake Nile river related projects without the consent of upper riparian states.
(f) Egypt assumed the right to veto any construction projects that would affect her interests adversely.

It is important to note that Egypt was still under British influence in 1929 and neither Sudan nor the remaining riparian states, aside from Ethiopia, were independent. The roles of both referee and player were taken, in the process of this agreement, by Britain in the name of its colonial territories in order to favor one, Egypt, by limiting the rights of the Sudan, and by rejecting those of the remaining riparian states. The agreement became the basis for the next agreement, called *The 1959 Nile Water Agreement*, which opened a door for Egypt and the Sudan to acquire rights to Nile water resources and for the full utilization of these waters by developing the Aswan high Dam, with its huge impact on the biophysical and social (disposition of human settlement) environment of the basin.

In May 1929, the Egyptian Prime Minister and the British High Commissioner to Egypt, signed the Nile River Treaty. Although the 1929 agreement was concluded between the British High Commission in Cairo and the Egyptian Government, while the 1959 treaty was between Egypt and the Sudan governments, *the pacts bind Uganda, Tanzania, and Kenya to date. They bar the three countries from using Lake Victoria waters, the mother of the Nile, without Egypt's permission*. The 1929 treaty was signed at a time when Lake Victoria was thought to be the source of river Nile. The treaty was a culmination of previous agreements made in 1889, 1891 and 1902—all which merely satiated British selfish interests against the Italian and (later) the Ethiopian Government. All ended up securing, for the 96% Egyptians crowded along the Nile, the use of 48 billion cubic meters of the water each year while Sudan got 4 billion cubic meters. These agreements were not sensitive to the water needs of the other eight countries in the basin (Tanzania, Uganda, Kenya, Rwanda, Burundi, Ethiopia, and the Democratic Republic of Congo).

The treaty acknowledged Egypt's natural and historical rights to the Nile waters. "Without the consent of the Egyptian Government, no irrigation or hydro-electric works can be established on the tributaries of the Nile or their Lakes, if such works can cause a drop in water level harmful to Egypt", so says a section of the treaty.

The original agreement concluded in 1929 heavily favored Egypt's "historic rights" and gave Egypt and Sudan 100% use and control of the waters from Lake Victoria. Even in the event of floods, other riparian states had no permission to use this water. Thus, in 1949, Owen Falls Treaty prohibited Uganda from using Lake Victoria water for irrigation in Karamoja.

2.4.3 *The 1959 Treaty: Shocking Clauses!*

The Nile Treaty of 1959: This was an agreement between the Sudan and Egypt for full utilization of Nile waters. In the 1950s, Egypt was planning the Aswan High Dam project to collect the entire annual flow of the Nile water. The objective of the 1959 Agreement was to gain full control and utilization of the annual Nile flow. One of the financiers of the project, the International Bank for Reconstruction and Development (IBRD) required a secure water allocation for Sudan and compensation for the population to be dislocated due to the project. In 1956 Sudan had become an independent country and wanted previous agreements, which it saw as being unfair, to be changed, so that it could pursue another agreement on the use of Nile water with Egypt. At the beginning of the talks, both Sudan and Egypt claimed large areas of irrigable land and amounts of Nile water: Sudan claimed 44 billion cubic meters of Nile water to irrigate 2.22 million hectares, while Egypt claimed even more water than Sudan that irrigates 7.1 million hectares. The debate over the claims delayed the agreement, but whether or not Sudan agreed, the construction of the Aswan High Dam was seen as a development priority for Egypt. Although neither the Sudan nor Egypt were contributors to the Nile water but only users, the agreement for the full utilization of the Nile Waters was signed in 1959, between Sudan and Egypt. The agreement contained the following main points:

1. The controversy on the quantity of average annual Nile flow was settled and agreed to be about 84 billion cubic meters measured at Aswan High Dam, in Egypt.
2. The agreement allowed the entire average annual flow of the Nile to be shared between the Sudan and Egypt at 18.5 and 55.5 billion cubic meters, respectively, with the annual lose due to evaporation and other factors pegged at 10 billion cubic meters.
3. Against this backdrop in the sharing of the waters of the Nile, which disregarded the contribution of the upper riparian states, changing the status quo has been an issue of hairsplitting interpretation. This is why sound water resources management is very crucial.

The general observation from the treaties discussed so far is that, countries have tried to make interpretations in their own words and contexts. It is, however, important for countries to adopt new ways that will help them forge ahead. This will be achieved through respect of the object and the purpose of treaties as established by the International Law. According to the Vienna Law of treaties, a treaty shall be interpreted in good faith in accordance with the ordinary meaning to be given to the terms of the treaty in that context and in the light of its object and purpose.

The object and the purpose of any agreement therefore must serve the interests of all the parties. In conclusion, as discussed above, none of the treaties and agreements dealing with the use of Nile waters signed during the colonial period involved all the riparian countries and they did not deal equitably with the interests of these riparian states. Also they did not take in to account the impact of water development on the basin's social and biophysical environment.

Some Important Terms of the 1959 Nile Treaty:

Egypt and Sudan signed a Treaty in Cairo on November 8, 1959. Its title was "*United Arab Republic and Sudan Agreement (with annexes) for the Full Utilization of the Nile Waters*". The agreement came into force on December 12, 1959. The 1959, agreement set Egypt's share at 55.5 bcm per year and Sudan's 18.5 bcm per year.

Agreement on a Unified View:

1. The Treaty said in part: "If it becomes necessary to hold any allegations with any riparian state, outside the boundaries of the two republics, the governments of the Sudan Republics and the United Arab Republic (Egypt) shall agree on a unified view after the subject is studied by the said technical commission. The said unified view shall be the basis of any negotiations by the commission with the said states".
2. The Sudan–Egypt treaty also notes: "If the negotiations result in an agreement to construct any works on the river, outside the boundaries of the two republics, the joint technical commission shall, after consulting the authorities in the governments of the States concerned, draw all the technical execution details and the working maintenance arrangements".
3. The treaty also gives Egypt and Sudan powers to "supervise the carrying out of technical agreements" by other riparian states.
4. According to the agreement, any other country that wishes to lay claims on the Nile waters, and that includes Lake Victoria, will only get a share, if Egypt and Sudan agree. The treaty reads: "As the riparian states, other than the two republics, claim a share in the Nile Waters, the two republics have agreed that they shall jointly consider and reach one unified view regarding the said claims".[4]

Deductions in Equal Parts:

1. The Treaty further says: "And if the said consideration results in the acceptance of allotting an amount of Nile water to one or the other of the said states, the acceptance amount shall be deducted from the shares of the two republics in equal parts, as calculated at Aswan. Even then, any water taken out will have to be monitored by a technical commission (which) shall make the necessary arrangement with the states concerned".
2. Little is known about an annex document in which Sudan gave a "water loan" to Egypt. The annex reads: "The Republic of the Sudan agree in principle to give water loan from the Sudan's share [to Egypt] in order to enable the latter to proceed with her planned programs for agricultural expansion".

Sharp focus has been put on the Nile water agreement but it is the 1959 treaty signed by Egypt and the Sudan that has shocking clauses. According to the treaty, none of the Lake Victoria nations can use the Lake's water without supervision by a "technical commission" appointed by Egypt and Sudan. Further, they can only be allocated a certain quota by the two. In one clause, the two nations agree, "if the other riparian states claim a share in the Nile waters, the two republics shall jointly consider and reach one unified view. And…the accepted amount shall be deducted from shares of the two". The treaty authorizes a technical commission to "make the necessary arrangement with the states concerned to ensure that their water consumption shall not exceed the amounts agreed upon".

[4]Daily Nation newspaper, March 28, 2002.

Consequences of the Nile Treaty: The implementation of the Nile Treaty means the following consequences:

According to the treaty, Egypt's approval must be obtained if any project is to be undertaken on the Nile River. Archival information shows that as early as 1949 when the Owen Dam was constructed, the British Government wrote to the Uganda Electricity Board telling it that the project would only go ahead "*If the Egyptian Government approved it*". To date, there is an Egyptian resident engineer at the dam.

A document titled "Exchange of Notes Constituting An Agreement Between the Government of Great Britain and Northern Ireland and the Government of Egypt regarding the Construction of the Owen Falls Dam, Uganda signed at Cairo on December 5, 1949" is one of the many agreements on the issue. The pack was signed by Britain and Egypt and it binds the former colonies.

Military Coup against Dissenting Foreign Governors: When President Gamal Abdel Nasser begun building the Aswan High Dam, contrary to the agreement, Sudan withdrew from the treaty. Later, there was a military coup in Sudan. The new Sudanese Government renegotiated the treaty resulting in the 1959 agreement, which increased Egypt's share to 55.5 billion cubic meters (or 82% of the annual flow), while Sudan's share was increased to 18.5 billion cubic meters (or 18%). Egypt is today accused of being behind the coup that followed the standoff and saw Mr. Jaffer Numeiri take power in Sudan. Once again, Egypt and Sudan ignored Kenya and all the other basin countries and went ahead to irrigate 6 million acres and 2.75 acres, respectively.

Threat of Usage of Force: To sustain this selfishness, the two countries have used the threat of force to intimidate other countries from utilizing the waters flowing into the Nile. For instance, in May 1978, the then Egyptian Minister of Irrigation and Land Reclamation, Mr. Abd a-Azim Abu al-Ata warned countries in the Horn of Africa that any attempt to control the source of the Nile was "an act of aggression".

Such has been the sensitivity with which Egypt treats any attempt by upstream countries to use the Nile waters for their needs. Even Kenya, which contributes two-thirds of the water flowing into Lake Victoria, is not spared from this threat. "The day that Kenya decides to use water from Lake Victoria, we'll have less water in Egypt. One litre of water used for their irrigation will be reduced from water received in Cairo", a senior official in Egypt was quoted by the Financial Times of London in January 1988. The then Egyptian Ambassador to Kenya, Dr Rifaat al-Ansary reacted that his country had nothing to do with the Financial Time's statement and that there was no controversy. But as the Ambassador denied any controversy, other countries were not happy with the fact that his country, and Sudan, in 2001 insisted that the treaty could only be renegotiated, if other Nile Basin countries first recognized it as binding.

Nevertheless, the agreement goes against the internationally recognized riparian states' rights of Kenya and the other nine Nile Basin countries. This law states that a riparian state has sovereignty over the stretch of any river within its international boundaries and may use such water as it deems fit so long as it does not infringe upon the rights of other riparian states on the same river. It could be with this law in mind that Tanzania disregarded the treaty in 1962 while Uganda constructed a second hydro-electric power generation dam on the Bujagari Falls on the Nile (Fig. 2.9a). Kenya seems to have settled for the renegotiation option.

The Near War with Ethiopia over the Use of Nile Water: In June 1980, Egypt nearly went to war with Ethiopia after the latter opposed moves by Egyptian President Anwar Sadat to divert the Nile waters to the Sinai Desert, which is outside the Nile Basin. After making peace with Israel, Mr Sadat had promised Israelis that he would irrigate the desert. On its part, Ethiopia had threatened to obstruct the Blue Nile, a notion that made Egypt prepare for war. Egypt has also placed its hydrologists along the 5,584 km-long course of the Nile ostensibly to monitor its volume. Nonetheless, Ethiopia is in the process of filling the biggest dam in the continent; the Grand Ethiopian Renaissance Dam (GERD; Fig. 2.9b),[5] which is currently (the year 2020) causing tension between it and Egypt.[6]

The nine countries need to disregard the treaty as already done by Uganda and Ethiopia if they are to meet their rising water needs. Today, they have a higher combined population as well as population growth compared to Sudanese and Egyptians (see Fig. 2.2c and d). Though a substantial section of the eight countries' population may have alternative sources of water, this has not been adequate for the rising population.

2.4.4 The Riparian States and Egyptian Concept on use of River Nile

Meanwhile countries sharing Lake Victoria cannot use its water for development projects. Egypt and Sudan must agree that Kenya and indeed the other basin countries have as much right over the use of the waters as they do. While there is merit in the Tanzanian President's statement in 1998 that the Nile resource should be "a formidable unifying factor instead of a source of division and conflict", the nine countries should handle the controversy with more boldness. Collectively, they should disregard the treaty so that Egypt and Sudan can approach the issue in a spirit of give-and-take. The nine Basin countries should not continue honoring a treaty that neglects others' interests.

Demands to Revoke the River Nile Treaty: Sudan was the first country to demand a review of the 1929 treaty. The action brought a brief standoff between Khartoum and Cairo after Sudan got independence from Britain in 1956. In April 2002, there was a heated debate in the Kenyan Parliament about the Nile Treaties signed in 1929 and 1959, respectively. The Members of Parliament (MPs) questioned the legality of those treaties. Consequently, there was a heated debate as to whether Kenya should or should not honor the treaties. Surprisingly, the debate, just like the treaty, was pegged on the wrong assumption that Lake Victoria is actually the only

[5] https://www.internationalrivers.org/campaigns/grand-ethiopian-renaissance-dam.

[6] https://www.bbc.com/news/world-africa-50328647.

source of river Nile and the MPs, during their contribution, argued from this point of ignorance.

As East African politicians question the legality of a treaty that gives Egypt and the Sudan absolute rights over river Nile waters, one issue is quickly coming to the fore: How will the two nations react given their enormous economic and agricultural interests? In Kenya, the then Energy Minister Raila Odinga and then minister for roads and public works and Kisumu East Town MP Gor Sungu questioned the legality of the 1929 and 1959 Nile waters Treaties while the same issue have cropped up at the new East African Legislative Assembly. Mr. Odinga threw the first salvo in Kenya's parliament, calling upon the government to review the treaty and denouncing it as "*obsolete*", accusing Egypt of planning to "Export" the water to Sinai via a tunnel [27]. The story of the tunnel is little known, although it is part of a huge land reclamation project in Sinai Desert called the North Sinai Agricultural Development project. Since 1987, the project has been diverting Nile water to agricultural developing plots west of Suez Canal,[7] with allegations that the water may end up in Israel.

Debate on a 1929 treaty that bound countries in the upstream of River Nile from "touching" its waters has been simmering for a while now. The treaty gave Egypt and Sudan absolute right to use 100% of the river's water. The debate re-surfaced in April 2002, when a legal consultant with the United Nations Environmental Program, Professor Charles Okidi, implored the three East African countries to ignore the treaty and go ahead and utilize the waters of Lake Victoria. He asked East African countries to ignore the 1929 treaty.[8] Yet, unlike this treaty, which was binding on British colonies, the 1959 one is between two independent nations—Egypt and Sudan—without reference to other concerned countries. The then Energy Minister Raila Odinga (later Kenya's prime minister 2007–2013) saying the treaty was "unfair" had earlier made a similar call. By and large, the Kenya government seems reluctant to be drawn into the controversy. It is not clear why Government officials have continued to refer to this flawed water use arrangement as an "Agreement".

For one, it was made before Kenya was a state, as we know it. "This treaty only benefits Egypt. We cannot sit back while we have water we cannot use to irrigate our land. Why should we preserve our water for Egypt"? Mr. Odinga asked Parliament? Experts predict minimal impact on the Nile if East African countries harvest Lake Victoria water for local use. Parliament was right in demanding a renegotiation of the 1929 Nile Water Agreement between Britain and Egypt over the use of Nile's waters. A treaty they did not sign in some distant past should not bind Kenyans.[9] In June 2003, members of the East African Legislative Assembly (EALA) called for a renegotiation on the Nile Treaty with a view to selling waters of the Nile to Sudan and Egypt.

---

[7] See https://earthobservatory.nasa.gov/images/1495/egypts-north-sinai-agricultural-development.

[8] Daily Nation newspaper, March 29, 2002.

[9] Daily Nation newspaper, October 19, 2001.

Source of Acrimony: With approximately 300 million people depending on the Nile for survival, the river is seen today as possible source of acrimony in East Africa. At the Regional Regulative Assembly, Uganda member Yona Kanyomozi questioned the importance of the 1949 Owen Fall Dam Agreement and wondered why Egypt was using more water than agreed. He also wonders why Egypt does not participate in conservation of Lake Victoria. He said, "What bothers me is that when Uganda developed a scheme to divert some of the Nile waters to Karamoja for irrigation, the plan was opposed by Egypt, yet for them they can do anything with the Nile waters".[10]

How the treaty binds other riparian countries is doubtful. But since then Egypt—the most powerful military power in the region - has been warning that it could go to war over Nile waters. In 1979, Egyptian President Anwar Sadat said: "The only matter that could take Egypt to war again is water". And in 1988, then Egyptian foreign minister Boutros-Ghali, who latter became UN Secretary-General, went a step further. He said: "The next war in our region will be over waters of the Nile, not politics".[11]

Such sentiments have for the years caused Nile nations eager to renegotiate the Nile water treaties to tread with caution. They are aware that for Egypt, in particular, Nile waters are a matter of life and death. From as early as 1898, when a French expedition tried to control the headwaters of the of the Nile, Egypt has discussed the Nile with passion. In 1958, it sent a military expedition to a disputed territory with the Sudan, pending negotiations over Nile waters. This eventually led to a coup and installation of a pro-Egyptian government in the country. From 1978 onwards tensions have persisted between Egypt and Ethiopia after the latter proposed to construct dams on headwaters of the Blue Nile. That was when President Sadat warned that Egypt's next war would be over the Nile. But can the region go to war over water? The only recorded incident of an outright war over water was 4,500 years ago in modern southern Iraq between two Mesopotamian city-states over the Tigris–Euphrates. Scholars say that 80% of water's war consists of verbal threats and posturing by state leaders, probably aimed at their own internal constituents.

In the past 50 years, there have been only 37-recorded events in which people actually shot at each other over water along international borders. Of those, 27 were between Israel and Syria over the Jordan and Yarmouk rivers. Is the Nile meandering towards that, or is this just a cold water war?

The New Nile River Treaty: In September 2002, the Coordinator of the Uganda-based inter-state organization, the Nile Basin Initiative, informed a workshop meeting in Kisumu, Kenya, that a new legal framework was to be signed in two months' time by 10 riparian countries, which include Burundi, Democratic Republic of Congo, Egypt, Ethiopia, Eritrea, Kenya, Rwanda, Sudan, Tanzania and Uganda.

The Nile Basin Initiative was launched in 1999 following consultations among the 10 riparian countries. Although Egypt is today backing a new initiative among the Nile Basin riparian states, which might redistribute more equitably the river's

[10] John Mbaria: Revoke obsolete Nile treaty. Daily Nation newspaper, March 29, 2002.

[11] See footnote 10.

water usage rights, it is not clear how far this will go. Note 2.1[12] below provides a historic perspective of the basin's initiatives.

Note 2.1 (*A Brief History of the Nile Basin Initiative (NBI), 1992–2001*) 1992: In 1992 the Council of Ministers (Nile-Com) of Water Affairs of the Nile Basin States launched an initiative to promote cooperation and development in the Basin. Six of the riparian countries—the Democratic Republic of Congo (D.R.C), Egypt, Rwanda, Sudan, Tanzania, and Uganda; formed the Technical Cooperation Committee for the Promotion of the Development and Environmental Protection of the Nile Basin (TECCONILE). The other four riparian states participated as observers. Within this framework, the Nile River Basin Action Plan (NRBAP) was prepared with support from Canadian International Development Agency (CIDA). One of the projects (Project 3), whose objective is to develop a co-operative framework for management of the Nile, was endorsed by all countries during the 3rd meeting of the Nile-COM (in Arusha, 9–11 February, 1995) and is being implemented with UNDP funding.

1995: The World Bank was asked by the COM to play a lead role in co-ordinating the inputs of external agencies to finance and implement the NRBAP.

March 1997: The request from COM to the World Bank was reiterated.

June 1997: The request was accepted by the World Bank. The World Bank proposed that it undertakes the task in partnership with UNDP and CIDA, and that a review and consultation process be launched prior to a Consultative Group-style donor meeting.

November 1997: A review of the Nile River Basin Action Plan was undertaken with the support of an International Advisory Group (IAG). The IAG meeting of international experts was held at Coolfont near Washington, D.C., USA.

January 1998: A Special Review Meeting was held in Cairo with senior officials attending from the riparian countries to discuss the Draft Review Report and move towards the definition of a priority, revised Action Program. Discussions converged on two complementary ideas, which provide a structure for the revised Action Program—a shared vision, and action on the ground.

March 1998: The 6th COM Meeting, held in Arusha, Tanzania, was attended by eight riparian countries (all except Eritrea and D.R. Congo). It was a major milestone in Nile co-operation. The meeting considered the Revised Action Plan.

July 1998: The 1st meeting of the Nile Technical Advisory Committee (Nile-TAC) was held in Dar es Salaam, Tanzania, under the chairman ship of Mr. Meraji Msuya.

21–22nd Sept 1998: The 2nd meeting of the Nile-TAC, held in Arusha, Tanzania. Meeting agrees on the Terms of Reference of the Nile-TAC, its Rules of Procedure, its Policy Guidelines and a Plan of Action.

23–24th Sept 1998: Extra-Ordinary meeting of the Council of Ministers (Nile-COM) was held in Arusha, Tanzania.

22nd Feb 1999: Extra-ordinary meeting of the Nile Basin Council of Ministers, was held in Dar es Salaam, United Republic of Tanzania. Agreed minutes were prepared which formally established the Nile Basin Initiative.

[12]Source: The Nile Basin Initiative Secretariat (Nile-SEC).

23–24th Feb 1999: 3rd meeting of Nile-TAC in Dar es Salaam, Tanzania.

4–7th May 1999: Nile-TAC held a Strategic Planning and training workshop in Sodere, Ethiopia, which initiated the preparation of projects within the Basin-wide Shared Vision Program.

10–14th May 1999: 4th meeting of the Nile-TAC in Addis Ababa, Ethiopia. The formation of Working Groups to develop Project Concept Documents for the Shared Vision Program.

12–13th May 1999: 7th meeting of the Council of Ministers for Water Affairs in the Nile Basin States (Nile-COM), Addis Ababa.

15th May 1999: First meeting of the Eastern Nile Council of Ministers.

1st June 1999: The Nile Basin Initiative Secretariat started operations in the former TECCONILE building in Entebbe, Uganda, with Mr. Meraj Msuya as Executive Director.

30th August–3rd Sept 1999: 5th meeting of the Nile-TAC in Entebbe and second meeting of the Working Groups to develop Project Concept Documents for the Shared Vision Program in preparation for the International Consortium for Cooperation on the Nile (ICCON) meeting.

3rd Sept 1999: The official opening of the Nile Basin Secretariat offices in Entebbe, Uganda.

13–18th Dec 1999: National experts from various sectors meet at the Nile Basin Secretariat in Entebbe to share ideas and work together in initiating project studies under the Shared Vision Program.

24–26th Jan 2000: 6th Nile-TAC meeting is held at the Nile Basin Secretariat in Entebbe to agree on the work plan for the priority project proposals for the Nile Basin.

18th March 2000: Senior officials from the Nile Basin Initiative present their "Shared Vision" for the time to the international community during the Second World Water World Forum in the Hague, the Netherlands, attended by over 4,000 participants from all over the world.

23rd March 2000: 7th Nile-TAC meeting is held in Deft, the Netherlands to review alternative work plans and meeting schedules towards the first ICCON.

31st July–3rd August 2000: 8th Nile-TAC meeting is held in Khartoum, Sudan, as a precursor to the 8th Nile-COM meeting, to finalize proposals for the priority projects under preparation.

4–5th Aug 2000: 8th Nile-COM meeting is held in Khartoum. It endorses the priority projects being prepared under the Shared Vision Program and instructs the Nile-TAC to complete preparation of full project documents to be submitted to the Nile-COM in the first week of December 2000. The Council confirms February 2001 and Geneva, Switzerland as the time and venue of the first ICCON

21–25th Aug 2000: National experts from the Nile Equatorial Lakes region meet in Entebbe to share ideas on a series of possible joint projects for the region, to be prepared under the Nile Equatorial Lakes Subsidiary Action Program (NELSAP).

20–29th January 2001: National experts from the Eastern Nile meet in Addis Ababa to share ideas about joint projects being prepared under the Eastern Nile Subsidiary Action Program (ENSAP).

6th September 2002: Mr. Antoine Sendama, Co-ordinator of Nile Basin Initiative, announces that a new legal framework—a revision of the Nile Treaty—is to be signed by 10 riparian countries (nations neigbouring water masses). The countries are Burundi, Congo, Egypt, Ethiopia, Eritrea, Kenya, Rwanda, Sudan, Tanzania and Uganda.

2.5 Concluding Remarks

One fact is outstanding. The Nile treaties of the last century (1929 and 1959) are obsolete! They have outlived their usefulness if they had any and as such, it is time that they are done away with. A probable way forward would require the 11 countries, all of which are stakeholder of the Nile River Basin to come up with a management framework of the Nile basin that will ensure equitable sharing of its resources. This would lead to a sustainable utilization of the Nile's precious resource for the benefit of its more than 300 million inhabitants.

References

1. Appelgren B, Klohn W, Alam U (2000) Water and agriculture in the Nile Basin. Nile Basin initiative report to ICCON, Food and Agriculture Organization (FAO) of the United Nations, Rome
2. Awange JL, Sharifi MA, Ogonda G, Wickert J, Grafarend E, Omulo M (2008) The falling lake victoria water levels: GRACE, TRIMM and CHAMP satellite analysis of the lake Basin. Water Resour Manag 22:775–796. https://doi.org/10.1007/s11269-007-9191-y
3. Agutu NO, Awange JL, Ndehedehe C, Kirimi F, Kuhn M (2019) GRACE-derived groundwater changes over greater horn of Africa: temporal variability and the potential for irrigated agriculture. Sci Total Environ. https://doi.org/10.1016/j.scitotenv.2019.07.273
4. Awange JL (2021) The greater horn of Africa's hydroclimate. Potential for agriculture amidst extreme drought. Springer Nature International, Berlin
5. Awange JL, Ogallo L, Kwang-Ho B, Were P, Omondi P, Omute P, Omulo M (2008) Falling lake victoria water levels: is climate a contribution factor? Clim Change 89:281–297. https://doi.org/10.1007/s10584-008-9409-x
6. Awange JL, Ong'ang'a O (2006) Lake Victoria-Ecology. Resource of the Lake Basin and Environment. Springer, Berlin
7. Awange JL, Forootan E, Kuhn M, Kusche J, Heck B (2014a) Water storage changes and climate variability within the Nile river basin between 2002 and 2011. Adv Water Resour 73:64–78. https://doi.org/10.1016/j.advwatres.2014.06.010
8. Awange JL (2021) Lake Victoria monitored from space. Springer Nature International, Berlin
9. Conway D, Hulme M (1993) Recent fluctuations in precipitation and runoff over the Nile subbasins and their impact on main Nile discharge. Clim Change 25:127–151. https://doi.org/10.1007/BF01661202
10. Beyene T, Lettenmaier DP, Kabat P (2010) Hydrologic impacts of climate change on the Nile river basin: implications of the 2007 IPCC scenarios. Clim Change 100:433–461. https://doi.org/10.1007/s10584-009-9693-0

11. Bonsor HC, Mansour MM, MacDonald AM, Hughes AG, Hipkin RG, Bedada T (2010) Interpretation of GRACE data of the Nile Basin using a groundwater recharge model. Hydrol Earth Syst Sci Discuss 7:4501–4533. https://doi.org/10.5194/hessd-7-4501-2010
12. Bullock A, et al (1995) Report on LVEMP tasks 11, 16 and 17, FAO
13. Conway D (2005) From headwater tributaries to international river: observing and adapting to climate variability and change in the Nile Basin. Global Environ Change 15:99–114. https://doi.org/10.1016/j.gloenvcha.2005.01.003
14. Hamouda MA, Nour El-Din MN, Moursy FI (2009) Vulnerability assessment of water resources systems in the Eastern Nile Basin. Water Resour Manag 23:2697–2725. https://doi.org/10.1007/s11269-009-9404-7
15. Martin W (2016) Nile Waters. Adventures of a geomorphologist at large. Springer International Publishing, Saharan Sands
16. Mohamed YA, Savenije HHG, Bastiaanssen WGM, van den Hurk BJJM (2006) New lessons on the Sudd hydrology learned from remote sensing and climate modeling. Hydrol Earth Syst Sci 10:507–18. https://doi.org/10.5194/hess-10-507-2006
17. Molen I, Hildering A (2005) Water: cause for conflict or co-operation? ISYP J Sci World Affairs 1(2):133–143
18. Owor M, Taylor RG, Tindimugaya C, Mwesigwa D (2009) Rainfall intensity and groundwater recharge: empirical evidence from the Upper Nile Basin. Environ Res Lett 4:035009. https://doi.org/10.1088/1748-9326/4/3/035009
19. Nile Basin Water Resources Atlas (2017). http://atlas.nilebasin.org/
20. Negm AM (2017) The Nile river. Springer International Publishing, Berlin
21. Sahin M (1985) Hydrology of the Nile Basin. Developments in water Science No. 21. Elsevier, New York 575 p
22. Samaan MM (2019) The Nile development game. Tug-of-war or benefits for all? Springer International Publishing, Berlin
23. Senay GB, Asante K, Artan G (2009) Water balance dynamics in the Nile Basin. Hydrol Process 23:3675–3681. https://doi.org/10.1002/hyp.7364
24. Wehling P (2020) Nile water rights. An international law perspective. Springer, Berlin
25. Whittington D, McClelland E (1992) Opportunities for regional and international cooperation in the Nile Basin. Water Int 17(3):144–154. https://doi.org/10.1080/02508069208686134
26. Yates DN, Strzepek KM (1998) Modelling the Nile Basin under climate change. J Hydrol Eng 3(2):98–108. https://doi.org/10.1061/(ASCE)1084-0699(1998)3:2(98)
27. Aseto O, Ong'ang'a O (2003) Lake Victoria (Kenya) and its environs: Resource. Opportunites and Challenges. Africa Herald Publishing House, Kendu Bay
28. Othieno H, Awange JL (2016) Energy resources in Africa. Distribution, opportunities and challenges. Springer International Publishing AG

Chapter 3
Weighing the Nile's Waters from Space

By far, the most abundant and available source of fresh water is underground water supplies or wellsprings known as aquifers. Therefore, scientists and natural resource managers are very interested in tracking how these underground reservoirs of fresh water are changing with time. Launched in 2002, a pair of identical satellites that make up NASA's Gravity Recovery And Climate Experiment (GRACE) are tackling the problem in an unexpected way: they are weighing Earth's fresh water from space. Serving as a sort of "divining rod" in space that moves in response to a powerful, fundamental force of nature—gravity—the satellites respond to changes in Earth's gravitation field that signal shifts in the movement of water across and under Earth's surface.

—NASA (https://earthobservatory.nasa.gov/features/WeighingWater)

3.1 Water Storage Changes: Space Measuring Techniques

Nile River Basin (NRB)'s overall water movement expressed through the hydrological cycle is characterised by a simple water balance equation: $\triangle S = P - E - Q$, where $\triangle S$ is the basin's total water storage (TWS representing the sum of groundwater, soil moisture, vegetation, and surface water), P the basin's precipitation, E its evapotranspiration, and Q its runoff. Due to its sheer size of area of 3,400,000 km^2 (see Sect. 2.3), monitoring changes in $\triangle S$, P, E, Q through "boots on the ground" ground-based (in-situ) observations is practically impossible and a daunting task indeed. On the one hand, the in-situ (ground-based products, e.g., [89]) data may be inaccessible while on the other hand, the mearger accessible data where available, might be inconsistent or suffer from missing data.

To supplement in-situ observations and water resources management over large and poorly gauged NRB, ***satellite remote sensing SRS*** offers the only possibility

J. Awange, *The Nile Waters*,
https://doi.org/10.1007/978-3-030-64756-8_3

of rigorously monitoring changes in these components (ΔS, P, E, Q) of the hydrological cycle within the NRB as will be shown in the Chapters ahead. To this end, the basin's total water storage changes $\Delta S/dt$, where dt is the temporal differences, e.g., changes between consecutive months can now accurately be measured using the Gravity Recovery And Climate Experiment (GRACE, discussed in Sect. 3.3.3) mission launched in 2002, see e.g., [4]. It enables the closure of NRB terrestrial water budget by providing large-scale quantitative estimates of changes in TWS fields ΔS on an approximately monthly basis.

NRB's precipitation P can be remotely sensed using both infra-red (IR) and microwave (MW) sensors on board geostationary satellites and low-Earth orbiting satellites (e.g., Tropical Rainfall Measuring Mission, TRMM, Kummerow et al. [22]). TRMM satellite, jointly managed by National Aeronautics and Space Administration (NASA) of the United States and Japan Aerospace Exploration Agency (JAXA) was designed to monitor and study tropical rainfall. It has the advantage of providing four dimensional (X, Y, Z, t) rainfall and latent heat data over inaccessible areas such as the ocean, un-sampled terrains, etc. The primary rainfall parameters on TRMM are the TRMM Microwave Imager (TMI), the Precipitation Radar (PR) and the Visible and Infrared Radiometer System (VIRS) [22]. Infrared-based rainfall estimates have a longer historical record and have higher spatio-temporal resolution than the more accurate (and shorter temporal record) Microwave-based estimates but suffer from higher uncertainty due to various problems [100, 101]. Awange et al. [5] discusses uncertainties in satellite based derived precipitation over Africa while Awange et al. [6] compares the performance of merged products in Africa and Australia. In this book, various satellite based precipitation will be employed and as such, they are discussed at their point of use. Evapotranspiration E on its part can be monitored using MODerate Resolution Imaging Spectroradiometer (MODIS) onboard NASA's Terra and Aqua satellites [27], which is also useful for monitoring vegetation. Details of MODIS will be covered in those chapters where they are used. For surfaces runoff Q, useful for measuring changes in levels of the lake and rivers, satellite altimetry missions [11, 12] such as Topex/Poseidon, Jason I and II, and the European Space Agency (ESA) discussed in Sect. 3.5.1 are employed.

Besides satellite data, ***reanalyses*** have made significant contributions to many global/regional hydrological and climatic studies. With the release of many new high-resolution reanalyses in the past decade, see e.g., [40–43], their application into regional- and basin-scale studies have become increasingly valuable [100]. Yet certain elements of the analyzed fields (e.g., precipitation) remain highly uncertain at global and regional scale both in terms of trends and interannual variabilities. The reliability of reanalysis fields can considerably vary in space and time due to lack of adequate observational data, instrumental changes, changing mix of observations, biases in observations, etc., which can introduce spurious variability and trends into reanalysis fields [100]. Khandu [100] opines that since reanalysis products are increasingly used as regional climate forcing data and hydrological model inputs, it is vital to estimate their skills in a region. A reanalysis system consists of (i) a "data assimilation system" that combines available observations from various data sources (e.g., SRS-based observations), and (ii), a "forecast model" consisting of a atmospheric model at its core, which is often coupled to a land surface model and/or ocean model [40, 41, 43]. The performance of reanalyses have significantly improved with the assimilation of additional data from SRS-based observations, e.g., [42, 43, 45–47]. Reanalysis are employed in various chapters of this book to complement satellite-based remotely sensed data. In this regard, the individual reanalysis product used will be discussed in the respective chapter.

Climate models and hydrological models add knowledge to the key climate and hydrologic process and has been rigorously applied to understand the contributions of climate change (both natural and man-made) and other drivers (e.g., land use change, water use) on water resources [47, 100]. SRS-based observations are also assimilated into hydrological models that are widely used to quantify climate change impacts on various hydrological components as will be demonstrated for NRB. Climate models and hydrological models are also employed in subsequent chapters to complement satellite remote sensing and help in understanding of the NRB hydrological process.

3.2 Remotely Sensed Optical and Microwave Products

The multi-spectral Landsat, MODIS and Sentinel-2 data (see Sect. 4.2.3) have a wide range of spectrum bands that includes, visible to Near-Infrared (VNIR) and short-wave Infrared (SWIR) wavelengths. Landsat satellite system collect multi-spectral images that are useful for monitoring and understanding human activities that could be beneficial for the NRB, e.g., detecting, measuring and highlighting changes in landscape patterns over time including those of coastal studies [9, 10]. Since changes in the landscape are dynamic, these imagery can be used to study and capture extent of changes at large scale within short time frames and at no cost since the images are free. Without remotely sensed data, the task of capturing and mapping changes in landscape within NRB would be much more difficult and time consuming. The imagery from Landsat platform, which can be accessed freely from USGS website (https://earthexplorer.usgs.gov/) are powerful data source providing accurate and timely spatial information on changes in the Earth's landscape. Because of its long-term digital archive since 1972, it has been used in many land use and land cover change studies over different parts of the world, see e.g., [13].

Over the past 46 years, there have been numerous studies that have used Landsat imagery for agriculture forecasting and management (production and conservation), see e.g., [14–16]. Its other applications are reported, e.g., in [17–21, 23–26]. Landsat data, therefore, are important when there are no other accurate way of mapping and detecting changes due to cost, time, and access constraints. Landsat datasets obtained from the USGS are Level 1 products, which have been corrected for terrain effects, calibrated radiometrically, and geographically georeferenced [10].

To validate the used Landsat images whose spatial resolution is 30 m, the Sentinel-2 (MSI) with 10 m spatial resolution, officially launched by ESA on June 23, 2015, with a 5-day temporal resolution or repeat cycle [28] are employed. The corporation between ESA and the USGS provides Sentinel-2 (MSI) data in Level-1C product to end users and can be accessed from the USGS website and the official website of the Copernicus Open Access Hub for the European Space Agency (ESA) (https://scihub.copernicus.eu/). Sentinel-2 data has been processed by ESA and USGS to include radiometric and geometric corrections along with ortho-rectification to generate highly accurate geo-located products for researchers [29]. Optical and microwave (MODIS, Landsat and Sentinel-2) products are thus useful for monitoring land use land cover (LULC) changes within the NRB at a high spatio-temporal resolution that will enhance management of its water resource. Optical and microwave techniques are discussed in detail in Awange and Kiema [8, 9].

3.3 Remote Sensing of Gravity Variations

In the subsections that follow, it is explained how satellites (particularly the low earth observing-LEO satellites) are used to monitor variations in gravity field, which are in turn used to remotely sense the changes in stored waters of the Nile River Basin (NRB). The most significant success of a LEO satellite is evidenced in the Gravity Recovery and Climate Experiment (GRACE) satellites discussed in Sect. 3.3.3. A possible use of the global navigation satellite systems (GNSS, see [30]) to measure variations in water mass is illustrated by Tregoning et al. [31] whose predictions derived from GRACE measured fields show a correlation with GNSS measured deformations, suggesting the possible use of such deformations to infer changes in stored water potential on much shorter temporal and spatial scales than GRACE provides (and with low-latency), while averaging over much larger spatial scales than afforded by multipath amplitude measurements [32]. In this book, GRACE satellites are employed to monitor changes in the NRB's stored water (see Sect. 3.1). To understand how GRACE satellites work, let us first discuss the relationship between mass variation and gravity next.

3.3.1 Mass Variation and Gravity

Two types of gravity field variation exists. The *first* is the long-term, also known as mean gravity field or static gravity field, which is due to the static part of the gravity field. The variation is constant over a very long time interval. To simplify the understanding of the mean gravity field, consider a case where you frequently go to a given destination and along the way, you pass a given mountain. Every time you pass by, the mountain is stationary at the same place. Come months and years, the mountain, which represents concentration of mass is stationary and does not change its position. In simple terms, one can visualize this mountain as representing the long term changes in gravity field as its mass concentration does not seem to change. Study of long-term (mean gravity field) is useful in understanding the solid structure of the Earth, ocean circulation, and in achieving a universal height measuring system.

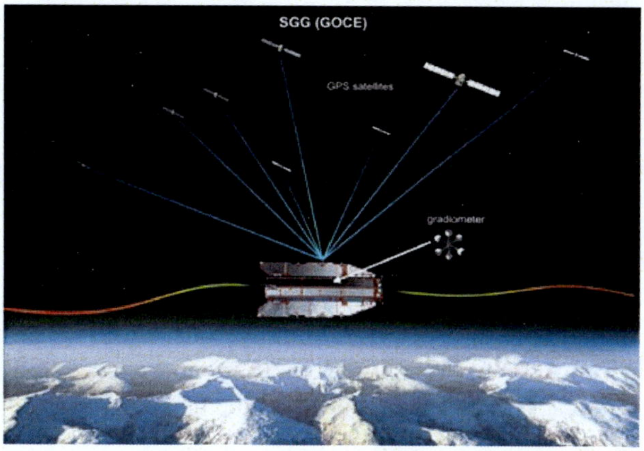

Fig. 3.1 Global Navigation Satellite System (GNSS) satellites track the GOCE satellite in space, thus contributing to the determination of its position (©ESA). The GOCE satellite's accurate determination of the static gravity field is expected to contribute towards studies of changes in sea level, earthquakes, and volcanoes. Figure modified by Rieser [38]

In this respect, the GOCE (Gravity field and the steady state-of-the ocean circulation explorer, Fig. 3.1) satellite[1] that completed its mission in mid-October 2011 used the state-of-the-art gradiometer with improved accuracy to provide data that are useful in mapping changes in gravity, see e.g., Hirt [33]. GOCE data is expected to benefit other studies such as those concerned with earthquakes, changes in sea level, and volcanoes.[2]

The *second* type of variation of the Earth's gravity field is associated with those processes that occur over shorter time scales, such as atmospheric circulation or the hydrological cycle. In the illustration of the mean gravity field above, we used the mountain as an example. To visualize time variable gravity field, instead of a mountain, consider now a small pond along your frequently used route to a given destination. As you frequently pass this pond, you will notice that its water level changes depending on climatic or human induced impacts. If it rains heavily and over a long period of time, the level rises while if there is a prolonged drought or massive use by humans, the level of the pond falls. This change in pond level over time symbolizes changes in mass with time. This is known as the *time-varying gravity field* and is the component, which enables the monitoring of, for example, changes in water level of the NRB as will be demonstrated in the chapters ahead.

By removing the effects of the other processes that cause changes in the time variable gravity field, changes in *terrestrial water storage* can thus be inferred from the observed temporal changes in the terrestrial gravity field. By assuming the density of water as 1.00 g/cm^3, and following the relation of [72], Ellet et al. [34] present the relationship between changes in stored water and gravity as

$$\Delta S = 0.419 \Delta g, \tag{3.1}$$

where water storage change ΔS is given in units of cm of water and gravity change Δg is in units of microGal (10^{-6} cm/s^2). From equation (3.1), it is seen therefore,

[1] http://www.esa.int/Our_Activities/Observing_the_Earth/GOCE.

[2] see, e.g., http://www.esa.int/esaCP/SEMV3FO4KKF_Germany_0.html.

that monitoring variations in the gravity field can enable hydrological changes to be monitored.

3.3.2 *High and Low Earth Orbiting Satellites*

At the broadest conceptual level, Low Earth Observing (LEO) satellites' gravity field missions observe (either directly or indirectly) gradients in the Earth's external gravitational field. This is essentially done through differential measurements between two or more points, thus largely eliminating spatially correlated errors. When done from space, two approaches can be used, e.g., [35, 36]:

1. Satellite-to-satellite tracking (SST), or
2. A dedicated gravity gradiometer on board a satellite, coupled with SST.

The SST methods can use either low-low inter-satellite tracking (ll-SST, see Fig. 3.2, right), where two LEO satellites track one another and additional observations in terms of high precision ranges and range rates between the two satellites are taken, or high-low inter-satellite tracking (hl-SST, see Fig. 3.2, left), where high-Earth orbiting satellites (notably GPS) track a LEO satellite. The low-low mode, compared to the high-low mode, has the advantage of signal amplification leading to a higher resolution of the obtained gravity variations, up to the medium wavelength spectrum of a few hundred km in spatial extent [35]. Taking this further, a combination of ll-SST and hl-SST is conceptually better still, as demonstrated by the GRACE mission (Fig. 3.2, right) with a baseline length between the two satellites of about 220 km. This is treated in detail in the next section.

In order to detect temporal gravity field variations at smaller spatial scales, the satellite(s) being tracked must be in as-low-as-possible orbits (close to the mass source), with the satellites being as free as possible from the perturbing effects of atmospheric drag [35]. In addition, so-called de-aliasing models (for correcting short-term—6 h—variations due to atmosphere and ocean mass variations) have to be used to mitigate the propagation of unwanted signals (e.g., leakage from the oceans) into the derived gravity solutions, e.g., [37].

3.3.3 *Gravity Recovery and Climate Experiment*

The GRACE mission, launched on 17th of March 2002, and which ended its mission on mid-October 2017, consisted of two near-identical satellites following one another in nearly the same orbital plane (about 400 km altitude) separated by a distance of 220 km; the so-called tandem formation (see Fig. 3.2, right). The ll-SST was measured using K-band ranging, coupled with hl-SST tracking of both satellites by GNSS (Fig. 3.2, right). Gobal Navigation Satellite System (GNSS) receivers were placed on GRACE satellites to measure occulted signals, see [30], and also to

GNSS tracking a low satellite, e.g., CHAMP.	GNSS tracking 2-low satellites, which are tracking each other, e.g., GRACE.

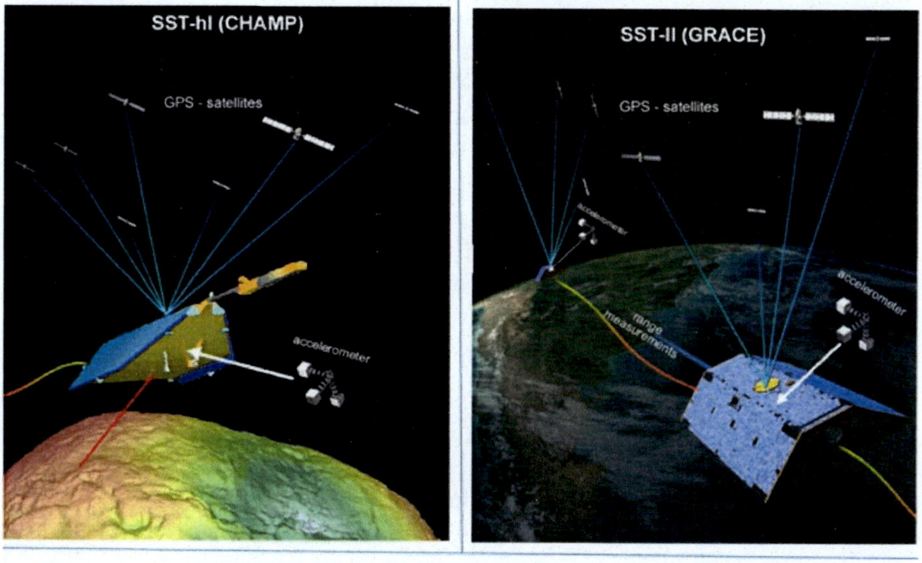

Fig. 3.2 Left: SST-hl realized with CHAMP (©GFZ Potsdam ([2.2]). Right: A combination of ll-SST and hl-SST realized with GRACE and GNSS satellites(©GRACE—CSR Texas ([2.2]). Figures modified by Rieser [38]). GNSS satellites are used in determining the positions of these satellites in space. For the GRACE satellites (right) inter-satellite distances can be computed from these positions and compared to the measured K-band distances, thus providing additional independent information

determined the orbital parameters of GRACE satellites required in order to determine gravity changes. On-board accelerometers monitored orbital perturbations of non-gravitational origin, see, e.g., [30].

GRACE mission processes GNSS data to contribute to the recovery of long-wavelength gravity field, remove errors due to long-term onboard oscillator drift, and aligns measurements between the two spacecraft [39, p. 200]. The timing function of GNSS for precision orbit determination, in terms of position and velocity as a function of time, enabled orbits to be determined within an accuracy better than 2 cm in each coordinate [39, p. 200]. The precise locations of the two satellites in orbit allowed for the creation of gravity maps approximately once a month.[3] These gravity maps, when converted to total water storage maps, are useful for monitoring changes in stored water potential, e.g., those of NRB as demonstrated in chapters to come.

The Earth's gravity field is mapped by making accurate measurements of changes in the distance between the satellites, using GNSS and a microwave ranging system. These changes in the distances between the two satellites occur due to the effect of the gravity (mass concentration) of the Earth. As the lead satellite passes through a region of mass concentration, it is pulled away from the trailing satellite (Fig. 3.2,

[3]http://www.csr.utexas.edu/grace/publications/brochure/page11.html.

right). As the trailing satellite passes over the same point, it is pulled towards the lead satellite thus *changing the distance between the satellites*.

Time-variable gravity field solutions are obtained by the exploitation of GRACE observation data over certain time intervals, i.e., every month [44, 48], or less, e.g., [49, 50, 102]. There are a number of institutions delivering GRACE products, each applying their own processing methodologies and, often, different background models. GRACE mission, which lasted for 15 years provide scientists with an efficient and cost-effective way to monitor time-varying component of the gravity field with unprecedented accuracy and in the process yield crucial information about the distribution and flow of mass within the Earth's system. The process causing gravity variations that are currently being studied by GRACE include [51];

- changes due to surface and deep currents in the ocean leading to more information about ocean circulation, e.g., [52, 53],
- changes in groundwater storage on land masses such as that of Nile River Basin (NRB), relevant to water resource managers, e.g., [2, 3, 51, 54, 55, 77, 88, 103],
- exchanges between ice sheets or glaciers and the oceans, needed for constraining the mass balance of the global ice regime and sea level change, e.g., [56, 57],
- air and water vapour mass change within the atmosphere, vital for atmospheric studies, e.g., [58, 59], and
- variations of mass distribution within the Earth arising from, e.g., on-going glacial-isostatic adjustments and earthquakes, e.g., [31, 60].

Currently, river and lake basins of the order of 200,000 km^2 and above in area can be successfully studied using the GRACE products [61]. In general, to understand how the GRACE satellites monitor changes in freshwater (all groundwater, soil moisture, snow, ice, and surface waters), first, the larger effect of the mass of the Earth, i.e, the static gravity field discussed in Sect. 3.3.1, which is always a constant G_0 corresponding to nearly 99% of the total field, is computed from a static model (e.g., *GGSM01S* [48]) and removed by subtracting it from the monthly gravity field ($G(t)$) measured by GRACE at a time t [62], i.e.,

$$\Delta G(t) = G(t) - G_0, \tag{3.2}$$

to give the monthly time-variable gravity field $\Delta G(t)$. Alternatively, instead of computing the static gravity field G_0, mean gravity field over a given number of years can also be used. Changes mostly related to the atmosphere and ocean, which occur over time scales shorter than one month, are then removed using models, see e.g., Wahr et al. [63]. Remnant atmospheric and oceanographic effects that last for more than one month can be removed using atmospheric and ocean circulation models before water storage change can be analyzed. The resulting difference in Eq. (3.2), which is called the gravity field anomaly is usually due to changes in stored water. If we consider $\overline{\Delta C_{nm}}(t)$ and $\overline{\Delta S_{nm}}(t)$ to be the normalized Stoke's coefficients expressed in terms of millimeters of geoid height, with n and m being degree and order respectively, the time-variable geoid in (3.2) is then expanded in-terms of spherical harmonic coefficients (see [64]) as

$$\Delta G(t) = \sum_{n=1}^{N}\sum_{m=0}^{n}(\overline{\Delta C}_{nm}(t)cos(m\lambda) + \overline{\Delta S}_{nm}(t)sin(m\lambda))\overline{P}_{nm}\left(cos(\theta)\right), \quad (3.3)$$

where N is the maximum degree of expansion, θ is the co-latitude, λ the longitude and $\overline{P}_{nm}$ the fully normalized Legendre polynomial [64]. From the gravitational spherical harmonic coefficients (3.3), the equivalent water thickness is computed using the following steps:

1. The gravitational residual coefficients are converted into the surface density coefficient differences by [63]

$$\begin{pmatrix}\Delta\check{C}_{lm}(\mathrm{M}_j)\\ \Delta\check{S}_{lm}(\mathrm{M}_j)\end{pmatrix} = \frac{\rho_{\mathrm{avg}}}{3\rho_{\mathrm{w}}}\frac{2l+1}{1+k_l'}\begin{pmatrix}\Delta\bar{C}_{lm}(\mathrm{M}_j)\\ \Delta\bar{S}_{lm}(\mathrm{M}_j)\end{pmatrix}, \quad (3.4)$$

   where k_l' is the *load Love* number of degree l, $\rho_{\mathrm{avg}} = 5517\,\mathrm{kg/m^3}$ the average density of the Earth, and $\rho_{\mathrm{w}} = 1000\,\mathrm{kg/m^3}$ the density of water.
2. The spatial variation of the surface density is then computed through

$$\Delta\sigma(\theta,\lambda,\mathrm{M}_j) = R\rho_{\mathrm{w}}\sum_{l=1}^{l_{\max}}\sum_{m=0}^{l}[\Delta\check{C}_{lm}(\mathrm{M}_j)\cos m\lambda + \Delta\check{S}_{lm}(\mathrm{M}_j)\sin m\lambda]\bar{P}_{lm}(\cos\theta), \quad (3.5)$$

   where $R = 6378137\,m$ is the radius of the Earth and $\Delta\sigma$ is in kg/m^2.
3. Finally, the changes in total water storage (TWS) are calculated by

$$\mathrm{TWS}(\phi,\lambda,\mathrm{M}_j) = \frac{\Delta\sigma(\theta,\lambda,\mathrm{M}_j)}{\rho_{\mathrm{w}}} = \frac{\Delta\sigma(\theta,\lambda,\mathrm{M}_j)}{1000}\quad [\text{meters}]. \quad (3.6)$$

The first steps in the analysis of GRACE data would provide an estimate of the changes in total water storage. In the second step, the changes can then be separated into their various components as discussed, e.g., in [51, 62] to obtain changes in the respective components (e.g., groundwater, surface water, soil moisture, and ice).

As stated previously, the GRACE satellites exceeded their planned 5 year life-span and ended their mission in mid-October 2017. However, given the excellent results that have been delivered so far, (see e.g., Fig. 1.2), GRACE follow-on mission (GRACE-FO) was launched on the 22nd of May 2018. Although GRACE-FO satellites, like their predecessor, will use the same kind of microwave ranging system giving a similar level of precision, they will also test an experimental instrument using lasers instead of microwaves, which promises to make the measurement of their separation distance at least 20 times more precise.[4] An example of GRACE's application is presented in Yang [65], where it is used to constrain freshwater flux from Greenland where the data show that Arctic freshwater flux started to increase rapidly in the mid-late 1990s, coincident with a decrease in the formation of dense Labrador Sea Water, a key component of the deep southward return flow of the Atlantic Merid-

[4] http://gracefo.jpl.nasa.gov/mission/.

ional Overturning Circulation (AMOC). The freshening of the polar oceans may be reducing formation of Labrador Sea Water and hence may be weakening the AMOC [65].

3.4 Gravity Field and Changes in Stored Water

In the discussion that follows, the concept of *gravity field variations* discussed in Sect. 3.3 is related to hydrological processes. Measurements of the time-varying gravity field by LEO satellites, e.g., GRACE discussed in Sect. 3.3.3 are the key to monitoring of changes in water levels at basin scales. Such techniques now enable the monitoring of groundwater recharge, which is the most important element in groundwater resources management and could also be applicable to monitoring changes in River Nile's water levels. For example, in 2009, GRACE satellites showed that northwest of India's aquifers had fallen at a rate of 0.3048 $m\,yr^{-1}$ (a loss of about 109 km^3 per year) between 2002 and 2008 (see Fig. 1.2.[5])

3.4.1 Gravity Field Changes and the Hydrological Processes

The hydrological cycle (Fig. 3.3) refers to the pathway of water in nature, as it moves in its different phases through the atmosphere, down over and through land, to the ocean and back to the atmosphere [66]. The associated variations in gravity field are therefore caused, e.g., by

- the redistribution of water in the oceans, including e.g., El Niño and Southern Oscillation (ENSO) events,
- movement of water vapour and other components in the atmosphere,
- seasonal rainfall; snow and subsequent drying and melting,
- groundwater extraction, or
- drying and filling of lakes, rivers, and reservoirs.

3.4.2 Monitoring Variation in Stored Water Using Temporal Gravity Field

The potential of using the relationship between temporal gravity changes and hydrology (Fig. 3.3) was first recognized by Montgomery [67] who estimated specific yield through a correlation between gravity and water-level changes [68]. In 1977, Lambert and Beaumont [69] used a gravity meter to correlate groundwater fluctuations and

[5]The Economist, September 12th 2009, pp. 27–29: Briefing India's water crisis.

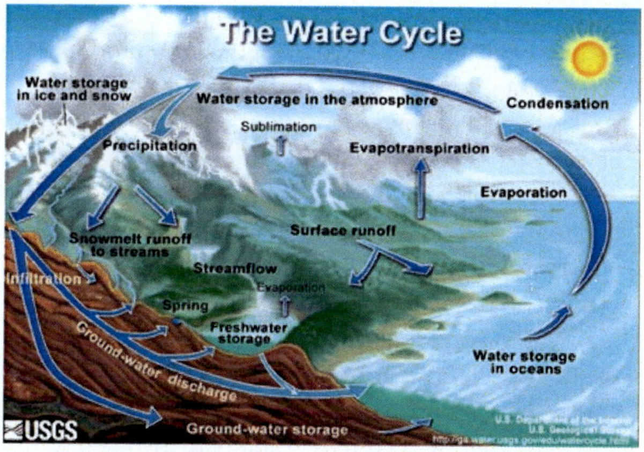

Fig. 3.3 Components of hydrological cycle that lead to temporal variations in the gravity field. *Source* US Geological Survey (USGS)

temporal changes in the Earth's gravity field. Goodkind [70] recorded observations from seven super conducting gravimetric stations to examine non-tidal variations in gravity and noted that at one of the stations (Geysers geothermal station), much of the variation could be correlated with rainfall and seismic activity. Such measurements had not been possible before the advent of super conducting gravimeters, thus providing evidence of the existence of temporal variation in gravity.

In 1995, while estimating the atmospheric effects on gravity observations around Kyoto, Mukai et al. [71] noted that changes in gravity around the station could have been caused by changes in underground water. In the same year, Pool and Eychaner [72] assessed the utility of temporal gravity-field surveys to directly measure *aquifer-storage* changes and reported gravity changes of around 100–134 μGal, equivalent to 2.4–3.2 m of water column, considering infinitely extended sheet approximation. Their results from the analysis of changes in stored water in the aquifer indicated an increase in the gravity field of 158 ± 6 μGal when the water table rose by about 17.7 m, providing further evidence of the possibility of using temporal gravity-field surveys to monitor changes in stored aquifer water. In fact, according to Bower and Courtier [73] who analyzed the effect of precipitation on gravity and well-levels at a Canadian absolute gravity site, 90% of the gravity variation was found to be due to the effects of precipitation, evapotranspiration and snow-melt.

The last decade has also recorded increased use of temporal gravity field studies in monitoring changes in stored water, see e.g., [34]. It saw the beginning of satellite missions dedicated to monitoring temporal variations in the gravity field. Smith et al. [74] investigated the ability of ground-based gravity meters to monitor changes in soil moisture storage.

Moving from local tests to regional, a different application of gravity surveys was investigated by Damiata and Lee [75], who simulated the gravitational response to aquifer hydraulic testing. The synthetic system was composed of an unconfined shallow aquifer and the purpose of the investigation was to assess the potential of the gravity measurements for detecting groundwater extraction. Drawdown due to pumping causes a decrease in mass and consequently in gravity measured at the surface. The results showed that the gravitational response to aquifer testing could be used to monitor the spatial development of the drawdown cone. For the configuration considered in the investigation, the signal was of the order of tens of μGals and could be detected up to several hundred meters away from the pumping well.

Water storage changes, such as changes in soil moisture, snow and ice cover, surface and groundwater, including deep aquifers, can be monitored either by in-situ observations or indirectly through changes in gravity [48]. While in-situ observations provide valuable localized information, they suffer from limited spatial coverage for regional to continent-wide studies [76]. Any change in water storage also manifests itself in a change in the gravity field. This property can be used to infer water-storage changes from time-variable gravity observations as demonstrated by Rodell and Famiglietti [77] for 20 globally distributed drainage basins of sizes varying from 130,000 km^2 to 5,782,000 km^2 to assess the detectability of hydrological signals with respect to temporal and spatial variations. Space borne techniques can provide time-variable gravity observations on a regional and global scale, thus allowing for large-scale water storage monitoring and the ability to close the 'gaps' between locally limited in-situ observations [78].

Since the launch of the GRACE satellite mission in 2002 (see Sect. 3.3.3) until the end of its mission in mid-October 2017, it provided a new and powerful tool for studying temporal gravity field changes as evidenced from numerous articles assessing the potential of GRACE recovering hydrological signals, see e.g., Awange et al. [35, and the references therein]. Tapley [48] provided early results of the application of the GRACE products for detecting hydrological signals in the Amazon-Orinoco basin. Following these results, numerous other authors have subsequently applied GRACE to detect hydrological signals in various situations and locations, see references in [35].

For instance, Ramillien et al. [51, 62] and Andersen et al. [80] investigated the potential of inferring inter-annual gravity field changes caused by continental water storage changes from GRACE observations between 2002 and 2003, and compared these changes to the output from four global hydrological models. It was possible to correlate large scale hydrologic events with the estimated change in the gravity field for certain areas of the world to an accuracy of 0.4 μGal, corresponding to 9 mm of water, see also [80–82].

Syed et al. [83] examined total basin discharge for the Amazon-Orinoco and Mississippi river basins from GRACE, while Rodell et al. [79] estimated groundwater storage changes in the Mississippi basin. Crowley et al. [84] estimated hydrological signals in the Congo basin, while Schmidt et al. [85] and Swenson et al. [61, 86] used GRACE to observe changes in continental water storage. Winsemius et al. [87] compared hydrological model outputs for the Zambezi River Basin with estimates derived from GRACE. Monthly storage depths produced by the hydrological model displayed larger amplitudes and were partly out of phase compared to the estimates based on GRACE data. Likely reasons included leakage produced by the spatial filtering used in the GRACE data, and the difficulty to identify the time of satellite overpass as opposed to simply averaging over the whole period.

Awange et al. [4, 7, 88] used GRACE to study the fall of Lake Victoria's water level in Africa while [1] employed it to study the impacts of global teleconnection on the continent's stored water. The Nile Basin and water changes in West Africa have been studied, e.g., by [91, 92] while its application to study agricultural droughts and changes in aquifer storage over the Greater Horn of Africa was pioneered in the study of Agutu et al. [93, 94].

As already discussed in Sect. 3.3.3, GRACE satellites detect changes in the Earth's gravity field by measuring changes in the distance between the two satellites at a 0.1 Hz sampling frequency. The variation in the distance between the two twin satellites caused by gravitational variations above, upon, and within the Earth all have an effect on the satellites. This variation in gravity could be due to *rapid* or *slow* changes caused, for example, by the redistribution of water in the oceans, the movement of water vapor and other components in the atmosphere, the tidal effect of the Sun and Moon, and the displacement of the material by earthquakes and glacial isostatic adjustment. The data therefore must be processed to isolate these effects so as to retain only those which correspond to the process of interest, in this case, terrestrial water storage changes, see e.g., [95]. Equation (3.6) is used to compute changes in stored water.

3.5 Satellite Altimetry

3.5.1 Remote Sensing with Satellite Altimetry

Satellites altimetry (Fig. 3.4) operates in two steps:

- *First*, the precise orbit of the satellite, i.e., its position, is determined. Through this, its *height* above the Earth is obtained.
- *Second*, range measurements are made by obtaining the time an emitted signal (radar or laser) travels to the Earth's surface and reflected back to the satellite.

GNSS satellites, see [30, 90], contribute to the *first step* where height is determined through GNSS receivers onboard the space satellites that enables monitoring of ranges and timing signals from GNSS satellites. The observed GNSS ranges provide precise and continuous tracking of the spacecraft, thereby delivering its position $\{\phi, \lambda, h\}$ at any time. The height component h is useful in determining the measured height (see Fig. 3.4). Besides GNSS tracking, other approaches such as satellite laser ranging (SLR) and DORIS (Doppler Orbitography and Radio positioning Integrated by Satellite) are also used to ensure that precise orbit determination is achieved.

In the *second step*, the Earth's surface heights (e.g., lake surface, ocean surface, glaciers, and ice sheets) are measured using ranges from the space altimetry satellite to the surface of interest. Radar altimeters send microwave signals to the Earth's surface and measures the time taken by the reflected signals to travel back. Using the velocity of light, the distance from the satellite to the Earth's surface is derived.

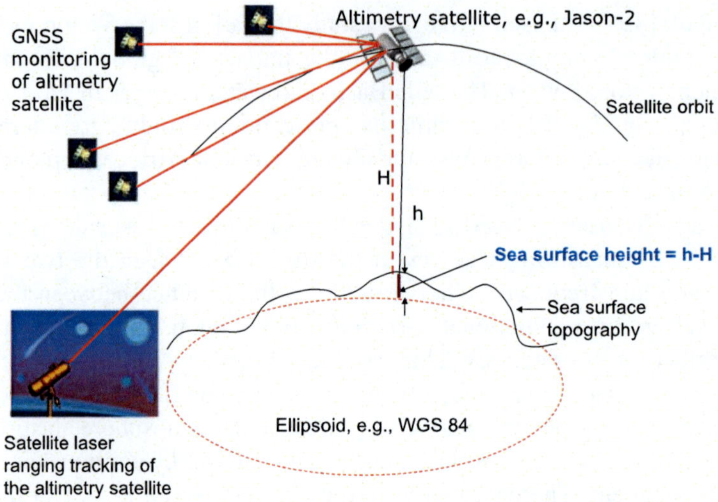

Fig. 3.4 Global Navigation Satellite System (GNSS) in support of monitoring changes in sea/lake level through the determination of the altimetry satellites' precise orbit. From the precise orbital parameters, the height component h is useful in determining changes in sea/lake level through the difference $\{h - H\}$, where H is measured by multiplying the speed of light with the time taken by the signals to travel from and to the satellite divided by 2, since the same distance is covered twice

Since the signals pass through the atmosphere from and to the satellites, they are affected by the atmosphere and as such, atmospheric corrections have to be made. The sea surface height is then obtained by subtracting the measured ranges in step 2 from the GNSS-derived satellite heights in step 1 (Fig. 3.4).

Example 3.1 (Satellite altimetry monitoring of Lake Victoria water level).

As an example, Lake Victoria water levels for the period 1993–2006 are plotted for both TOPEX/Poseidon altimetry-derived heights and tide gauge (in-situ) data in Fig. 3.5. The figure indicates a close relationship between the two data sets. Crétaux et al. [96] compares water levels of Lake Victoria from the Jinja tide gauge and those from Jason-1 altimetry satellite and obtain a coefficient of correlation value of 0.99 with a standard deviation of 2.7 cm for the period 2004–2007. This supports the fact that satellite altimetry provides useful information on changes in the lake level. Satellite altimetry are largely used for observing changes in sea level. By averaging the few-hundred thousand measurements collected by the satellite in the time it takes to cover the global oceans (i.e., 10 days for TOPEX/Poseidon), global mean sea level can be determined with a precision of several millimeters [97]. Such information is vital for mitigation of disasters related to sea level changes. Detailed satellite altimetry study on East African lakes (e.g., Fig. 3.5) are given e.g., in [96, 99].

End of Example 3.1.

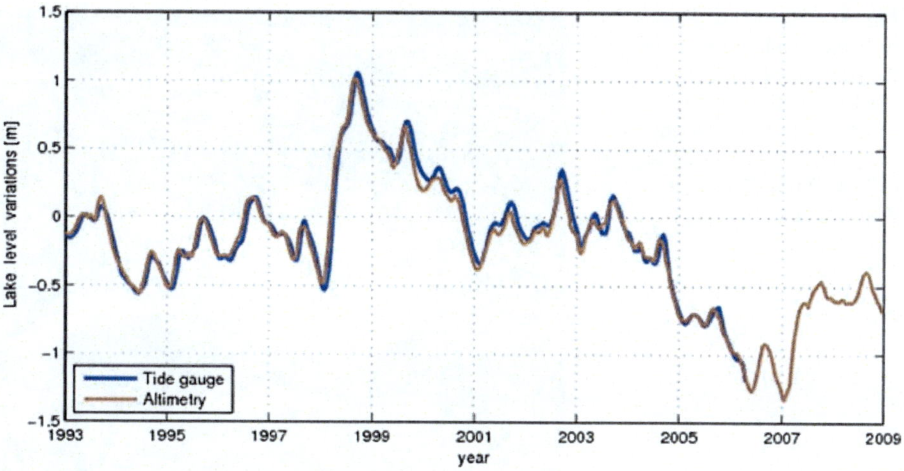

Fig. 3.5 A comparison of water gauge readings at Jinja station in Uganda (near Lake Victoria's outlet, see Fig. 2.4b) and water Levels from Topex/Poseidon and Jason-1 altimetry satellites. The figure shows a close match between the tide gauge and satellite altimetry data (cf. Fig. 4.6 obtained from GRACE)

3.5.2 Satellite Altimetry Missions

The first altimetry mission was TOPEX/Poseidon, developed by NASA and the Centre National d'Etudes Spatiales (CNES) and launched on 10th of 1992. Its mission ended in 2006 after 13 years of operation, providing 11 years of data. It was followed by Jason-series (Jason-1 was launched on 07/12/2001 and Jason-2 on 20/06/2008). Both TOPEX/Poseidon and Jason-1 were dedicated to measuring global mean sea level from space. TOPEX/Poseidon orbited at 1336 km above the Earth and covered the global oceans every 10 days, measuring the heights of the ocean surface directly underneath the satellite with an accuracy of 2–4 cm or better when averaging over several measurements [97]. Jason-2 is expected to be replaced by Jason-3 launched on 17th of January 2016, and subsequently Sentinel-6. Sentinel-6 will continue high precision ocean altimetry measurements in the 2020–2030 time-frame using two successive, identical satellites (Jason-Continuity of Service): Jason-CS-A and Jason-CS-B, and as a secondary objective, collect high resolution vertical profiles of temperature using the GNSS Radio-Occultation sounding technique discussed in Awange [30, 90].[6] Combined, all these satellites will provide long-term series of data capable of undertaking *climatological studies* resulting from changes in lakes and sea level.

ICESat (launched on January 12, 2003) uses a 1064 nm-laser operating 40 Hz to make measurements at 172-m intervals over ice, ocean, and land [98]. It combines state-of-the-art laser ranging capabilities with precise orbit and attitude control and

[6] https://eospso.nasa.gov/missions/sentinel-6.

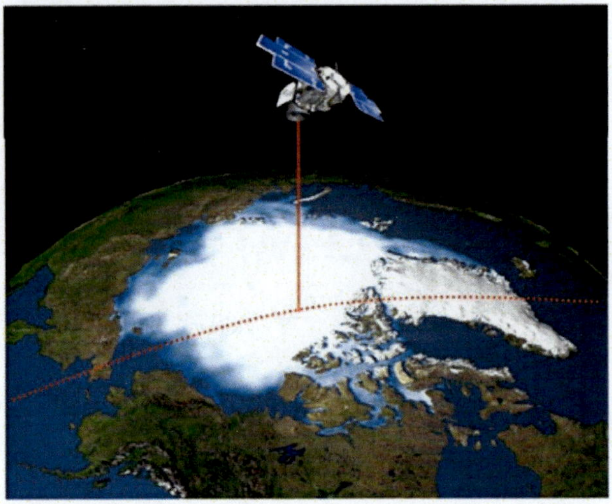

Fig. 3.6 Schematic diagram of ICESat on a transect over the Arctic. ICESat uses a 1064 nm-laser operating 40 Hz to make measurements at 172-m intervals over ice, oceans, and land. *Source* Abdalati et al. [98]

knowledge to provide very accurate measurements of ice sheet topography and elevation changes along track. It has the specific objective of measuring changes in polar ice as part of NASA's Earth Observing System.

By observing changes in ice sheet elevation, it is possible to quantify the growth and shrinkage of parts of the ice sheets with great spatial detail, thus enabling an assessment of ice sheet mass balance and contributions to sea level. Moreover, because the mechanisms that control ice sheet mass loss and gain in accumulation, surface ablation, and discharge presumably have distinct topographic expressions, ice sheet elevation changes also provide important insights into the processes causing the observed changes [98]. ICESat-2[7] launched in September 15th, 2018 that uses uses lasers and a very precise detection instrument to measure the elevation of Earth's surface replaced ICESat (Fig. 3.6) with improved laser capability compared to ICESat and will have the objectives of measuring ice sheet changes, sea ice thickness, and vegetation biomass. Achieving these objectives will contribute to the following [98]:

- Contribute to the development of predictive models that capture both dynamic and surface processes.
- Since the thickness distribution of sea ice controls energy and mass exchanges between the ocean and atmosphere at the surface, and the fresh water fluxes associated with melting ice serve as stabilizing elements in the circulation of the North Atlantic waters, basin-scale fields of ice thickness are therefore essential to improve our estimates of the seasonal and interannual variability in regional mass balance, the freshwater budget of the polar oceans, and the representation of these processes in regional and climate models.
- The capability of producing a vegetation height surface with 3-m accuracy at 1-km spatial resolution, assuming that off-nadir pointing can be used to increase

[7]https://www.nasa.gov/content/goddard/icesat-2.

the spatial distribution of observations over terrestrial surfaces. This sampling, combined with a smaller footprint of 50 m or less, would allow characterization of vegetation at a higher spatial resolution than ICESat, and is expected to provide a new set of global ecosystem applications.

- In addition, the atmospheric measurement capability of ICESat-2, even at near-IR wavelengths, will enable global measurements of cloud and aerosol structure to extend the record of these observations beyond those provided by the current lasers onboard ICESat.

3.6 Concluding Remarks

Satellite remote sensing and its application to water resource monitoring, particularly GRACE, is an ongoing active area of research. The data that has been collected from GRACE so far has provided information that was hitherto difficult to fathom. This technique is currently contributing significantly to water resource studies of the Nile River Basin and other freshwater bodies the world over. When the life spans of the various missions (e.g., GRACE) are reached, thousands of data sets will have been collected that will help to unravel some of the complex nature of the hydrological cycle and climate variation/change studies.

References

1. Anyah RO, Forootan E, Awange JL, Khaki M (2018) Understanding linkages between global climate indices and terrestrial water storage changes over Africa using GRACE products. Sci Total Environ 635:1405–1416. https://doi.org/10.1016/j.scitotenv.2018.04.159
2. Awange JL, Saleem A, Sukhadiya RM, Ouma YO, Kexiang H (2019) Physical dynamics of Lake Victoria over the past 34 years (1984–2018): Is the lake dying? Sci Total Environ 658:199–218. https://doi.org/10.1016/j.scitotenv.2018.12.051
3. Awange JL, Forootan E, Kuhn M, Kusche J, Heck B (2014) Water storage changes and climate variability within the Nile Basin between 2002 and 2011. Adv Water Res 73:1–25. https://doi.org/10.1016/j.advwatres.2014.06.010
4. Awange JL, Sharifi MA, Ogonda G et al (2008) The Falling Lake Victoria Water Level: GRACE, TRIMM and CHAMP Satellite Analysis of the Lake Basin. Water Resour Manag 22:775. https://doi.org/10.1007/s11269-007-9191-y
5. Awange JL, Ferreira VG, Forootan E, Khandu Andam-Akorful, SA, Agutu NO, He XF, (2016) Uncertainties in remotely sensed precipitation data over Africa. Inter J Climatolol 36(1):303–323. https://doi.org/10.1002/joc.4346
6. Awange JL, Hu KX, Khaki M (2019) The newly merged satellite remotely sensed, gauge and reanalysis-based Multi-Source Weighted-Ensemble Precipitation: Evaluation over Australia and Africa (1981–2016). Sci Total Environ 670:448–465
7. Awange J (2021) Lake Victoria monitored from space. Springer International Publishers, Berlin
8. Awange JL, Kiema JBK (2013) Environmental Geoinformatics. Monitoring and Management. Springer, Heidelberg

9. Awange JL, Kiema JBK (2019) Environmental Geoinformatics. Exploiting the Big Data, Extreme Hydro-Climatic and Food Security Challenges. Springer Nature, Heidelberg
10. USGS (2015) Landsat data. USGS, United States
11. Berry PAM, Garlick JD, Freeman JA, Mathers EL (2005) Global inland water monitoring from multi-mission altimetry. Geophys Res Lett 32(16). https://doi.org/10.1029/2005GL022814
12. Shum CK, Ries JC, Tapley BD (1995) The accuracy and applications of satellite altimetry. Geophys J Int 121:321–336. https://doi.org/10.1111/j.1365-246X.1995.tb05714.x
13. Jensen JR, Rutchey K, Koch MS, Narumalani S (1995) Inland wetland change detection in the Everglades Water Conservation Area 2A using a time series of normalized remotely sensed data. Photogrammetr Eng Remote Sens 61:199–209
14. Kauth R, Lambeck P, Richardson W, Thomas G, Pentland A (1979) Feature extraction applied to agricultural crops as seen by Landsat
15. Stow D, Tinney L, Estes J (1980) Deriving land use/land cover change statistics from Landsat-A study of prime agricultural land
16. Peterson U, Aunap R (1998) Changes in agricultural land use in Estonia in the 1990s detected with multitemporal Landsat MSS imagery. Landscape Urban Plan 41:193–201
17. Cihlar J (2000) Land cover mapping of large areas from satellites: status and research priorities. Inter J Remote Sens 21:1093–1114
18. Jusoff K, Senthavy S (2003) Land use change detection using remote sensing and geographical information system (GIS) in Gua Musang district, Kelantan, Malaysia. J Trop Forest Sci 15:303–312
19. Porter-Bolland L, Ellis EA, Gholz HL (2007) Land use dynamics and landscape history in La Montaña, Campeche, Mexico. Landscape Urban Plan 82:198–207
20. Dewan AM, Yamaguchi Y (2009) Land use and land cover change in Greater Dhaka, Bangladesh: Using remote sensing to promote sustainable urbanization. Appl Geography 29:390–401. https://doi.org/10.1016/j.apgeog.2008.12.005
21. Kesgin B, Nurlu E (2009) Land cover changes on the coastal zone of Candarli Bay, Turkey using remotely sensed data. Environ Monitor Assess 157:89–96
22. Kummerow C, Barnes W (1998) The tropical rainfall measuring mission (TRMM) sensor package. J Atmos Ocean Technol 15:809-817
23. Vittek M, Brink A, Donnay F, Simonetti D, Desclée B (2014) Land cover change monitoring using landsat MSS/TM satellite image data over West Africa between 1975 and 1990. Remote Sens 6:658–676
24. Yadav P, Kapoor M, Sarma K (2012) Land use land cover mapping, change detection and conflict analysis of Nagzira-Navegaon corridor, Central India using geospatial technology. Int J Remote Sens GIS 1:90–98
25. Otukei JR, Blaschke T (2010) Land cover change assessment using decision trees, support vector machines and maximum likelihood classification algorithms. Int J Appl Earth Obser Geoinf 12:S27–S31
26. Moran EF (2010) Land cover classification in a complex urban-rural landscape with QuickBird imagery. Photogrammetr Eng Remote Sens 76:1159
27. Justice CO, Townshend JRG, Vermote EF, Masuoka E, Wolfe RE, Saleous N, Roy DP, Morisette JT (2002) An overview of MODIS Land data processing and product status. Remote Sens Environ 83:3–15. https://doi.org/10.1016/S0034-4257(02)00084-6
28. ESA (2017) European Space Agency, Copernicus Open Access Hub. In: SERCO
29. Drusch M, Del Bello U, Carlier S, Colin O, Fernandez V, Gascon F, Hoersch B, Isola C, Laberinti P, Martimort P (2012) Sentinel-2: ESA's optical high-resolution mission for GMES operational services. Remote Sens Environ 120:25–36
30. Awange JL (2018) GNSS Environmental Sensing. Revolutionizing Environmental Monitoring. Springer, Berlin
31. Tregoning P, Watson C, Ramillien G, McQueen H, Zhang J (2009) Detecting hydrologic deformation using GRACE and GPS. Geophys Res Lett 36:L15401. https://doi.org/10.1029/2009GL038718

32. Hammond WC, Brooks BA, Bürgmann R, Heaton T, Jackson M, Lowry AR, Anandakrishnan S (2010) The scientific value of high-rate, low-latency GPS data, a white paper
33. Hirt C, Gruber T, Featherstone WE (2011) Evaluation of the first GOCE static gravity field models using terrestrial gravity, vertical deflections and EGM2008. Quasigeoid heights. J Geodesy 85:723–740. https://doi.org/10.1007/s00190-011-0482-y
34. Ellett KM, Walker JP, Western AW, Rodell M (2006) A framework for assessing the potential of remote sensed gravity to provide new insight on the hydrology of the Murray-Darling Basin. Aust J Water Resour 10(2):89–101
35. Awange JL, Sharifi MA, Baur O, Keller W, Featherstone WE, Kuhn M (2009) GRACE hydrological monitoring of Australia. Current limitations and future prospects. J Spatial Sci 54(1):23–36. https://doi.org/10.1080/14498596.2009.9635164
36. Rummel R, Balmino G, Johannessen J, Visser P, Woodworth P (2002) Dedicated gravity field missions - principles and aims. J Geodynam 33(1):3–20. https://doi.org/10.1016/S0264-3707(01)00050-3
37. Schrama EJO, Visser PNAM (2007) Accuracy assessment of the monthly GRACE geoids based upon a simulation. J Geodesy 81(1):67–80. https://doi.org/10.1007/s00190-006-0085-1
38. Rieser D (2008) Comparison of GRACE-derived monthly surface mass variations with rainfall data in Australia. MSc Thesis. Graz University of Technology
39. Prasad R, Ruggieri M (2005) Applied satellite navigation using GPS. GALILEO and Augmentation Systems. Artech House, Boston/London
40. Kalnay E, Kanamitsu M, Kistler R, Collins W, Deaven D, Gandin L, Iredell M, Saha S, White G, Woollen J, Zhu Y, Chelliah M, Ebisuzaki W, Higgins W, Janowiak J, Mo KC, Ropelewski C, Wang J, Leetmaa A, Reynolds R, Jenne R, Joseph D (1996) The NCEP/NCAR 40-year reanalysis project. Bull Amer Meteorol Soc 77(3):437–470. https://doi.org/10.1175/1520-0477(1996)077<0437:TNYRP>2.0.CO;2
41. Onogi K, Tsutsui J, Koide H, Sakamoto M, Kobayashi S, Hatsushika H, Matsumoto T, Yamazaki N, Kamahori H, Takahashi K, Kadokura S, Wada K, Kato K, Oyama R, Ose T, Mannoji N, Taira R (2007) The JRA-25 reanalysis. Quart J R Meteorol Soc 85(3):369–432. https://doi.org/10.2151/jmsj.85.369
42. Saha S, Moorthi S, Pan H-L, Wu X, Wang J, Nadiga S, Tripp P, Kistler R, Woollen J, Behringer D, Liu H, Stokes D, Grumbine R, Gayno G, Wang J, Hou YT, Chuang HY, Juang H-MH, Sela J, Iredell M, Treadon R, Kleist D, Delst PV, Keyser D, Derber J, Ek M, Meng J, Wei H, Yang R, Lord S, Dool HVD, Kumar A, Wang W, Long C, Chelliah M, Xue Y, Huang B, Schemm J-K, Ebisuzaki W, Lin R, Xie P, Chen M, Zhou S, Higgins W, Zou C-Z, Liu Q, Chen Y, Han Y, Cucurull L, Reynolds RW, Rutledge G, Goldberg M (2010) The NCEP climate forecast system reanalysis. Bull Amer Meteorol Soc 91(8):1015–1057. https://doi.org/10.1175/2010BAMS3001.1
43. Dee DP, Uppala SM, Simmons AJ, Berrisford P, Poli P, Kobayashi S, Andrae U, Balmaseda MA, Balsamo G, Bauer P, Bechtold P, Beljaars ACM, van de Berg L, Bidlot J, Bormann N, Delsol C, Dragani R, Fuentes M, Geer AJ, Haimbergere L, Healy SB, Hersbach H, Holm EV, Isaksen L, Kållberg P, Köhler M, Matricardi M, McNally AP, Monge-Sanz BM, Morcrette J-J, Park B-K, Peubey C, de Rosnay P, Tavolato C, Thépaut J-N, Vitarta F (2011) The ERA-Interim reanalysis: configuration and performance of the data assimilation system. Q J R Meteorolog Soc 137:553–597. https://doi.org/10.1002/qj.828
44. Luthcke S, Rowlands D, Lemoine F, Klosko S, Chinn D, McCarthy J (2006) Monthly spherical harmonic gravity field solutions determined from GRACE inter-satellite range-rate data alone. Geophys Res Lett 33:L02402. https://doi.org/10.1029/2005GL024846
45. Rodell M, Houserand PR, Jambor U, Gottschalck J, Mitchell K, Meng C-J, Arsenault K, Cosgrove B, Radakovich J, Bosilovich M, Entin JK, Walker JP, Lohmann D, Toll D (2004) The global land data assimilation system. Bull Amer Meteorol Soc 85:381–394. https://doi.org/10.1175/BAMS-85-3-381
46. Rienecker MM, Suarez MJ, Gelaro R, Todling R, Bacmeister J, Liu E, Bosilovich MG, Schubert SD, Takacs L, Kim GK, Bloom S, Chen J, Collins D, Conaty A, da Silva A, Gu

W, Joiner J, Koster RD, Lucchesi R, Molod A, Owens T, Pawson S, Pegion P, Redder CR, Reichle R, Robertson FR, Ruddick AG, Sienkiewicz M, Woollen J (2011) MERRA: NASA's Modern-Era retrospective analysis for research and applications. J Clim 24(14):3624–3648. https://doi.org/10.1175/JCLI-D-11-00015.1
47. Döll P, Müller Schmied H, Schuh C, Portmann FT, Eicker A (2014) Global-scale assessment of groundwater depletion and related groundwater abstractions: combining hydrological modeling with information from well observations and GRACE satellites. Water Resour Res 50(7):5698–5720. https://doi.org/10.1002/2014WR015595
48. Tapley BD, Bettadpur S, Ries JC, Thompson PF, Watkins MM (2004) GRACE measurements of mass variability in the Earth system. Science 305:503–505. https://doi.org/10.1126/science.1099192
49. Bruinsma S, Lemoine J, Biancale R, Valès N (2010) CNES/GRGS 10-day gravity field models (release 2) and their evaluation. Adv Space Res 45(4):587–601. https://doi.org/10.1016/j.asr.2009.10.012
50. Lemoine F, Luthcke S, Rowlands D, Chinn D, Klosko S, Cox C (2007) The use of mascons to resolve time-variable gravity from GRACE. In: Tregoning P, Rizos C (eds) Dynamic Planet. Springer, Berlin, pp 231–236
51. Ramillien G, Cazenave A, Brunau O (2004) Global time variations of hydrological signals from GRACE satellite gravimetry. Geophys J Int 158(3):813-826. https://doi.org/10.1111/j.1365-246X.2004.02328.x
52. Chambers D, Wahr J, Nerem R (2004) Preliminary observations of global ocean mass variations with GRACE. Geophys Res Lett 31(L13310). https://doi.org/10.1029/2004GL020461
53. Wahr J, Jayne S, Bryan F (2002) A method of inferring changes in deep ocean currents from satellite measurements of time-variable gravity. J Geophys Res 107(C12):3218. https://doi.org/10.1029/2002JC001274
54. Tiwari V, Wahr J, Swenson S (2009) Dwindling groundwater resources in northern India, from satellite gravity observations. Geophys Res Lett 36:L18401. https://doi.org/10.1029/2009GL039401
55. Werth S, Güntner A, Petrovic S, Schmidt R (2009) Integration of GRACE mass variations into a global hydrological model. Earth Planet Sci Lett 27(1–2):166–173. https://doi.org/10.1016/j.epsl.2008.10.021
56. Baur O, Kuhn M, Featherstone W (2009) GRACE-derived ice-mass variations over Greenland by acocunting for leakage effects. J Geophys Res 114(B06407). https://doi.org/10.1029/2008JB006239
57. Velicogna I (2009) Increasing rates of ice mass loss from the Greenland and Antarctic ice sheets revealed by GRACE. Geophys Res Lett 36:L19503. https://doi.org/10.1029/2009GL040222
58. Boy J-P, Chao B (2005) Precise evaluation of atmospheric loading effects on Earth's time-variable gravity field. J Geophys Res - Solid Earth 110(B8):4–12. https://doi.org/10.1029/2002JB002333
59. Swenson S, Wahr J (2002) Estimated effects of the vertical structure of atmospheric mass on the time-variable geoid. J Geophys Res 107(B9):2194. https://doi.org/10.1029/2000JB000024
60. Barletta V, Sabadini R, Bordoni A (2008) Isolating the PGR signal in the GRACE data: impact on mass balance estimates in Antarctica and Greenland. Geophys J Int 172(1):18–30. https://doi.org/10.1111/j.1365-246X.2007.03630.x
61. Swenson S, Wahr J, Milly PCD (2003) Estimated accuracies of regional water storage variations inferred from the Gravity Recovery and Climate Experiment (GRACE). Water Resour Res 39(8):1223. https://doi.org/10.1029/2002WR001736
62. Ramillien G, Frappart F, Cazenave A, Güntner A (2005) Time variations of land water storage from an inversion of two years of GRACE geoids [rapid communication]. Earth Planet Sci Lett 235(1–2):283–301. https://doi.org/10.1016/j.epsl.2005.04.005
63. Wahr J, Molenaar M, Bryan F (1998) Time variability of the Earth's gravity field: Hydrological and oceanic effects and their possible detection using GRACE. J Geophys Res (Solid Earth) 103(B12):30205–30230. https://doi.org/10.1029/98JB02844

64. Heiskanen WA, Moritz H (1967) Physical geodesy. W.H. Freeman and Company, San Francisco
65. Yang Q (2016) Applications of satellite geodesy in environmental and climate change. Graduate Theses and Dissertations. http://scholarcommons.usf.edu/etd/6440 Accessed 26 Jan 2017
66. Brutsaert W (2005) Hydrology. An introduction, 4th edn. Cambridge University Press, New York
67. Montgomery EL (1971) Determination of coefficient of storage by use of gravity measurements. PhD Thesis, University of Arizona, Tucson
68. Leirião S, He X, Christiansen L, Andersen OB, Bauer-Gottwein P (2009) Calculation of the temporal gravity variation from spatially variable water storage change in soils and aquifers. J Hydrol 365(3–4):302–309. https://doi.org/10.1016/j.jhydrol.2008.11.040
69. Lambert A, Beaumont C (1977) Nano variation in gravity due to seasonal groundwater movements: implications for the gravitational detection of tectonic movements. J Geophys Res 82(2):297–306. https://doi.org/10.1029/JB082i002p00297
70. Goodkind JM (1986) Continuous measurement of nontidal variations of gravity. J Geophys Res 91(B9):9125–9134
71. Mukai A, Higashi T, Takemoto S, Nakagawa I, Naito I (1995) Accurate estimation of atmospheric effects on gravity observations made with a superconducting gravity meter at Kyoto. Phys Earth Planet Inter 91(1–3):149–159
72. Pool DR, Eychaner JH (1995) Measurements of aquifer-storage change and specific yield using gravity surveys, groundwater 33(3):425–432. https://doi.org/10.1111/j.1745-6584.1995.tb00299.x
73. Bower DR, Courtier N (1998) Precipitation effects on gravity measurements at the Canadian absolute gravity site. Phys Earth Planet Inter 106:353–369. https://doi.org/10.1016/S0031-9201(97)00101-5
74. Smith AB, Walker JP, Western AW, Ellett KM (2005) Using ground based measurements to monitor changes in terrestrial water storage. In: 29th Hydrology and water resources symposium (CD Rom), Institute of Engineers Australia
75. Damiata BN, Lee TC (2002) Gravitational attraction of solids of revolution - part 1: vertical circular cylinder with radial variation of density. J Appl Geophys 50(3):333–349. https://doi.org/10.1016/S0926-9851(02)00151-9
76. Rodell M, Famiglietti JS, Chen J, Seneviratne SI, Viterbo P, Holl S, Wilson CR (2004) Basin scale estimates of evapotranspiration using GRACE and other observations. Geophys Res Lett 31:L20504. https://doi.org/10.1029/2004GL020873
77. Rodell M, Famiglietti JS (1999) Detectability of variations in continental water storage from satellite observations of the time dependent gravity field. Water Resour Res 35(9):2705–2724. https://doi.org/10.1029/1999WR900141
78. Ellett KM, Walker JP, Rodell M, Chen JL, Western AW (2005) GRACE gravity fields as a new measure for assessing large-scale hydrological models. In: Zerger A and Argent RM (Eds). MODSIM 2005 International Congress on Modelling and Simulation. The Modelling and Simulation Society of Australia and New Zealand Inc, December 2005, pp 2911–2917, ISBN: 0-9758400-2-9
79. Rodell M, Chen J, Kato H, Famiglietti JS, Nigro J, Wilson CR (2006) Estimating groundwater storage changes in the Mississippi River basin (USA) using GRACE. Hydrogeol J 15(1):159–166. https://doi.org/10.1007/s10040-006-0103-7
80. Andersen OB, Seneviratne SI, Hinderer J, Viterbo P (2005) GRACE-derived terrestrial water storage depletion associated with the 2003 European heat wave. Geophys Res Lett 32(L18405):2–5. https://doi.org/10.1029/2005GL023574
81. Neumeyer J, Barthelmes F, Dierks O, Flechtner F, Harnisch M, Harnisch G, Hinderer J, Imanishi Y, Kroner C, Meurers B, Petrovic S, Reigber C, Schmidt R, Schwintzer P, Sun HP, Virtanen H (2006) Combination of temporal gravity variations resulting from superconducting gravimeter (SG) recordings, GRACE satellite observations and global hydrology models. J Geodesy 79(10–11):573–585. https://doi.org/10.1007/s00190-005-0014-8

82. Yan JP, Hinderer M, Einsele G (2002) Geochemical evolution of closed-basin lakes: general model and application to Lakes Qinghai and Turkana. Sedimentary Geology 148(1–2):105–122. https://doi.org/10.1016/S0037-0738(01)00212-3
83. Syed T, Famiglietti J, Rodell M, Chen J, Wilson C (2008) Total basin discharge for the Amazon and Mississippi River basins from GRACE and a land-atmosphere water balance. Water Resour Res 44:W02433. https://doi.org/10.1029/2006WR005779
84. Crowley JW, Mitrovica JX, Bailey RC, Tamisiea ME, Davis JL (2006) Land water storage within the Congo Basin inferred from GRACE satellite gravity data. Geophys Res Lett 33:L19402. https://doi.org/10.1029/2006GL027070
85. Schmidt R, Schwintzer P, Flechtner F, Reigber C, Güntner A, Döll P, Ramillien G, Cazenave A, Petrovic S, Jochmann H, Wünsch J (2006) GRACE observations of changes in continental water storage. Global Planet Change 50(1–2):112–126. https://doi.org/10.1016/j.gloplacha.2004.11.018
86. Swenson S, Yeh PJ-F, Wahr J, Famiglietti J (2006) A comparison of terrestrial water storage variations from GRACE with in situ measurements from Illinois. Geophys Res Lett 33:L16401. https://doi.org/10.1029/2006GL026962
87. Winsemius HC, Savenije HHG, van de Giesen NC, van den Hurk B, Zapreeva EA, Klees R (2006) Assessment of Gravity Recovery and Climate Experiment (GRACE) temporal signature over the upper Zambezi. Water Resour Res 42:W12201. https://doi.org/10.1029/2006WR005192
88. Awange JL, Anyah R, Agola N, Forootan E, Omondi P (2013) Potential impacts of climate and environmental change on the stored water of Lake Victoria Basin and economic implications. Water Resour Res 49:8160–8173
89. Awange JL, Ogallo L, Kwang-Ho B, Were P, Omondi P, Omute P, Omulo M (2008) Falling lake victoria water levels: is climate a contribution factor? J Climat Change 89:287–297. https://doi.org/10.1007/s10584-008-9409-x
90. Awange JL (2012) Environmental monitoring using GNSS global navigations satellite systems. Springer, Heidelberg
91. Awange JL, Forootan E, Kuhn M, Kusche J, Heck B (2014) Water storage changes and climate variability within the Nile Basin between 2002 and 2011. Adv Water Resour 73:1–25. https://doi.org/10.1016/j.advwatres.2014.06.010
92. Ndehendehe C, Awange J, Agutu N, Kuhn M, Heck B (2016) Understanding changes in terrestrial water storage over West Africa between 2002 and 2014. Adv Water Resour 88:211–230. https://doi.org/10.1016/j.advwatres.2015.12.009
93. Agutu N, Awange JL, Zerihun A, Ndehedehe C, Kuhn M, Fukuda Y (2017) Assessing multi-satellite remote sensing, reanalysis, and land surface models' products in characterizing agricultural drought in East Africa. Remote Sens Environ 194:287–302. https://doi.org/10.1016/j.rse.2017.03.041
94. Agutu NO, Awange JL, Ndehedehe C, Kirimi F, Kuhn M (2019) GRACE-derived groundwater changes over greater horn of Africa: temporal variability and the potential for irrigated agriculture. Sci Total Environ. https://doi.org/10.1016/j.scitotenv.2019.07.273
95. Bettadpur S (2007) UTCSR Level-2 Processing Standards Document for Level-2 Product Release 0004. Gravity Recovery and Climate Experiment (GRACE) Rev 3.1, GRACE 327-742 (CSR-GR-03-03). Center for Space Research, The University of Texas at Austin
96. Crétaux J-F, Jelinski W, Calmant S, Kouraev A, Vuglinski V, Bergé-Nguyen M, Gennero M-C, Nino F, Abarca Del Rio R, Cazenave A, Maisongrande P (2011) SOLS: a lake database to monitor in the Near real-time water level and storage variations from remote sensing data. Adv Space Res 47:1497–1507. https://doi.org/10.1016/j.asr.2011.01.004
97. Pugh D (2004) Changing sea levels. Effect of tides, weather and climate. Cambridge University Press, Cambridge
98. Abdalati W, Zwally HJ, Bindschadler B, Csatho B, Farrell SL, Fricker HA, Harding D, Kwok R, Lefsky M, Markus T, Marshak A, Neumann T, Palm S, Schutz B, Smith B, Spinhirne J, Webb C (2010) The ICESat-2 laser altimetry mission. Proceed IEEE 98(5):735–751. https://doi.org/10.1109/JPROC.2009.2034765

99. Becker M, Llowel W, Cazenave A, Güntner A, Crétaux J-F (2010) Recent hydrological behaviour of the East African Great Lakes region inferred from GRACE, satellite altimetry and rainfall observations. Comptes Rendus Geosci 342(3):223–233. https://doi.org/10.1016/j.crte.2009.12.010
100. Khandu (2016) Assessing climate change impacts on water resources in the Ganges-Brahmaputra-Meghna River Basin. PhD thesis, Curtin University
101. Ebert EE, Janowiak JE, Kidd C (2007) Comparison of near-real-time precipitation estimates from satellite observations and numerical models. Bull Amer Meteorol Soc 88(1):47–64. https://doi.org/10.1175/BAMS-88-1-47
102. Awange JL, Fleming KM, Kuhn M, Featherstone WE, Heck B, Anjasmara I (2011) On the suitability of the 4° × 4° GRACE mascon solutions for remote sensing Australian hydrology. Remote Sens Environ 115:864–875. https://doi.org/10.1016/j.rse.2010.11.014
103. Khaki M, Awange J, Forootan E, Kuhn M (2018) Understanding the association between climate variability and the Nile's water level fluctuations and water storage changes during 1992–2016. Sci Total Environ 645: 1509–1521, https://doi.org/10.1016/j.scitotenv.2018.07.212

Part II
The Headwaters: The White and Blue Niles

Chapter 4
Lake Victoria: The Mother of the Nile

Lake Victoria is a living Lake. From Kisumu (Kenya) to Mediterranean, millions of people are being fed by it!—Pastor Martin Mbandu, former pastor, Kisumu Pentecostal Church.

4.1 Summary

Lake Victoria (Fig. 4.1a), the world's third largest lake and the largest in the developing world, is ideally the "*mother of the Nile*" given the seasonality characteristics of the Blue Nile. It is the source of the White Nile (see Fig. 2.4), which flows from Jinja in Uganda (Fig. 4.1b) to join the Blue Nile (see Chap. 2) in Khartoum (Sudan) to form one Nile and provides water for irrigation, transport, domestic and livestock uses, and supports the livelihood of more than 76 million people who live around it [6, 12, 17]. The presentation of this book, therefore, would not be complete without a special coverage of the Lake, which is so vital to the entire Nile River Basin (NRB). Nicholson [14, 15] documents its significance as an indicator of environmental and climate change over long-term scales. Since the 1960s, the lake level has experienced significant fluctuation, see e.g., [14, 15], which impacts on the overall water budget of the Nile. From 2001 to 2006, for example, its water level showed a dramatic fall that alarmed water resource managers as to whether the lake was actually drying up [21]. Kull [13] reported that the lake's levels fell by more than 1.1 m below the 10 year average. This chapter presents the Lake and its challenges, which are significant for the overal Nile waters. A more detailed elaborate coverage of this precious lake, its potentials and challenges are presented in the books of Awange [6, 12].

J. Awange, *The Nile Waters*,
https://doi.org/10.1007/978-3-030-64756-8_4

4.2 Features of the Lake and Its Environs

4.2.1 The Origin

It is important to know, for historical and scientific purposes, the possible origins and age of Lake Victoria (Fig. 4.1). How old is Lake Victoria? How did it come about? Has it changed much during its existence? If so, how? Why should we be interested in knowing the age of Lake Victoria? These are some of the pertinent questions, which have occupied the attention of scientists for centuries; and rightly so. The origins and age of the Lake are still shrouded in deep mystery. Although scientists have grappled with these questions, no definitive answers have been forthcoming. As it is, the origins of Lake Victoria are still the subject of scientific dispute [3]. It seems likely that it is much more recent in its origins than the other great Lakes of Eastern Africa. But there is an interesting aspect of the direction of flow of waters into the Lake. There is a scientific contention that many of the rivers now flowing east into Lake Victoria (including Kagera) once flowed west.

Hickman et al. [2, p. 36] suggests that Lake Victoria was formed due to warping and tilting of the land. They note that lower reaches of some river valleys have been

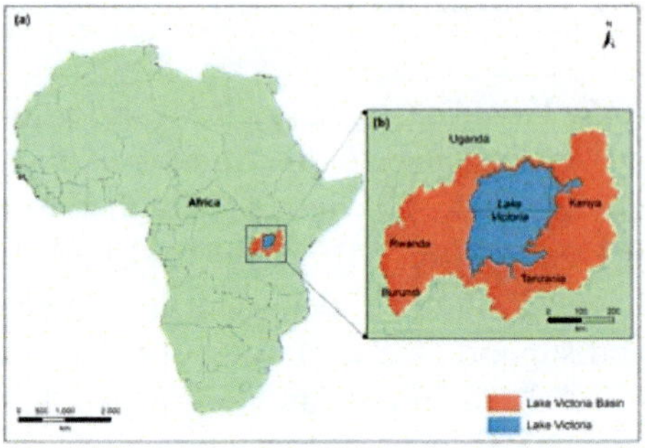

(a)

(b)

Fig. 4.1 **a** Lake Victoria Basin (LVB). *Source* Morgan et al. [1], and **b** The White Nile flowing from lake Victoria through Nalubale dam (Uganda) (Fig. 2.4b) raised, so that they flooded, and the directions of the rivers changed at least in the Miocene, Pliocene, and part of the Pleistocene eras; within the past 2 million years, possibly eventually into the Nile system. A more recent upthrust of the western side of the basin is thought to have however reversed these rivers, and caused Lake Victoria to form by flowing eastwards [3]. Although Fuggle [4] suggest that the Lake formed sometimes during the last 400,000 years, Aseto and Ong'ang'a [3, 8] suggest the possibility of the Lake having formed as recently as 25,000 to 35,000 years ago and dried up completely between 10,000 and 14,000 years. Samples of Earth taken from the bottom of the Lake corroborate this suggestion by indicating that the Lake may have temporarily but completely dried up 12,000 to 14,000 years ago. But this suggestion raises a host of other questions. For instance, *what happened to make the Lake dry up? What happened to make the dry Lake become a freshwater Lake? Could it dry up again in the future?* In attempting to address the issue of the drying of the Lake, Johnson et al. [7] suggest that the Lake probably dried up entirely about 15,000 years ago due to climatic phase of extreme dryness. Be that as it may, some scientists maintain that tectonic movement that swept across the Eastern African plateau about 14,600 years ago caused the formation of shallow basins, which filled with waters to become what is today ***Lake Victoria***. However, Some unresolved issues requiring further research include:

- Resolving the age of Lake Victoria since the conflicting explanations as to the age of the Lake still persists.
- Finding out whether the Lake can die under stress, as was supposed to have happened thousands of years ago?
- Continuous monitoring of fluctuation of its water level to establish the trend as to whether the lake is dying.
- Establishing whether the Lake is getting deeper or shallower with the passage of time and also whether it is expanding or decreasing, thus increasing or reducing its area of coverage.
- The Lake's parameters, e.g., length, area, etc., vary according to various authors as will be seen in Sect. 4.2.3. Use of satellite techniques discussed in Chap. 3) can resolve this once and for all as demonstrated by Awange et al. [5].

4.2.2 The Name "Lake Victoria"

We elaborate in this section how the name Lake Victoria was coined. Is the name Lake Victoria a proper name for the Lake? What is in a name? Does a name matter? If so, how and why? From time immemorial, these questions have been asked over and over again, and in different contexts. The general answer is that giving a name to an entity is an extremely important decision that people can make, in that, it is a means of providing an identity. A good name is given to a good thing; and a bad name is usually given to a bad thing. In the Luo language in Kenya, it is taken axiomatically that "Nying' ema ichiemogo", which is translated to mean "one eats by the fame of one's name". It is for this reason that human beings name their children after important persons, or give their children the names of their parents and fore parents to maintain their genealogy, or after important events, or in honour of important events or entities. It is also the reason why human beings give names to animals, hills, valleys, and stars. And even at an advanced age, individuals can acquire new names. Some are given names such as "sibuor" (Luo name for a lion) in recognition of one's bravery, etc.

The inhabitants around the Lake have followed suit; they always had different names for the Lake. The Lake is also known as Victoria Nyanza. This appears to have been derived from the Basuba (a tribe in Kenya) name for the Lake: "Nyancha" or "Nyanja" by certain communities in Uganda. The Luos on the Kenya side call it "Nam (Namb) Lolwe" (endless waters). The Baganda in Uganda name it "Nalubaale", the Sukuma in Tanzania call it "Sukuma Lake". This being the case, why then is the Lake called Lake Victoria?

The name *Victoria* is "foreign" in that it was named in honour of an individual in a foreign land. "Foreigners" had different names for the Lake. The Arabs called it *"Ukerewe"*. But an Englishman, explorer John H. Speke, stumbled onto its southern shore and proclaimed he had discovered the fabled source of River Nile in 1858. He built a pillar at the site (Fig. 2.4) and being the first European to sight it, he decided to honour Queen Victoria of England by naming it after her. One recalls that during 1890s, Great Britain colonized Kenya and Uganda while the Germans had a stronghold of Tanzania's mainland in 1899. In 1875, an American journalist, Henry Morton Stanley, seeking to confirm Speke's claim and looking for the British missionary Dr. David Livingston, circumnavigated the Lake. He spent two weeks spinning tales of God and England to curry favour with Mtesa, king of Buganda and ruler of the northern Lake region. Then Stanley sent word back to England, calling for missionaries. They came with soldiers and traders. He popularized the name Lake Victoria in the Western world. And the name Lake Victoria has continued to-date.

Nevertheless, African countries around the Lake have not been content with the present name of Lake Victoria. Thus, soon after the independence of the three surrounding East African states of Uganda, Kenya and Tanzania, there were attempts to rename the Lake so as to give it a proper African name; but such attempts have not yet succeeded, and the name Lake Victoria has been retained. It may be just a matter of time before the Lake is given its rightful appropriate African name that would give its true African identity and character. After all, changing the name of the Lake from a 'foreign' tag to an African name should not be a problem. Several examples of name change abound in East Africa and elsewhere in the continent and in the world. In the case of Kenya, African names have replaced foreign names. For instance, upon the country's independence on December 12, 1963, all landmarks, monuments and streets in Kenya that bore foreign names were given local, or to put it bluntly, African names. Lord Delamere Avenue in Nairobi was renamed Kenyatta Avenue. Queen's Way was renamed Mama Ngina Street. Government Road became Moi Avenue. And Duke Street became Ronald Ngala Street. Similar examples abound in the country. Thus changing the name of the Lake from a foreign one to an African name is in order and is consistent with the tradition of name changing initiated since independence.

Could the name "***Nyanza***" be a good candidate for the changed name, given that originally, the natives in the Suba District called it by that name? Already, the communities around the Lake have recently expressed their desire to rename the Lake. At a cultural conference held in Mwanza, Tanzania, in 2001, the cultural leaders selected from major ethnic communities around Lake Victoria questioned the rationale for retaining the name Victoria. Following extensive discussions on the matter, the leaders recommended that the Lake be renamed. Their argument was that the current name makes their Lake sound foreign and detached from their cultures.

4.2.3 Lake Victoria Basin: Physical Description

Here, only a brief background of lake Victoria is presented. For more details, we refer the reader to the books and article of [3, 5, 12]. Lake Victoria is the largest freshwater Lake in Africa (also largest tropical Lake) and the second largest in the world after Lake Superior in the United States. The Lake is also the largest in the developing world. It is located at Longitude 31° 39'E–34° 53' E and Latitude 0° 20'N–3° S, and has a surface area of c.a. 69,295 km^2 [5]. Its greatest length is about 400 km and its breadth is about 320 km.[1] Using a high spatio-temporal resolution Sentinel-2 remotely sensed images (see Sect. 3.2), Awange et al. [5] established the 2018 values to be 386 km (length) and 362 km (width).[2] It contains about 2,760 cubic kilometers of water and is situated at an altitude of 1135 m above mean sea level. The shoreline is very irregular and totals some 3,300 km in length. Hughes and Hughes [9] have put the shoreline close to 3,500 km in length, which is less than the 2018 value of 4.572 km established by Awange et al. [5]. The shoreline on the Kenyan side is 760 km. Much of the Lake is relatively shallow, reaching a maximum depth of about 80 m (Fuggle [4] puts the figure to 85 m), and an average depth of about 40 m; the deepest zone (60–90 m)[3] lies toward the shore. There is little annual variation in water temperatures, the mean surface being about 24°C, and that of deeper waters to be 23°C.

Its basin (Fig. 4.1) spreads from Longitude 30° E–35° 50' E to Latitude 0° N–3° 20'S. The ***shoreline*** is comprised as follows: 17% in Kenya, 33% in Tanzania and 50% in Uganda. Of the total catchments area of 193,000 km^2, Burundi accounts for 7.2%, Kenya 21.5%, Rwanda 11.4%, Tanzania 44.0% and Uganda 15.9%. The entire drainage basin covers an area of 258,000 km^2. Awange et al. [5] show the lake's mean surface area to be 69,295 km^2 (i.e., 812 km^2 or 1.2% more than that of the 37 publications that they reviewed) and its 2018 value to be 69,216 km^2 (i.e., $\sim$733 km^2 (1.1%) more than the mean of the 37 publications). Details of this study are presented in [5, 6]. The Lake is also the source of the river Nile (i.e., the source of

---

[1] Other sources, e.g., Britannica Concise put the length at 337 km and width at 240 km.

[2] The mean values of Awange et al. [5] from the 1984, 2002, 2017 and 2018 images were 388 km for length and 364 km for width respectively.

[3] 82 m—Britannica Concise.

Table 4.1 Morphoedaphic characteristics of Lake Victoria [5, 6, 10]. The mean width*, length* and area* are with respect to the Landsat (1984, 2002, 2017) and Sentinel-2 (2018) images

Characteristic	Measure
Position: Latitude	00° 20' N to 03° 00' S
Longitude	31° 39' E to 34° 53' E
Altitude (m above mean sea level)	1134
Catchment area (km^2)	184,000
Lake surface area in 2018 (km^2)	69,216
Mean area* (km^2)	69,295
Lake area as percentage of catchment	37
Shoreline in 2018 (km)	4,572
Maximum length in 2018, North-South (km)	386
Maximum width in 2018, East-West (km)	362
Mean width* (km)	364
Mean length* (km)	388
Maximum depth (m)	84
Mean depth (m)	40
Volume (km^3)	2,760
Inflow (km^3/yr)	20
Outflow (km^3/yr)	20
Precipitation (km^3/yr)	114
Annual fluctuations in level (m)	0.4–1.5
Flushing time (yrs)	138
Residence time (yrs)	21

the White Nile while the Blue Nile originates from Ethiopia), arguably the world's longest river. The Lake's main physical parameters have been summarized by [5, 6, 10] in Table 4.1.

Because the Lake is shallow, its volume is substantially less than that of other Eastern African lakes with much smaller surface area. The volume of Lake Victoria's water is only 15% of the volume of Lake Tanganyika, even though the latter has less than half the surface area. The Lake straddles the Equator and touches the Equator in its northern reaches. It is shared by Kenya (6%), Uganda (45%) and Tanzania (49%). The Lake's shoreline is long and convoluted, enclosing innumerable small, shallow bays and inlets, many of which include swamps and wetlands, which differ a great deal from one another and from the Lake itself.

Lake Victoria contributes much to the regional rainfall and subsequently the hydrological cycle. Of particular importance to Lake Victoria are the South-East trade winds, which picks moisture from the lake while passing. They move to the northern and western shores of Lake Victoria where they eventually condense to give heavy rainfall. Uganda benefits more due to the fact that winds carry more moisture to the *Northwest* shores of Lake Victoria, with major rainfall season normally between March, April and May (MAM), with low rainfall season occurring between September, October and November (SON). There exist no remarkable dry months.

The climate of Lake Victoria catchment area is mild with small variations in monthly average air temperature between 19°C and 25°C throughout the year. Daily temperature fluctuates more widely, ranging from 15°C to 30°C. Rainfall in this catchment area averages about 1,300 mm annually, varying from 2,000 mm in highlands to 1,000 mm in the north, southwest and lowlands along the Lakeshore.

4.3 Population and Demographic Features

Lake Victoria Basin (LVB) had a population of approximately 21 million in 1997, 26 million in 1999 (above 30 million if Rwanda and Burundi are considered), with a growth rate of around 3% per annum with considerably higher rates in urban areas. Interesting results are presented by [18] who puts the 2010 population to over 42 million and projects the population to increase to over 60 million in 2020, over 76 million in 2030 and tripple to over 113 million in 2050. Indeed the population estimate of Lake Victoria basin inhabitants by 2017 is placed at above 76 million, with a population growth above 4% as shown in Fig. 2.2d. On the commonly used figure of 30 million population of LVB, [18] writes:

> *Interestingly, our population estimates derived for the LVB reveal that the total population that most stakeholders for the basin, including the Lake Victoria Basin Commission, were commonly citing—about 30 million inhabitants likely underestimated by more than 10 million people. While the number of 30 million people was being used in reports and other communications about the threat of population growth, we could identify no original source for this estimate.*

The region is increasingly characterized by growing variability of ethnic composition. However the larger ethnic groups dwelling around Lake Victoria are the Luo, Samia, and the Suba people in Kenya; the Buganda and Busoga in Uganda; In Tanzania the distribution is as follow; in Mwanza region are the Sukuma, Kerewe, Jita, Kara, and Zinza. Sukuma constitute the overwhelming majority in this region. In Mara region are the Jita, Luo, Kuria, Zanaki, and Ruli. In Kagera there are Haya, Hangaza, Nyambo, Subi, and Sukuma. This means that there is a very great cultural diversity in the basin requiring a complex analytical framework as regards harmonization towards better and sustainable use of the natural resources of the basin for improving people's livelihoods.

The five countries in the Lake Region (i.e., Kenya, Uganda, Tanzania, Rwanda and Burundi) cover a total land area of 170,270 km^2, supporting one of the densest and poorest rural populations in the world. Women constitute just over 50% of the population in the region. In the Lake Victoria region as a whole, compared to the 2010 population of over 42 million, Bremner et al. [18] estimate the population increase by 43% in 2020, 80% in 2030 and 167% in 2050 (see also Fig. 2.2d for the projected growth rates). About 50% of the population in the region comprise youth who need more gainful employment opportunities to avoid the risks of social insecurity. The region already experiences some insecurity because of internally displaced persons as well as refugees from neighbouring countries.

In practically all the East African countries, increasing attention is being directed toward the status of women in society. Programmes aimed at mainstreaming gender issues in development activities are at various stages of development. To date, however, the socio-economic status of women is lower than that of men in a number of ways. For instance, women have limited access to productive resources such as land and capital. The literacy levels are lower among women than among men and the number of girls in secondary schools is considerably lower than that of boys. At the primary school level, the numbers are about the same. Women bear a heavier burden of the widespread poverty. Commercial fishing and processing factories employs more men than women and has impacted negatively on women who were previously engaged in fish processing and marketing. Moreover, when men migrate from rural to urban areas in search of employment, they leave behind women and children to bear the brunt of agricultural activities. The heavy work load women bear on farms, in their reproductive and community roles is a major source of poor health. Due to lack of money, women have limited access to health facilities [19, p. 8].

There is a strong cultural dimension to environmental degradation in the region. The dictates of cultural practices of sons inheriting their fathers' lands and wives owning lands to cultivate are reinforcing the need to subdivide land into small units, which are uneconomic for meaningful farming. Such practices continue to generate a population of landless youth who must migrate elsewhere to earn a living and the cycle of poverty created continues to cause further environmental degradation. Population pressure on limited land leads to rapid land degradation. It is, therefore, imperative that conservation measures are adopted on a massive scale, if not then, it may not be possible to control the rate of environmental degradation.

Livelihood standards of the area have deteriorated as already noted by [20], not just because of the consequences of the population increases, but through a host of limiting factors and livestock production that lead to declining land productivity. There are also the people of Rwanda and Burundi in the wider Lake catchment region. Population density in the Lake basin (see e.g., Fig. 2.5b) is above the national averages found in all three riparian countries and the populations within these riparian communities grow at rates that are among the highest in the world. Population density in the region ranks among the highest in the world for rural areas where the majority of the people live (Fig. 2.5b).

4.3.1 Historical Perspective of Early Settlements

East Africa is largely considered the cradle of mankind, since the first paleontologic traces of humans have been found there. Most of the present Bantu-speaking inhabitants of the LVB started entering the basin around the beginning of the second century. The communities presently settled around Lake Victoria were immigrants

Fig. 4.2 Kabaka palace. **a** main office, **b** the compound with Kabaka's flag (**c**) Kabaka's lake where he could go and bathe, and **d**, remnant of Kabaka's vehicle when he fled during President Obote's take over in 1966

from other parts of Africa. As early as the 13th and 14th centuries, Lake Victoria was well surrounded by organized and settled communities, which had and still enjoy a lot of interactions. The Luos arrived via the River Nile from the North of Africa and settled in the Lake Victoria basin in Uganda before moving to Kenya. When the Luos arrived, they forced out other tribes like Elgon Masai and Bantus that had earlier settled in the Lake basin. They interacted with ethnic groups that lived in Uganda. Some of the Luos later moved to North Mara in Tanzania in search of grazing ground when drought destroyed their crops. The Bantus who now occupy the Lake Victoria basin in Uganda and some parts of Tanzania trace their origins in Central Africa. Their settlement developed into indigenous kingdoms in the 14th century. Among them were the Baganda, Banyoro, Toro, Ankole and Basoga. The Baganda people created a dominant kingdom that could not be penetrated until the 19th century (see Fig. 4.2 for the Buganda Kingdom's palace[4]). The Kabaka king Mutesa II reigned until his overthrowal in 1966 by Milton Obote who himself was overthrown in 1971 by the

[4] http://www.buganda.com/crisis66.htm.

Fig. 4.3 Historical sites within the lake Victoria surrounding. **a** Path to Uganda's torture chambers during Idi Amin's regime, **b**, **c** and **d** inside the torture chambers

butcher of Uganda (Idi Amin, see Fig. 4.3). The other most important group was the Basoga who later established their kingdom around Jinja. Before colonization, the larger tribes that occupied the southern shores of Lake Victoria in Tanzania were the Wasukuma and Wanyamwezi. Currently their population runs into millions. Other smaller tribes like the Wanazaki, Wajita, Wakerewe of Tanzania; the Bunyala, Samia and Suba of Kenya and the Samia of Uganda are some of the most important ethnic groups that live on the shores of Lake Victoria. We have among the communities the Asians who, although constituting a minority, were brought by the colonialists to build the then Kenya-Uganda railway. The group is among town dwellers and still controls major businesses in the Lake Victoria port towns and cities (Kampala and Kisumu).

The earlier settlers brought with them distinct cultures into the basin. Through interactions, intermarriages amongst the different cultures became imminent. The cross transfers of communities were enhanced among the neighbourhood communities, through trade. The barter system amongst the communities became prominent as canoes and dhows transported goods. Trade and market centers were developed and later small towns were established in the Lake ports. This was the origin of the municipalities and cities that are now well established. The major towns include Kampala, Entebbe and Jinja in Uganda, with Kampala being the capital city of Uganda; Kisumu, Homa Bay in Kenya, with Kisumu being the third largest city in Kenya after Nairobi and Mombasa; and Mwanza, Musoma and Bukoba in Tanzania. These towns' populations run into millions. The peoples of the LVB share a relatively common history characterized by intensive interaction, extensive trading, intermarriage and welfare. In other cases, assimilation occurred, e.g., the Kenyan Suba were assimilated into Luo and now speak and practise Luo culture. Young Subas can hardly speak their native Suba language but Dholuo (Luo language).[5] In essence, the Subas are Luos for all intent and purpose. The regime of former president Moi attempted to divide the two communities through the 1988 census but the Subas and Luos remained a single entity. Furthermore, the communities within LVB also share the historic experience of external interventions in the area, such as the slave trade and colonialism.

In 1885, Tanzania, Rwanda, and Burundi were placed under the rule of the Imperial German Government. However, after the first World War, Tanzania was placed under the British mandate until the establishment of the Trusteeship system under the Charter of the United Nations, while Rwanda and Burundi came under Belgium rule. In 1887 Kenya was placed under the Imperial British East Africa Company, but the territory was transferred in 1895 to Britain. Likewise, the British established the Uganda Protectorate in 1894. However, after 1945, various nationalist movements emerged in this region, and began a sustained and successful effort to emancipate themselves from European domination. In 1961, Tanzania became independent followed by Uganda, Rwanda, and Burundi in 1962 while Kenya gained her independence in 1963. The delay in Kenya's independence is argued to be due to Jaramogi Oginga Odinga's (a Luo from Siaya County within Lake Victoria Basin) refusal to assent to the throne unless Jomo Kenyatta (Kenya's first president) was released from prison.

4.3.2 Impacts of Colonialism

The establishment of colonial rule over the people of the region brought with it resource management structures that removed the power from traditional leaders to the central governments of larger territories of Kenya, Uganda, and Tanganyika. This meant that people who had no interest in the Lake could be given the responsibility to manage resources such as fisheries. The ownership of the resources shifted with time. The population influx and lack of properly planned infrastructural development has brought many problems to the urban centers. These problems include non-functional sewerage systems, industrial pollution and shortage of water supply. These have become a serious headache for the citizens. County governments alone cannot address these problems. The citizens have to join them with 'own key' solu-

[5] https://www.standardmedia.co.ke/the-standard-insider/article/2001386038/how-abasuba-accepted-luo-love-and-lost-their-culture.

tions through organized groups such as Non-Governmental Organizations (NGOs) and Community-Based Organizations (CBOs). Kenya changed its constitution in 2010 to allow for the devolved system and is proposing the current building bridges initiative (BBI) to increase county revenue from the current 15% to 35%. This if properly managed could see huge development within the Kenyan side of the LVB. However, rampant corruption and mismanagement of the resources that bedevilled the first county governments of Kisumu, Siaya, Migori, Homabay and Busia in the Kenyan side of LVB could not let this happen. Hopefully, the future county governments will learn from the mistakes of their predecessors. The current (2020) Kisumu county government under Professor Anyang Nyongo seems to have hit the right pedal.

4.4 Lake Victoria: Benefits and Challenges

Of all the tropical's Lakes, Lake Victoria stands tall as the greatest fresh water body. In the entire world, it comes only second to Lake Superior. Within its surrounding, it directly supports a population of more than 76 million East Africa's inhabitants (i.e., $\frac{1}{3}$ of the combined population) [5, 17], while globally it is the source of Tilapia (*Oreochromis niloticus*) and Nile Perch (*Lates niloticus*) [3]. So important is the Lake such that Egypt and Sudan entered the Nile treaties of 1929 and 1959 for exclusive use of its waters (see e.g., Sect. 2.4). Just as oil is an important resource of the world, Lake Victoria and its basin plays a major role in the world such that it appears at the top of a number of lists as pointed out by [11]. Specific roles played by this Lake are [3, 6, 12]:

- Agricultural (rain-fed as well as irrigated) potential. Its surface water could be used for irrigation whereas its modulating of the regional hydroclimate contributes to rainfall that is essential for rain-fed agriculture.
- It is a source of fresh water that supports livelihood of people living within its shores. The water is used for drinking, domestic use, in industries, for transportation and sports among others.
- It contributes to the river Nile, the lifeline of Egypt and Sudan.
- It is the habitat of Tilapia (*Oreochromis niloticus*) and Nile Perch (*Lates niloticus*), the two types of fish that are in great demand the world over together with other species of fish who reside in its waters.
- Its wetlands provide homage to various bird species, insects and thus preserving biodiversity.
- It helps fight poverty by creating employment in fish industry and transportation for people living within its environs.
- It acts as eco-tourism center.

The list could go on and on but we have chosen to mention only but a few.

Fig. 4.4 A huge floating island in Lake Victoria clogging a turbine in a hydroelectric power station. *Source* https://www.bbc.com/news/world-africa-52286296

Although the Lake and its basin boasts of such rich endowment, the fact of the matter is that it is a sick giant [4]. This has necessitated the establishment of several measures to help address the challenges faced by this Lake. Such measures include the establishment of Lake Victoria Management Environmental Programme (LVEMP) and other institutions. The challenges facing the Lake include:

- Having as its dependants people who are faced with the health threat such as HIV/AIDs and presently (2020–2021), the COVID-19 pandemic.
- Being infested with the deadly water hyacinth, which is depriving it of oxygen and thus suffocating it, and posing threats to hydropower plants (see Fig. 4.4).
- Faced with people living in its surrounding languish in abject poverty although it is endowed with great resources.
- Consequences of decision and policies made in far parts of the world, i.e., ecological power, and by global economic structure [4].

In general, the Lake faces four critical and mutually reinforcing problems. The *first* is the *obsolete Nile treaties* that lock up its potential for irrigated agriculture. If these treaties could be done away with, and the lake used sustainably, then its contribution to irrigated agriculture could go along way in alleviating food insecurity over the entire Greater Horn of Africa (GHA, see e.g., [23]). *Second* is *productivity*. The most important commercial species caught is Nile Perch. A predatory species that has over time eaten to near extinction many species endemic to the Lake that were once important sources of food and livelihood for the local communities. The scramble for the catch has become the source of conflict between fishers from the different countries. Overfishing is attributed to the use of inappropriate gear and insufficient enforcement capacity. Open access and inappropriate exploitation increases the fishing effort against the diminishing reproduction rate. In addition, territorial boundaries tend to concentrate the fishing effort on a limited area especially in Kenya. Authorities lack the resources nor the will to effectively police the Lake.

The *third* set of challenges are *environmental*, namely resource degradation emanating from catchment degradation and discharge of human, industrial and agricultural effluent. The *fourth* set of challenges is *market related* and includes price competitiveness, quality assurance and reliability of supply. The principal export product Nile Perch, is under intense price pressure from aquaculture, in particular the new tilapia varieties. While the fishery has made substantial progress in meeting European (EU) standards, it has not yet attained the eco-labelling certification that is increasingly required by the retailers [16].

Owing to these overwhelming evidence of the importance of Lake Victoria together with its emerging threats, a continuous monitoring of its status is inevitable. A comprehensive coverage of Lake Victoria's resources and related challenges are presented in Awange and Ong'ang'a [12]. Awange [23] presents the hydroclimate of Greater Horn of Africa and considers the potential of irrigated agriculture, e.g., from Lake Victoria's surface water. The current book looks at the feasibility of weighing the Nile River Basin from space while.

4.5 Fluctuations: Climatic or Anthropogenic Induced?

The lake's level has undergone fluctuations in the past. For example, with the receding of the lake waters in 2002–2006, acres of land that were lost to the floods of the 1960s were fast being reclaimed, creating sources of conflicts between man and wildlife. In some beaches, e.g., Usoma in Kenya, wetlands that were once breeding places for fish were dying up, leaving areas of land as playing fields for children and farmland. Ships were now forced to dock deep inside the lake, while the landing bays needed to be extended. Those who directly depended on the lake waters for domestic use were forced to go deeper into the lake to draw water, thus exposing women and children to water-borne diseases and risks of snakes and crocodiles. Water intakes that supplied major towns and cities had to be extended deeper into the lake, thus causing more financial burden to the municipalities that were already strained financially [21]. In 2020, the floods were back leading to speculations that the lake was retracing its 1960s level.

With 80% of Lake Victoria water coming from direct rainfall, changes in the lake level are directly related to the variation in the water stored in its basin, which contributes around 20% in the form of river discharge. A decrease in stored basin water was therefore suspected to contribute to the drop in the lake level. An analysis of the stored water in the Lake Victoria Basin (LVB) in relation to rainfall and evaporation was therefore necessary as a first diagnosis. This would provide water resource managers and planners with information on the state and changing trend of the stored water within the basin. Such basin scale observations could only be achieved through the use of satellites such as Gravity Recovery And Climate Experiment (GRACE) discussed in Chap. 3. Conventional methods for studying variations in stored water such as the Artificial Neural Network, Geographical Information System (GIS) and remote sensing could not diagnose the problem, see e.g., [21].

Having been motivated by the potential of the GRACE satellites, Awange et al. [21] undertook a satellite analysis of the entire LVB in an attempt to establish the cause of

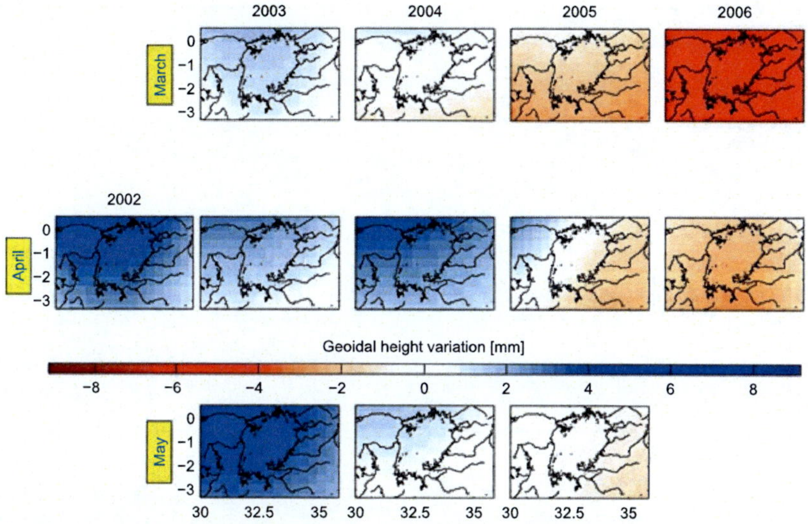

Fig. 4.5 Inter-annual comparison in the geoid during the high rainy season of MAM from 2002–2006. The figure indicates a decline in total water storage in the Lake Victoria basin during this period. *Source* Awange et al. [21]

the decline in Lake Victoria's water levels. The GRACE and Challenging Minisatellite Payload (CHAMP) satellites (Fig. 3.2) together with data from the The Tropical Rainfall Measuring Mission (TRMM) satellite were employed in the analysis. Using 45 months of data spanning a period of 4 years (2002–2006), the GRACE satellite data were used to analyze the gravity field variation caused by changes in the stored waters within the lake basin. Figure 4.5 presents the annual variation of the geoid in the lake's basin during the high rain season months of March, April and May (MAM) for the period 2002–2006. The GRACE results indicated that the basin's total water storage dramatically decreased at a rate of 6.20 mm/month. These changes are expressed in equivalent water thickness (also known as total water storage (TWS)) in Fig. 4.6. For the period 2002–2006, the results indicate a general decline in the lake basin's water level at a rate of 1.83 km^3/month [21]).

To validate the GRACE results, TRMM Level 3 monthly data for the same period of time were used to compute mean rainfall at a spatial resolution of $0.25° \times 0.25°$ (25×25 km), as shown in Fig. 4.7, from which the rainfall trends were analyzed (Fig. 4.8). To assess the effect of evaporation, Global Navigation Satellite System (GNSS) remote sensing data (59 CHAMP satellite occultations) for the period 2001 to 2006 were analyzed to define if tropopause warming took place [21]. The results indicated that the tropopause temperature fell in 2002 by about 3.9 K and increased by 2.2 K in 2003 and remained above the 189.5 K value of 2002. The tropopause heights showed a steady increase from a height of 16.72 m in 2001 and remained above that value reaching a maximum of 17.59 km in 2005, an increase in height by 0.87 m. Temperatures did not, therefore, increase drastically to cause massive evaporation. TRMM results indicated the rainfall over the basin (and directly over

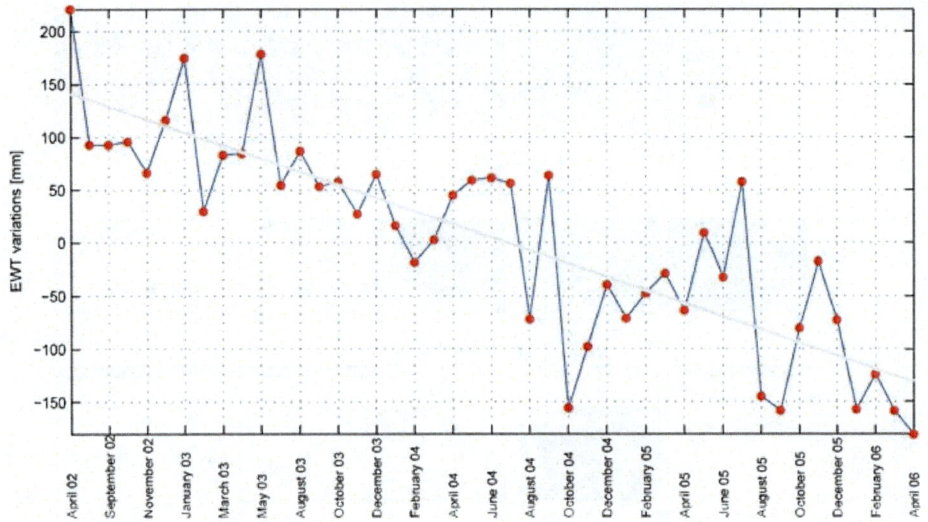

Fig. 4.6 Lake Victoria basin total water storage (equivalent water thickness) changes between 2002–2006, as seen by the GRACE satellite. The figure indicates that the GRACE satellites observed a general decline in the lake's basin waters over this period (cf. Fig. 3.5 obtained from satellite altimetry). *Source* Awange et al. [21]

the lake) to have been stable during this period (see Figs. 4.7 and 4.8). Since rainfall over the period remained stable, and temperatures did not increase drastically to cause increased evaporation, the remaining major contributor during the period 2002–2006 was suspected to be discharge from the expanded Owen Falls (Nalubaale) dam. Awange et al. [21] concluded, thanks to the GRACE and GNSS satellites, that the fall in Lake Victoria's water level between 2002 and 2006, was due to human impact on the basin's environment (i.e., expanded dam) as opposed to natural factors.

In a related work, Swenson and Wahr [22] used satellite gravimetric and altimetric data to study trends in water storage and lake levels of multiple lakes in the Great Rift Valley region of East Africa for the years 2003–2008. GRACE total water storage estimated by Swenson and Wahr [22] corroborated the findings of Awange et al. [21] that the lake's water level had declined by as much as 60 mm/year, while their altimetric data indicated that levels in some large lakes in the East African region dropped by as much as 1–2 m. Swenson and Wahr [22] concluded that the largest decline occurred in Lake Victoria and, like Awange et al. [21], attributed this to the role of human activities.

Both the findings of Awange et al. [21] and Swenson and Wahr [22] provide evidence that the GRACE satellites could be used to provide independent means of assessing the relative impacts of climate and human activities on the balance of stored water that does not depend on in-situ observations, such as dam discharge values, which may not be available to the public domain.

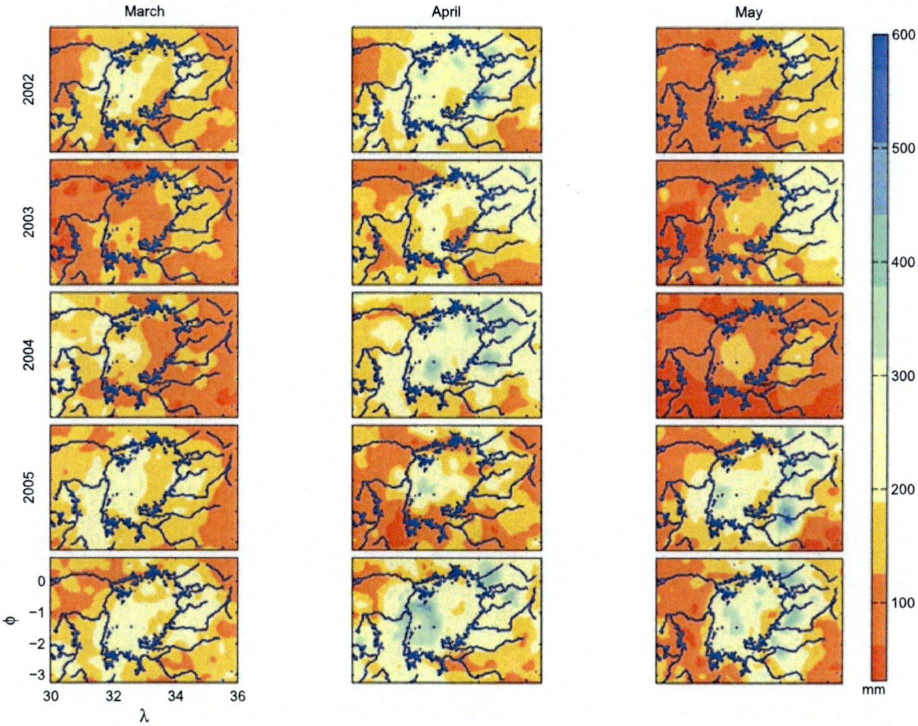

Fig. 4.7 Inter-annual comparison of rainfall over the Lake Victoria basin during the high rainy season of MAM from 2002–2006 using TRMM results. *Source* Awange et al. [21]

4.6 Concluding Remarks

This chapter has discussed the origin of Lake Victoria, its name and some elements of demographic features of the region. It is a region where the population growth rate is very high and average population density also high. These two phenomena have a direct bearing on the environment, leading to serious environmental degradation, soil erosion, pollution of the Lake and the decline in soil fertility. It is therefore imperative that measures be taken to address the issues concerning population pressure and environmental degradation. Whereas these measures are necessary, it is more important that the lake waters themselves be continuously monitored for the benefit of NRB at large. However due to its size, ground-based observation methods by themselves are insufficient hence the need of space-based methods discussed in Awange [6].

Lake Victoria is a riparian resource shared by 3 East African countries: Kenya, Uganda and Tanzania. It serves as a source of water for irrigation, which if properly utilized, could significantly enhance GHA's food security. As argued previously, however, the caveats imposed by obsolete Nile treaties of the 1920s and 1960s have to be removed. Furthermore, its modulation of the regions climate leading to rainfall

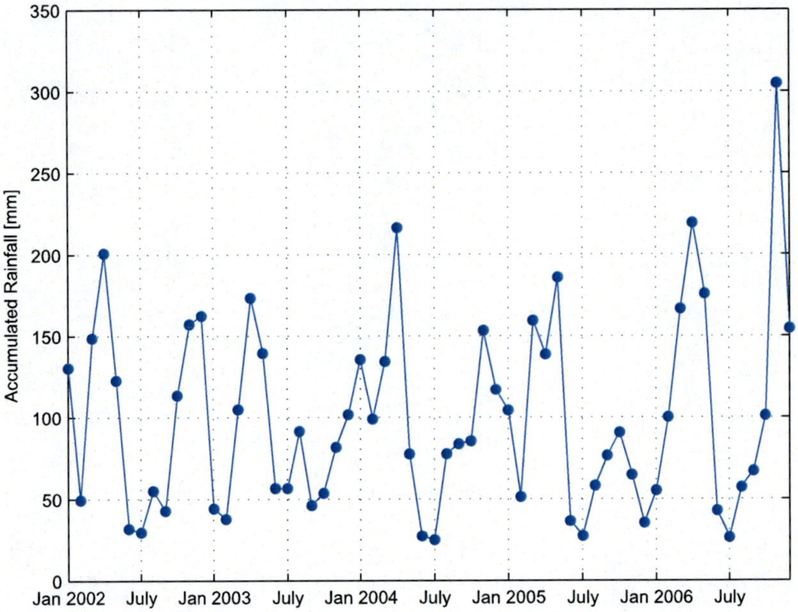

Fig. 4.8 Time series of rainfall 2002–2006 for the lake Victoria basin as observed by the TRMM satellite. *Source* Awange et al. [21]

within its basin is significant for rain-fed agriculture. Moreover, it is vital for transport, domestic and livestock uses and a source of livelihood for over 42 million people living around the lake. Its fish (tilapia and Nile perch) are exported the world over. Its role as an indicator of environmental and climate change on long-term scales has been documented in [14]. The vital role of this lake and its surroundings as a source of food has been documented e.g. in Awange and Ong'ng'a [6, 12]. The lake and its environment is under threat from environmental pollution and more recently from declining water levels. In 2020, the floods were back leading to speculations that the lake was retracing its 1960s level.

References

1. Morgan B, Awange JL, Saleem A, Hu K (2020) Understanding vegetation variability and their "hotspots" within Lake Victoria Basin (LVB: 2003-2018), 122. https://doi.org/10.1016/j.apgeog.2020.102238.
2. Hickman GM, Dickins WHG, Woods E (1973) The lands and people of East Africa. Longman, Essex
3. Aseto O, Ong'ang'a O (2003) Lake Victoria (Kenya) and its environs: resource, opportunites and challenges. Africa Herald Publishing House, Kendu Bay, Kenya
4. Fuggle RF (2004) Lake Victoria: a case study of complex interrelationship. In: United Nations environmental program: Africa environmental outlook case studies. Physical dynamics of

5. Awange JL, Saleem A, Sukhadiya RO, Ouma YO, Hu K (2019) Physical dynamics of Lake Victoria over the past 34 years (1984–2018): is the lake dying? Sci Total Environ 658:199–218
6. Awange JL (2021) Lake Victoria monitored from space. Springer Nature International
7. Johnson TC, Kelts K, Odada EO (2000) The holocene history of Lake Victoria. Ambio 29(1)
8. Ong'ang'a O (2002) Poverty and Wealth of fisherfolks in the Lake Victoria basin of Kenya. Africa Herald Publishing House, Kendu Bay
9. Hughes RH, Hughes JS (1992) A directory of African Wildlife. With a chapter on Madagascar, IUCN/UNEP/WCMC
10. Balirwa SJ (1998) Lake Victoria Wetlands and the Ecology of the Nile Tilapia Oreochromis niloticus. Linne. Ph.D. Dissertation, Balkema Publishers, Rotterdam, Netherlands
11. Pitcher TJ, Hart PJB (1995) The impact of species changes in African Lakes. Fish and fisheries series 18. Chapman and Hall, London, p 601
12. Awange JL, Ong'ang'a O (2006) Lake Victoria-ecology. Resource of the lake basin and environment. Springer, Berlin
13. Kull D (2006) Connections between recent water level drops in Lake Victoria, dam operations and drought. Available at: http://www.irn.org/programs/nile/pdf/060208vic.pdf. Accessed 25 Sep 2011
14. Nicholson SE (1998) Historical fluctuations of Lake Victoria and other lakes in the Northern Rift Valley of East Africa. In: Lehman JT (ed) Environmental change and response in East African lakes. Kluwer, Dordrecht, pp 7–35
15. Nicholson SE (1999) Historical and modern fluctuations of lakes Tanganyika and Rukwa and their relationship to rainfall variability. Clim Change 41:53–71. https://doi.org/10.1023/A:1005424619718
16. KLI (2004) Market based fisheries regulation strategy for Lake Victoria. A report of the regional inception workshop, 6–8 June 2004, Lake Naivasha Country Club, Kenya, 50pp
17. Nile basin water resources atlas (2017). http://atlas.nilebasin.org/
18. Bremner J, Lopez-Carr D, Zvoleff A (2013) Using new methods and data to assess and address population, fertility, and environment links in the Lake Victoria Basin, population and the environment, Session 247. https://www.iussp.org/en/event/17/programme/paper/6136
19. Ehlin U (1997) Lake Victoria basin: natural resource under stress. Stockholm, Department of natural resource and environment, Sida
20. Aseto O, Ong'ang'a O, Awange JL (2003) Poverty. A challenge for the Lake Victoria basin. OSIENALA Series 5, Printed by Africa Herald Publishing House, Kendu Bay, Kenya
21. Awange JL, Sharifi M, Ogonda G, Wickert J, Grafarend EW, Omulo M (2008) The Falling Lake Victoria Water Level: GRACE. Water Resource Management, TRIMM and CHAMP Satellite Analysis. https://doi.org/10.1007/s11269-007-9191-y
22. Swenson S, Wahr J (2009) Monitoring the water balance of Lake Victoria, East Africa, from space. J Hydrol 370:163–176. https://doi.org/10.1016/j.jhydrol.2009.03.008
23. Awange JL (2021) The greater horn of Africa's hydroclimate. Potential for agriculture amidst extreme drought. Springer Nature International, Berlin

Chapter 5
Ethiopian Highlands: The Water Tower of Nile

Water towers are areas that generate high stream flow compared with other areas. For the Nile River Basin (NRB), Ethiopian Highlands stand tall amongst its competitors and as such, understanding the spatio-temporal characteristics of its water storage changes is crucial for not only Ethiopia, a country that is facing a range of challenges in water management caused by anthropogenic impacts as well as climate variability but also for the overall management of the entire NRB's water resource.

—Awange et al. [90]

5.1 Summary

Water towers are areas that generate high stream flow compared with other areas [89]. For the Nile River Basin (NRB), Ethiopian Highlands stand tall amongst its competitors and as such, understanding the spatio-temporal characteristics of its water storage changes is crucial for not only Ethiopia, a country that is facing a range of challenges in water management caused by anthropogenic impacts as well as climate variability but also for the overall management of the entire NRB's water resource. In addition to this, the scarcity of in situ measurements of soil moisture and groundwater, combined with intrinsic "scale limitations" of traditional methods used in hydrological characterization are further limiting the ability to assess water resource distribution in the region. This chapter presents the study of [90] who applied remotely sensed and model data over Ethiopia for a period of 8 years (2003–2011) to (i) test the performance of models and remotely sensed data in modeling water resources distribution in un-gauged arid regions of Ethiopia, (ii) analyze the inter-annual and seasonal variability as well as changes in total water storage (TWS; surface, groundwater, soil moisture and vegetation) over Ethiopia, (iii) understand the relationship between TWS changes, rainfall, and soil moisture anomalies over the region, and (iv), identify the relationship between the characteristics of aquifers

J. Awange, *The Nile Waters*,
https://doi.org/10.1007/978-3-030-64756-8_5

and TWS changes. The data used included (see Chap. 3 for a discussion); monthly gravity field data from the Gravity Recovery And Climate Experiment (GRACE) mission, rainfall data from the Tropical Rainfall Measuring Mission (TRMM), and soil moisture from the Global Land Data Assimilation System (GLDAS) model. The results showed that the western part and the north-eastern lowlands of Ethiopia experienced decrease in TWS water between 2003–2011, whereas all the other regions gained water during the same period. The impact of rainfall seasonality was also seen in the TWS changes. Applying the statistical method of Principal Component Analysis (PCA) to TWS, soil moisture and rainfall variations identified the dominant annual water variability in the western, north-western, northern, and central regions, and the dominant seasonal variability in the western, north-western, and the eastern regions. A correlation analysis between TWS and rainfall indicated a minimum time lag of zero to a maximum of six months, whereas no lag was noticeable between soil moisture anomalies and TWS changes. The delay response and correlation coefficient between rainfall and TWS appeared to be related to recharge mechanisms, revealing that most regions of Ethiopia received indirect recharge. The results further showed that the magnitude of TWS changes was higher in the western region and lower in the north-eastern region, and that the elevation influenced soil moisture as well as TWS.

5.2 Ethiopian Hydrogeological Regimes: Characterization

Ethiopia's hydrology plays a significant international role, being the headwaters of the Blue Nile Basin, where it contributes about 86% of the total annual flow of the Nile [52, 77] and also approximately 90% of inflow into Lake Turkana [32], a lake situated in the arid area of northern Kenya. In recent decades, however, extreme hydrological variability, seasonality, and anthropogenic factors are posing challenges to the region's water resource management. For example, due to the large and increasing population pressure, insufficient agricultural production, a low number of developed energy sources, and drought episodes, Ethiopia, which has almost 94% of its population depending on wood fuel, is undertaking major hydropower and irrigation development schemes [12, 83]. In addition to irrigation and hydroelectric dams, land degradation and changes in land cover in Ethiopia where forests are being converted to agricultural land are having impact on the Nile flow, see, e.g., [63, 74].

Characterizing water-storage in all its forms (surface, soil moisture, and groundwater) and their responses to the incoming (precipitation) and outgoing (evaporation and discharge) water masses, therefore, is of great importance in terms of understanding extreme changes in stored water triggered by natural and anthropogenic factors. This is of interest especially in areas where the seasonality of rainfall is strong and the welfare of the society relies on the availability of water, such as in Ethiopia. To quantify the stored water resource of a region, soil moisture and groundwater have been documented to play a significant role, e.g., [64, 65, 78]. For developing countries such as Ethiopia, however, in situ networks of soil moisture and groundwater measurements are sparse. Therefore, these components are among the most difficult water budget parameters to obtain. To circumvent this shortfall, characterizing the hydrogeological regimes has been undertaken by considering their hydrogeological environments, see e.g., [11, 37, 43, 44, 88]. Most of the previous studies have concentrated on the climatic characteristics, e.g., [21, 58–60, 72, 73, 75, 82] and the use of environmental isotopes and hydrochemicals, e.g., [12] to trace the water availability in Ethiopia. For example, Berhane et al. [12] found the environmental isotopes and hydrochemicals approach to be the most effective tool for differentiating various forms of geochemical reaction, and to infer the environmental factors affecting groundwater quality and its flow in the region.

Besides concentrating on climatic characteristics, most of the studies above are also restricted to small scale hydrological characterizations, which do not reflect the large-scale water storage variability over Ethiopia. Moreover, isotopes and hydrochemical-based methods are costly, require skilled experts, and are often difficult to apply over large areas, which are not easy to venture into, particularly in developing countries such as Ethiopia. Furthermore, most of the hydrological studies over Ethiopia focused only on regional characterizations. Generalizing such local outputs to the whole of Ethiopia is extremely difficult due to its vast range of climatic and topographic conditions.

Total water storage (TWS; surface, groundwater, soil moisture and vegetation water), which is defined as the sum of all forms of water stored above and underneath the Earth's surface, is a key component of the terrestrial and global hydrological cycles that exerts important control over water, energy and biogeochemical fluxes, and thus plays a major role in the Earth's climate system [66, 80, 81], see Sect. 3.3.3. The Gravity Recovery And Climate Experiment (GRACE) mission offers the possibility of remotely sensing global and regional TWS changes. Launched in 2002 as a joint project of the US and Germany, GRACE products have contributed enormously to the study of changes in total water storage. With the help of complementary data sets, GRACE-derived products offer the possibility of monitoring groundwater depletion in data poor regions of the world [29, 36, 64]. To date, however, except for studies that have been done at continental or global scales [62, 70] and those in connection with the Nile Basin [4–9, 17, 51, 91–94], GRACE satellite products have not been applied specifically to the whole Ethiopian Basins at a local level. Within Ethiopia, for example, the application of GRACE products are reported, e.g., in Bonsor et al. and Melesse et al. [17, 51] with respect to the study of the Blue Nile. For instance, Melesse et al. [51] presents the low and high flow characteristics of the Blue Nile River using wavelets and applies GRACE products to analyse moisture flux. One of the major challenges in applying the GRACE products to estimate TWS changes over Ethiopian Basins is the fact that only 2 out of the 12 basins (Abbay and Wabishebele) can barely fulfill the requirement for the smallest resolvable basin area of 200,000 km^2, see also [81]. Furthermore, Longuevergne et al. [49] have pointed out that an accurate estimate of TWS changes in small basins using GRACE-derived products requires a compromise between competing needs for noise suppression and spatial resolution. To overcome the spatial limitations, in [90], Ethiopia was divided into ten regions of equal sizes of $4^\circ \times 4^\circ$. For each region, TWS were computed from GRACE data using the approaches presented in Wahr et al. [84].

Awange et al. [90]'s contribution focused on the remotely-sensed TWS changes over Ethiopia using GRACE products. For the purpose of evaluating GRACE products, [90] also uses rainfall and soil moisture data based on products from the Tropical Rainfall Measuring Mission (TRMM) [45] and the Global Land Data Assimilation System (GLDAS) [66], respectively. The major aims of [90] were (i) understanding the response of Ethiopian aquifers to TWS changes from information obtained following the analysis of inter-annual (i.e., between years variability), and intra-annual (i.e., processes that occur on a time scale of less than one year, but more than one month) GRACE products, (ii) depicting the reaction of each region to hydrological input (rainfall) including any time delays, and (iii), identifying the dominant pattern of intra-annual, annual, and seasonal variability over Ethiopia by applying the statistical method of Principal Component Analysis (PCA) on GRACE-derived TWS, soil moisture, and rainfall patterns. Knowledge of the lag time is important for understanding the longest period over which the available stored groundwater can be sustainably exploited after rainy seasons. The remainder of this chapter is organized as follows; in Sect. 5.3, the background of Ethiopian Highlands and its characteristics are discussed. Section 5.4 presents the remotely sensed/model products used and the employed analysis methods. The water tower is then characterized in Sects. 5.5 and finally, Sect. 5.6 summarizes the major findings of this chapter.

5.3 Ethiopian Highlands: Background

5.3.1 Location

Ethiopia, located between latitudes 3°15' N to 15°N and longitudes 33°E to 48°E, is a landlocked country bounded by Eritrea (North), South Sudan (South West), Sudan (North West), Kenya (South), Somalia (East) and Djibouti (North East), with a surface area of 1,127,127 km^2. Its altitude ranges from nearly 120 m below mean sea level in the Dallol depression to about 4620 m above mean sea level at Mount Ras Dashen. It contains three major physiographic regions that include the western highlands and associated lowlands, the eastern highlands and associated lowlands, and the Rift Valley in between them, running from north-east to south-west, separating the eastern and western highlands (Fig. 5.1). It is because of these physiographic influences on the drainage systems that Ethiopia is counted as the water tower of NRB, with twelve major basins; eight of which are river basins, one is a lake basin, and the remaining three are dry basins with no or insignificant out flow [31].

Ethiopia contributes to three major drainage systems (Fig. 5.1), the Mediterranean Sea drainage system (Abbay, Blue Nile, Baro-Akobo, Mereb and Tekeze), the Great East African Rift-Valley drainage system (Omo-Ghibe, Awash, Rift-Valley Lakes, Danakil and Aysha) and the Indian Ocean drainage system (Genale-Dawa, Wabishe-

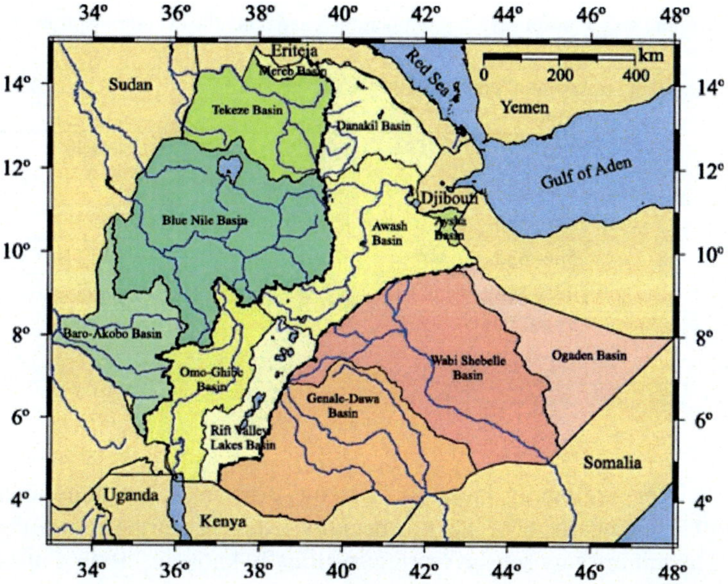

Fig. 5.1 Major rivers and basins in Ethiopia. The black dashed line represents the major water divides between the Mediterranean Sea Basin (West), the Rift Valley endorheic Basins (Center), and Indian Ocean Basin (East). *Source* Modified from Nyssen et al. [56]. The different colors mark out the separate basins. *Source* [90]

belle and Ogaden). The groundwater resources of Ethiopia and their distribution vary depending on their respective geological, structural, and climatic conditions. The near-surface geological pattern that mainly govern the hydrogeological characteristics of Ethiopia constitutes the region's oldest basement rocks (Precambrian basement) (18%), Paleozoic and Mesozoic sedimentary rocks (25%), Tertiary volcanic (40%), and Quaternary sediments and volcanics (17%) [2]. It should be pointed out that there are also large areas with Tertiary sediments occurring mainly in the Rift Valley.

5.3.2 Climate

The climate of Ethiopia ranges from equatorial rainforest in the south and southwest, which is characterized by high rainfall and humidity; afro-alpine conditions on the summits of the Semien (western highlands) and Bale (eastern highlands) mountains, to the desert-like conditions of the Northeast, East and Southeast lowlands (e.g., Fig. 2.3a). The temperatures range from 60 °C at the Dallol depression, to freezing temperatures on the Mount Ras Dashen Plateau [54]. The mean annual rainfall varies from 3000 mm at Masha in the western highlands to barely 200 mm in the eastern lowlands [28, 68, 73]. Ethiopia experiences two rainfall seasons. The major rainy

Table 5.1 Annual precipitation, average temperature, and potential evapotranspiration (ET). *Source* [90]

Climatic zones	Annual precipitation (mm)	Temperature (°)	Annual ET (mm)
Arid	<302	>27.5	2159
Semi-arid	302–350	27.5–21	1737–2159
Sub-moist	350–566	21–16	1431–1737
Moist	566–835	16–11	1124–1431
Sub-humid	835–1189	11–7.2	895–1124
Humid	1189–1711	<7.5	<895
Per-humid	>1711	–	–

season (summer, regionally know as "Kiremt") extends from June to September, and accounts for nearly 60% [71], especially over the northern two thirds of the country. The minor rainy season (spring, regionally know as "Belg") usually begins in January/ February and ends in April/May [20]. In general rainfall over Ethiopia is influenced by both local scale forcing mechanisms associated with the Ethiopian Highlands as well as heterogeneous land surface characteristics. In addition, the rainfall variability is significantly influenced by large (global) atmospheric circulation and sea surface temperatures. These large scale forcing mechanisms are normally expressed through El Niño Southern Oscillation (ENSO) induced anomalies, Quasi-Biennial Oscillation (QBO), as well as west-east sea surface temperature gradients over the equatorial Indian Ocean (e.g., [15, 71, 95]). Table 5.1 shows the annual precipitation and potential evapotranspiration (ET) over the seven climatic zones over Ethiopia (Table 3 of Berhanu et al. [13]).

The common indicators of climate variability and changes are trends in precipitation and the maximum and minimum temperatures. Mengistu et al. [53] reported a consistent warming trend in the maximum and minimum temperatures over the past few decades in Ethiopia. However, rainfall shows a declining trend for the east, south and southwest parts of Ethiopia [73]. Additionally, for the central highlands, there is no evidence of trend in rainfall [19]. In general, there is no trend in the extremes of seasonal rainfall in Kiremt and Belg over Ethiopia (cf., [72]). However, Marshall et al. [50] have reported a gradual decline in evapotranspiration on the order of 5 mm/year over the coast of West Africa, the Sahel, and the western Ethiopian highlands, over a 31 years study period.

5.4 GRACE/Model Products and Analysis Method

5.4.1 Data

5.4.1.1 Gravity Recovery and Climate Experiment (GRACE) Products

GRACE, a US/German satellite gravimetry mission that tracks changes in the Earth's gravity field [81] has been the most commonly exploited satellite in the past decade for computing total water storage (TWS). GRACE products have been used to quantify water storage changes at regional as well as global scales as reported for different regions, e.g., [9, 34, 36, 64]. The GRACE Release-05 (RL05) Level-2 (L2) dataset provides the processed time-variable gravity field products applied in this chapter. The products are provided as sets of spherical harmonic coefficients (Stokes's coefficients) averaged over certain time periods (usually monthly). Spherical harmonic coefficients at high degrees are affected by correlated noise [46, 47] that need to be smoothed before being used for the hydrological analysis. These correlated noise can be reduced, using either an empirical method based on a polynomial fit (e.g., [79]), or an a priori synthetic model of the observation geometry, e.g., [46]. In this regard, the [90] employed the DDK3 [47] filtered monthly spherical harmonic coefficients dataset available from the official website of the International Center for Global Earth Models (ICGEM) (http://icgem.gfz-potsdam.de/ICGEM/TimeSeries.html). The suitability of the employed DDK filter is discussed, e.g., in Werth et al. [85]. The investigations of [90] covered the period from January 2003 to December 2011, and included 96 months of the German GeoForschungsZentrum (GFZ), Potsdam, products.

5.4.1.2 Global Land Data Assimilation System (GLDAS)

GLDAS model is developed jointly by the National Aeronautics and Space Administration (NASA), the National Oceanic and Atmospheric Administration (NOAA), and the National Centers for Environmental Prediction (NCEP) [66]. It drives multiple offline (not coupled to the atmosphere) land surface models that integrate huge quantities of observation-based data. GLDAS executes its outputs globally at a relatively high spatial and temporal resolution, enabled by the Land Information System (LIS) [30]. The parameters in the GLDAS-monthly data fall into three main categories: water balance, energy balance, and forcing parameters. Water balance includes soil moisture parameters and other variables such as rainfall rate and surface runoff. To investigate the spatial and temporal variation of soil moisture over Ethiopia, the water balance monthly soil moisture data generated by NOAH LSM at a spatial resolution of $1^\circ \times 1^\circ$ was used. The data was obtained freely from NASA, Goddard Earth Science Data and Information Services Center (GES DISC) (http://disc.sci.gsfc.nasa.gov/hydrology/dataholdings).

5.4.1.3 Tropical Rainfall Measuring Mission (TRMM) Satellite Rainfall

TRMM, a joint USA/Japan satellite mission, is designed to survey the rain structure, rate, and distribution in tropical and subtropical regions (latitude range ±50°) [45]. Amongst the TRMM precipitation products are those derived from integrated rainfall estimates from various sensors on-board satellites such as Precipitation Radar (PR), Special Sensor Microwave Imager (SSM/I), infra-red (IR) data from geostationary satellites, and additional merged rain gauge observations [40, 48]. TRMM-3B42 version 7 [39], which was made available to the public on 22 May, 2012 was used. It has a spatial resolution of 0.25° × 0.25° and temporal resolution of 3-hour. The major difference between the previous versions and the one used by [90] is that the later has an improved scheme for weighting the incorporated rain-gauge data [33, 39]. The data was obtained from NASA GES DISC (http://mirador.gsfc.nasa.gov/).

The motivation for selecting TRMM satellite data in this chapter is informed by the works of [16, 23–25] in terms of the validation of satellite products in the region, which includes nearly half of Ethiopia (i.e., the Blue Nile portion). In particular, Dinku et al. [25] reported the fairly good performance of the TRMM- 3B42 rainfall product compared to locally available rain gauge data. It should be pointed out that although another product, TRMM-3B43, provides monthly data, [90] used the shorter daily TRMM-3B42 product mainly due to the fact that these have been validated in the region, see, e.g., [25].

5.4.2 Analysis Methods

As already stated in Sect. 3.3.3, GRACE products are limited in their usefulness to a spatial extent of greater than 200,000 km^2. On the one hand, only two of the Ethiopian Basins (Abbay and Wabishebele) can fulfill this criterion. On the other hand, Ethiopian catchments exhibit very different characteristics as they are climatically and environmentally extremely heterogeneous (see Sect. 5.3). In this respect, computing the TWS over Ethiopia as a whole might lead to wrong conclusions, since the basins' averages will follow the general variability of the area with the dominant signal. To overcome this problem, Ethiopia was subdivided into 4° × 4° (approximately 440 km × 440 km, i.e., 193,000 km^2) blocks that enable (i) the spatial extent required for the valid consideration of the GRACE products, and (ii), minimizing the errors that might result from generalizing products for Ethiopia as a whole. This way, ten regions are obtained, whose minimum/maximum latitude and longitude, and designation to be used in this chapter from now on are shown in Fig. 5.2. These subdivisions are made in such a way that larger physiographic regions of Ethiopia are represented as close as possible. Since GRACE data sets have already been smoothed using the DDK3 filter [47], the computation of GRACE-derived TWS is undertaken following the approach presented in Wahr et al. [84]. For all 10 regions (Fig. 5.2), TWS values were generated with an approximate monthly temporal resolution over the study period (2003–2011), where the average field from the study period (2003–

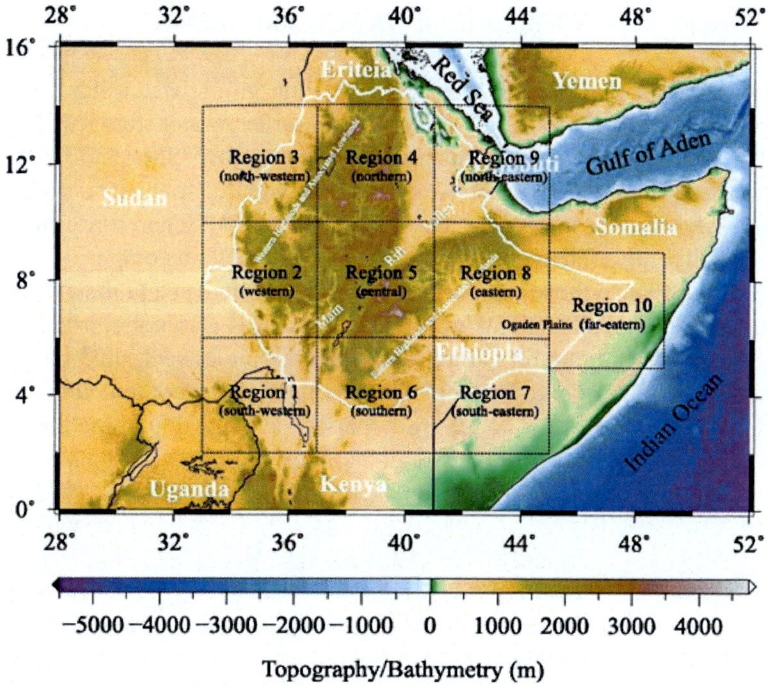

Fig. 5.2 Ethiopia is divided into ten study regions of 4° × 4° to meet the spatial extent requirement of the GRACE products of 200,000 km^2 and the physiographic regions. *Source* [90]

2011) was removed from the monthly TWS. The missing data from June 2003 was filled using a simple linear interpolation. The average error of the derived TWS over the whole region of the study was estimated to be less than 5 mm of water columns. This error was estimated using the formal errors of the Stokes coefficients smoothed using DDK3 filter and converted to TWS cf. [84].

Monthly soil moisture down to 2 m beneath the surface has been computed from NOAA Land Surface Model (LSM) and water budget data obtained for four layers (0–10, 10–40, 40–100 and 100–200 cm), with spatial resolutions of 1° × 1° and aggregated to a spatial domain of 4° × 4°. The average rainfall over each study region obtained from TRMM-3B42 was also rescaled to 4° × 4° and a monthly temporal resolution. Since the TRMM 3B42-V7 rainfall rate are 3-hourly averages centered at the middle of each 3-hour period (i.e., 0, 3, 6, 9, 12, 15, 18, and 21 h), they are converted to total daily rainfall by first multiplying each 3-hourly rainfall rate by 3 to get the total rainfall for each 3-hour period, then added to obtain the desired monthly data, the sum of all the 3-hourly total rainfalls in within 24-hour period is multiplied by the number of the days for a specific month. Both TRMM and GLDAS data were filtered using a Gaussian filter of 300 km radius. The radius of 300 km was selected to be consistent with the smoothing impact of the DDK3 filter applied to the GRACE products. This value has been tested, e.g., in Werth et al. [85] and was found to be applicable to nearly all river basins.

5.4.2.1 Groundwater Changes from GRACE and GLDAS

The most important variables to use when deriving groundwater storage changes (GW) from GRACE-TWS data are; (i) changes in surface water storage (SWS) associated with lakes and reservoirs, (ii) changes in soil moisture (SM), and (iii), changes in snow water equivalent (SWE). However, when compared to the land surface, only about 0.7% of Ethiopia is covered by water bodies [54]. Therefore, this chapter does not consider the contribution of surface water storage in computing TWS over Ethiopian regions since the areas of the existing water bodies and their fluctuations levels were quite small. The snow water equivalent was also not considered in the computation since its temporal variability over Ethiopia is negligible compared to the other contributors of TWS. Soil moisture anomalies were computed by removing the mean value over the extent of the study period (2003–2011) of the monthly soil moisture values. The monthly groundwater anomaly (ΔGWS) was then obtained by subtracting changes in soil moisture from TWS changes as:

$$\Delta GWS = \Delta TWS - \Delta SM, \tag{5.1}$$

assuming that the surface water and snow are insignificant. This approach has been previously applied by Rodell et al. [67] in Northern India with the aim of estimating groundwater depletion.

In association with the GRACE-derived TWS estimates over the Ethiopian Highland, Bonsor et al. [17] used a recharge model for the Nile Basin, Zoomable Object Oriented Distributed Recharge Model (ZOODRM), to interpret the seasonal variation in TWS. They found that the simulated annual variation in groundwater storage and soil moisture accounts for 50–90% of the variations in the GRACE-derived TWS.

5.4.2.2 Duration Curve (TDC) and Total Storage Deficit (TSD)

To simplify the information available for judging the characteristics of TWS, this chapter expresses the TWS changes as a single value through the use of the TWS duration curve (TDC). The TDC is derived through the principle of the Flow Duration Curve (FDC). The FDC is a frequency distribution formed from daily/monthly stream flow data and their exceedence probability [86]. In this approach, all the available data are first listed in descending order and given a rank (the maximum will take the value 1 and the minimum will take the last rank). Using the ranks m, the probability of exceedence (in percentage) is computed using the Weibull method [38, p. 23]:

$$\%Exceedence = \frac{m}{(N+1)} \times 100\%, \tag{5.2}$$

where m is rank and N is the sample size. TDC is obtained by plotting the TWS against its probability of exceedence. The slope STDC of the TDC is then calculated between the 33rd and 66th TWS percentiles using

$$S_{TDC} = \frac{TWS(33\%) - TWS(66\%)}{(66 - 33)}, \tag{5.3}$$

modified for the TWS case from Sawicz et al. [69] as this portion represents a relatively linear part of the TDC. This slope reflects the aquifer storage size and the aquifer through flow property.

To assess the anthropogenic effect on TWS through soil moisture, total storage deficit (TSD) in each regions is computed from GRACE derived TWS anomalies using the approach stated by Yirdaw et al. [87] as

$$TSD_i = \frac{TWS_i - Mean(TWS_i)}{Max(TWS_i) - Min(TWS_i)} \times 100\%, \tag{5.4}$$

where TSD_i is the Total Storage Deficit (%); TWS_j is the monthly total storage anomaly as derived from GRACE-measurements, Mean (TWS_i), Max (TWS_j), and Min (TWS_j) are the long term mean, maximum, and minimum TWS, respectively, for each month j over the study period (2003–2011).

5.4.2.3 Correlation Analysis

To study the lag time of water storage variations (TWS and soil moisture) in response to rainfall, lagged correlation analyses at 95% confidence level were carried out between (a) TWS and rainfall, (b) soil moisture variation and rainfall, and (c), soil moisture anomalies and TWS.

5.4.2.4 Principal Component Analysis (PCA)

Principal Component Analysis (PCA) is a statistical technique that can be used to extract dominant uncorrelated patterns from spatio-temporal observations [61, 96]. In principle, PCA expands the derived TWS changes within the 10 defined regions in terms of new sets of orthogonal vectors known as empirical orthogonal functions (EOFs) associated with their uncorrelated temporal evolutions known as principal components (PCs) [35]. The PCA method used in this chapter is based on the eigenvalue decomposition of the data-derived auto-covariance matrix [61]. Here, only the dominant components of TWS behavior over Ethiopia are shown. The less dominant components that most often correspond to lower correlations with overall TWS changes are not presented. To decide on the significant number of modes, North's rule of thumb, which states that if the sampling error of a particular eigenvalue is comparable to a nearby eigenvalue, then the sampling errors in the EOF will be comparable to the "nearby" EOF [55] was used (see [61] for details).

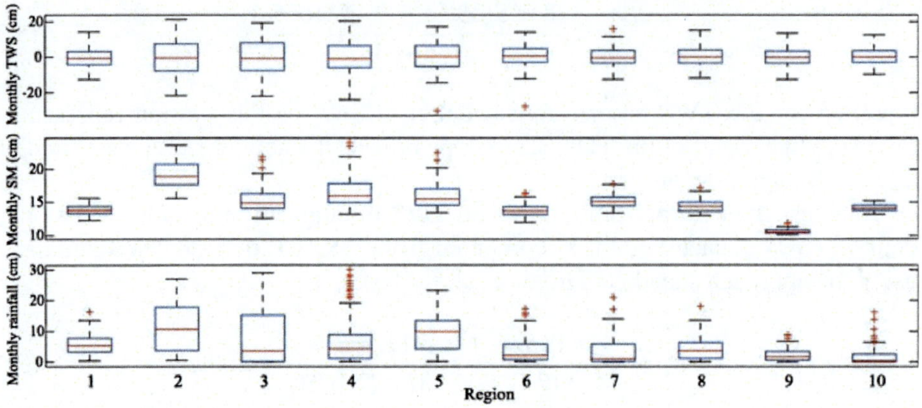

Fig. 5.3 Summary of TWS, soil moisture, and rainfall variability for each of the Ethiopian region shown in Fig. 5.2. The red crosses indicates outliers. TWS (top) is seen to be greatest in western (region 2), north-western (region 3) and northern (region 4) region and less so in the far east (region 10). Regions 2–5 (central) show a strong variation in soil moisture (middle) and rainfall (bottom). *Source* [90]

5.5 Hydrogeological Characterization

Figure 5.3 shows box plots used to highlight the overall patterns of response for the variations in water storage over Ethiopia computed from GRACE, soil moisture (from GLDAS), and rainfall (from TRMM) for the 10 sub-regions over the study period (2003–2011). The lower whisker of the box plot indicates the lowest observed value (sample minimum), the lower end of the box is the lower quartile (25%), the line across the box indicates the median, the upper end of the box specify the upper quartile (75%), and the upper whisker of the plot illustrates the highest observed value of the sample. The red crosses indicate outliers that represent cases that have values more than three times the height of the boxes. Overall, the results in Fig. 5.3 (top) show that the variation in TWS is greatest in the western highlands area (see Fig. 5.1), i.e., western (region 2), north-western (region 3), and northern (region 4) of Ethiopia, while the far eastern corner (region 10) shows relatively low variation. Highland dominated regions (Fig. 5.3, middle and bottom, regions 2, 3, 4 and 5) show strong variations in soil moisture and rainfall, while the lowland dominated segments display comparatively less variability (Fig. 5.3, middle and bottom).

5.5.1 Dominant Variability of Water-Storage Over Ethiopia

The PCA method was applied to the TWS, soil-moisture, and precipitation fields derived for the 10 sub-regions (Fig. 5.4) in order to identify patterns of simultaneous temporal variations in the whole country rather than any localization of the signal.

Applying North's rule of thumb shows that for the three data sets above, only the first two components are statistically significant, which are shown in Fig. 5.4. The first mode of PCA on TWS (EOF1 and PC1) is equivalent to 65.83% of total TWS variability, with PC1 showing a dominant annual variability and EOF1 showing that regions 2–5 are the dominant. Note that the PCA/EOF results are somewhat close to the regions that show predominant variability shown in Fig. 5.3a, as one could expect. The same statement is also true for the first mode of soil-moisture (67.4% of variance) and rainfall (70.1% of variance). The second modes of all three data sets extract mostly inter-annual variability in the data. There are also some annual and intra-annual variations detectable in the second modes, which are due to the incapability of the PCA method to perfectly separate such signals [35].

The second mode (EOF2 and PC2) of the TWS (Fig. 5.4) is equivalent to 14.9% of the total variance and shows that the regions 2, 3, 6–9 are dominant. The dominant inter-annual soil moisture variations are found in regions 1, 2, 5–8, and 10 (EOF2 and PC2 of soil-moisture in Fig. 5.4, middle). Finally, the same regions, i.e., 1, 2, 5–8, and 10 are also found to exhibit the most inter-annual precipitation (EOF2 and PC2 of rainfall in Fig. 5.4, right), again confirming the findings in Fig. 5.3a. It should be mentioned that the accuracy of the annual and semi-annual GRACE-TWS estimations are around 1 cm (in amplitude) for the selected regions. This has also been tested by considering the sampling errors originating from length-limited data sets [61, p. 199], as well as the coefficients errors, provided by the data producer, the GFZ Potsdam Center (the error-bars are not shown here in order to enhance the visual interpretation of the patterns).

5.5.2 *Annual and Seasonal Mean TWS Changes*

Figure 5.5 depicts the mean water mass lost or gained in cm/month aggregated over seasonal and annual time scales from monthly GRACE-derived TWS values over the 2003–2011 period. The results of the seasonal analysis of TWS show regions 1–5 having the same characteristics, where they experienced a decrease in TWS during winter and spring and gained it during summer and autumn. Regions 7 and 8 reflect the same characteristics as the first group, but differ in that their main water loss season is during winter and the water gain in summer (Fig. 5.5, top). The other three regions (region 6, 9, and 10) have different characteristics from all the others. The only characteristic that regions 6 and 10 have in common is that they gain water mass for 75% of the year (i.e., 3 seasons), while it is 50% for region 9. The results, from both GLDAS and GRACE products, confirm the fact that summer is the main water gain season for Ethiopia corresponding to the main rainy season.

The annual mean indicates that mass has been gained in the northern (region 4), central and eastern (regions 5 and 8 respectively), southern (regions 6 and 7) and far eastern (region 10) regions, while the south-western (regions 1), western (region 2), north-western (region 3), and north-eastern (region 9) regions showed that mass has been lost (Fig. 5.5, bottom). The maximum mean annual water mass

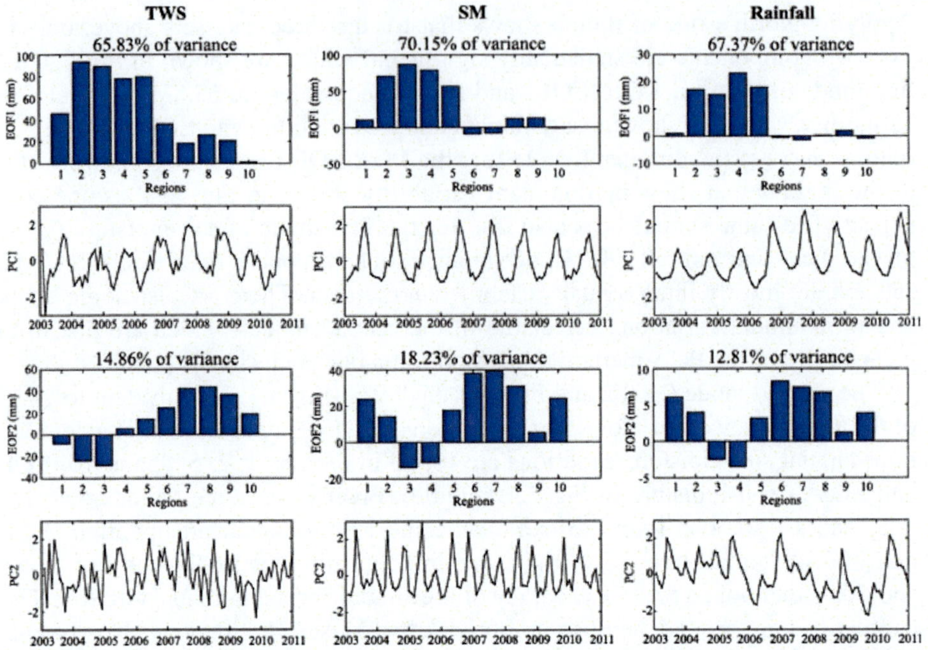

Fig. 5.4 Results of PCA, applied on the time series of 10 predefined regions; (left) TWS data from GRACE, (middle) soil moisture data from GLDAS and, (right) rainfall data from TRMM. *Source* [90]

gains were recorded in central Ethiopia (region 5) during summer (Fig. 5.5), which may be attributed to three main factors. First, the larger part of the Rift Valley Basin lakes lies in this region (see Figs. 5.1 and 5.2), and could be exerting a greater surface water influence. The Ethiopian Rift Valley lakes' basin (Fig. 5.1), which lies in the main Ethiopian Rift, is a collection of cascade lakes, which might have underground connections with other lakes. However, regarding the second factor, considering the amount of surface water, the Rift Valley lakes' basin is known to have a closed watershed, and as such there is water going into the lakes, but have no surface water outlets, e.g., Lake Naivasha [5]. Thirdly, the region is again the starting point of four of the six Ethiopian international basins that include the Awash, Omo-Ghibe, Ganale-Dawa and Wabishebele.

The GRACE-derived TWS duration curve (TDC, see Eq. (5.2)) is presented in Fig. 5.6. From the probability of exceedence at the no loss/no gain point (i.e., point at which the TWS anomaly equals zero), it can be seen that almost all regions gained water mass only 45% of the time, and lost it 55% of the time. The TDC curves in Fig. 5.6 also allow us to reflect upon the probability of change in amount of TWS. For, example, the maximum water mass gain, which has a probability of occurrence of less than 1% over the 2003–2011 period, ranges from 12 cm/month in region 10–21 cm/month in region 2, while the maximum water loss with a percentage

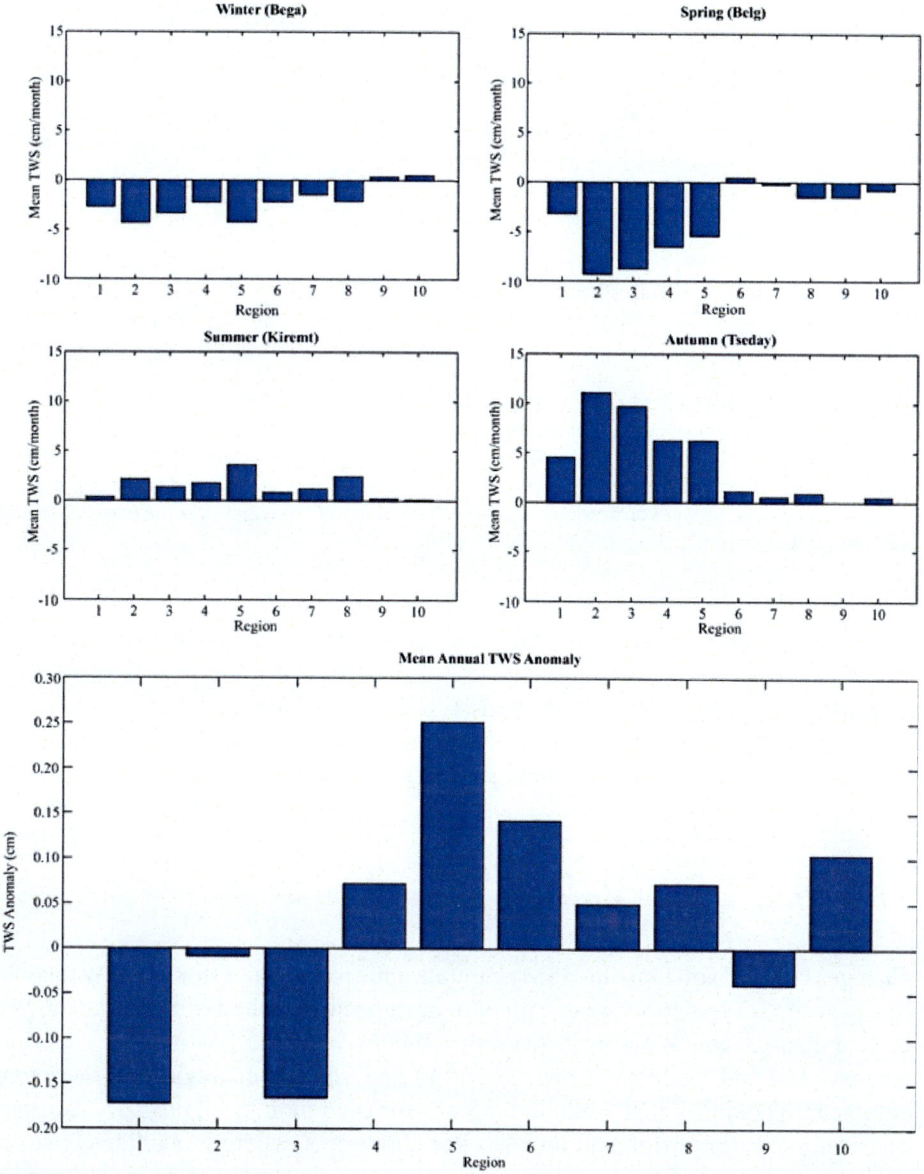

Fig. 5.5 Seasonal (top) and Annual (bottom) mean of TWS variation over Ethiopia derived from GRACE. *Source* [90]

exceedence of about 99% ranges from 30 cm/month in region 5–10 cm/month in region 10.

To quantify the TWS changes into a single value, the slope of the TDC was calculated using Eq. (5.4). A high slope value means rapid response and small storage potential, while a low slope value means slow response to incoming water mass and

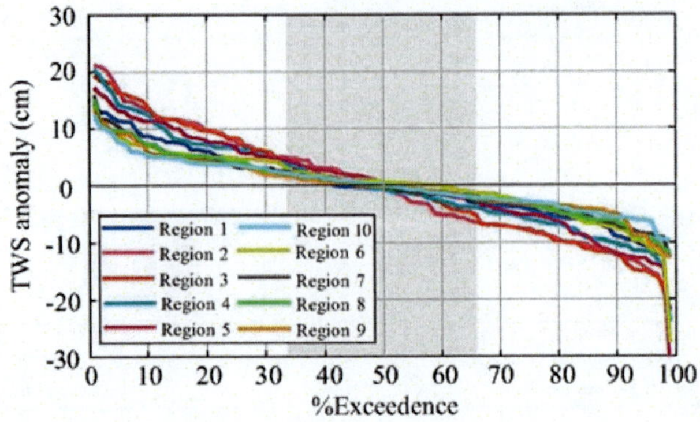

Fig. 5.6 Total Duration Curve (TDC) of the GRACE-based monthly TWS anomalies over the study period. The figure indicates that almost all regions gained water 45% of the time. Slope analysis of this figure show that over Ethiopia, TWS changes are high in the western region and low in the eastern region. The extreme fall at the end of the curves for regions 5 and 6 could be the consequence of the loss of a large amount of water in January and February 2003, which was the result of drought that propagated from the previous year. *Source* [90]

large storage potential. For simplicity, we can group the regions based on their TDC slopes as, (i) $S_{TDC} > 30\%$ (only region 2), (ii) $20\% < S_{TDC} < 30\%$ (regions 3–5), and (iii), $10\% < S_{TDC} < 20\%$ (regions 1, 6–10). This indicates that over Ethiopia, the TWS changes are highest in the western region and lowest in the north-eastern region, i.e., consistent with the findings in Figs. 5.3 and 5.4.

5.5.3 Inter-annual Variation

Changes in TWS, soil moisture (SM), rainfall, and groundwater storage (GW) anomalies over the study period for each month have been compiled in Fig. 5.7. The year to year changes in TWS (Fig. 5.7a) shows that except for a few months, e.g., June (regions 1, 4 and 9), July (regions, 7 and 8) and September (region 6), there exist significant variability in TWS in all regions for each month. Figure 5.7b also show considerable variations in soil moisture for all months, where it is relatively strong during rainy seasons. The inter-annual plot of rainfall illustrated in Fig. 5.7c describes the considerable variation that exists in all months in regions 1 and 2, while regions 3–5, 8 and 9 show relatively low variability except during the major rainy season (June–September; JJAS). Additionally, there is a low variability in June–September for regions 6, 7 and 10, i.e., south east and east. The GW plots (Fig. 5.7d) present similar characteristics as the soil moisture and TWS. From Fig. 5.7, the variations of TWS, SM and GW anomalies are seen to be less correlated to rainfall for all regions with the exception of region 2 (western), see also Fig. 5.11.

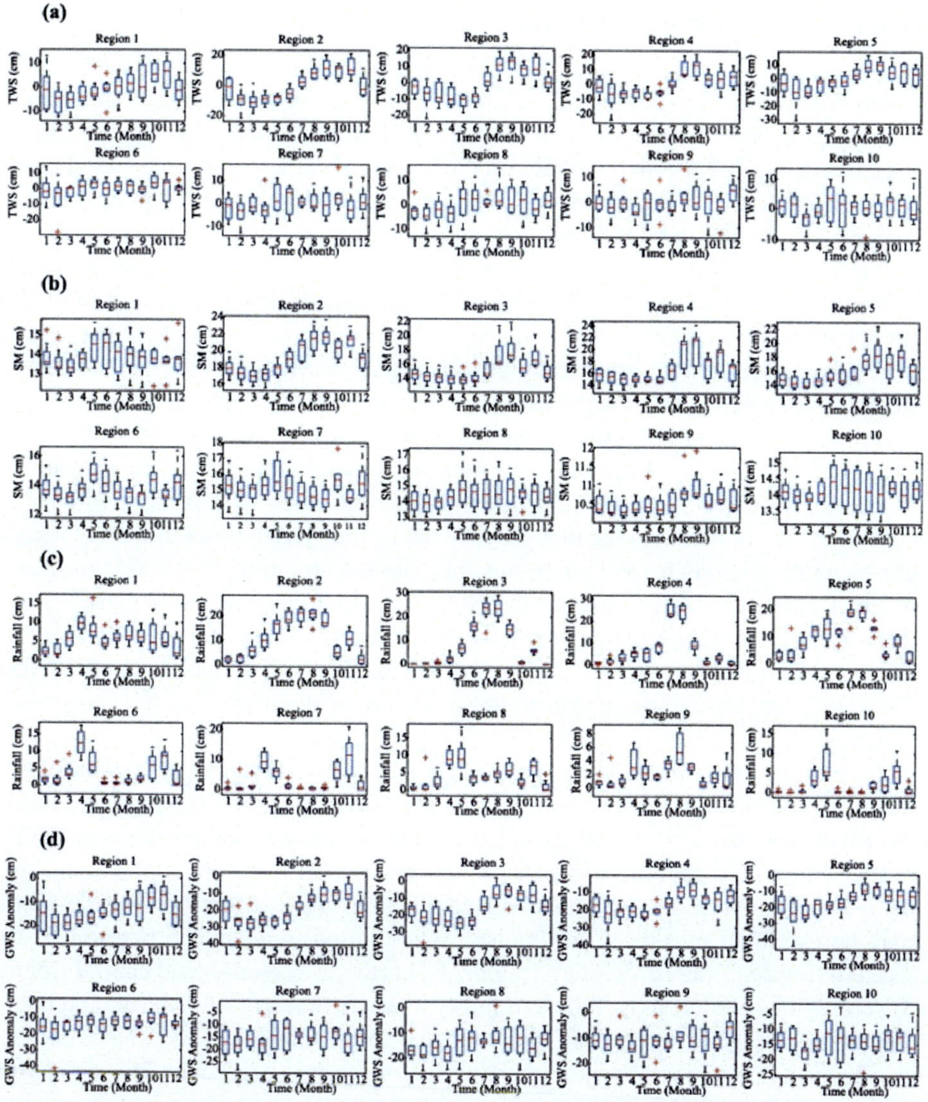

Fig. 5.7 Inter-annual variability of TWS (**a**), Soil Moisture (**b**), Rainfall (**c**) and GW anomaly (**d**) over Ethiopia. The variation in TWS (**a**), soil moisture (**b**), and GW (**d**) are unrelated to the rainfall (**c**), thus indicating a possible human influence on hydrological changes in Ethiopia. Note the difference in amplitude of the figures. Although **a** and **d** look similar, they differ considerably in amplitude. *Source* [90]

5.5.4 Intra-annual Variation

Figure 5.8 shows the intra-annual variability of TWS, soil moisture, GW and rainfall developed from the mean of each month over the study period as opposed to the red lines in Fig. 5.7. From this, one can clearly see two distinct groups based on the behavior of their TWS anomaly values over the months. The first group includes regions 2–5, where the TWS and GW anomalies shows three similar parts; (i) the rising limb, which occurs from June to September, indicating the recovery period of TWS, (ii) the peak, which occurs in August, and (iii), the recession or falling limb that covers a longer period than the previous ones, i.e., starting from a point at which the peak is attained and extending up to June. The results indicate that, like the rainfall patterns, TWS shows seasonality, having seasons showing mass gain (JJA and some parts of SON), while in the other two seasons, the mass loss exceeds the gain in most regions of Ethiopia. This happens since the mentioned periods are the main rainy seasons in Ethiopia. The second group includes all the other regions with characteristics opposite to the first group. That is, they have short duration recession curves with very long recovery or rising limb period, attaining their maximum peaks in the SON season. Despite the fact that SON is the long rainy period in these regions, the length of the rising period might be attributed to the aquifer type and size, as well as recharge mechanism. Compared to the rainfall, and the response of GW and TWS, one can see that there appears to be very little fluctuation in soil moisture in all cases.

Examining now the soil moisture in greater detail (Fig. 5.9), one can visually identify extreme regions that include (i) the western part of Ethiopia (i.e., region 2), which has the highest soil moisture content of all the studied regions, (ii) the north-eastern (i.e., region 9), which has the lowest soil moisture content, and (iii), the intermediate regions, which include all the other areas. Apart from this, Ethiopia may be classified into three soil moisture variability regions. The first covers the western (region 2), north-western (region 3), northern (region 4) and central (region 5) sub-regions, which show little variability over the first half of the year, and then a significant rise in the second half, attaining their peaks in September. This is because the second half of the year is the main rainy season for most segments. The second soil moisture region (southern; region 6 and south-eastern; region 7) is characterized by the occurrence of two small peaks, one in May and the other in November, while the third soil moisture region includes the south-western (region 1) and eastern (region 10), displaying very little variability in soil moisture throughout the year compared to others.

The soil moisture variation over Ethiopia bears a resemblance to the agro-ecological zonation described in Hurni [41]. Depending on the dominant agro-ecological zone in each region, three sub-divisions may exist. These are; (i) Dega/Woyna Dega consisting of regions 2, 4 and 5, (ii) the Bereha made up of regions 1, 6, 7, 8, 9 and 10, and (iii), the transitional zone Kolla/Dega, comprised only of region 3. The Dega/Woyna Dega group displays high soil moisture levels with strong seasonality, while the Bereha group shows little seasonality behavior as they receive

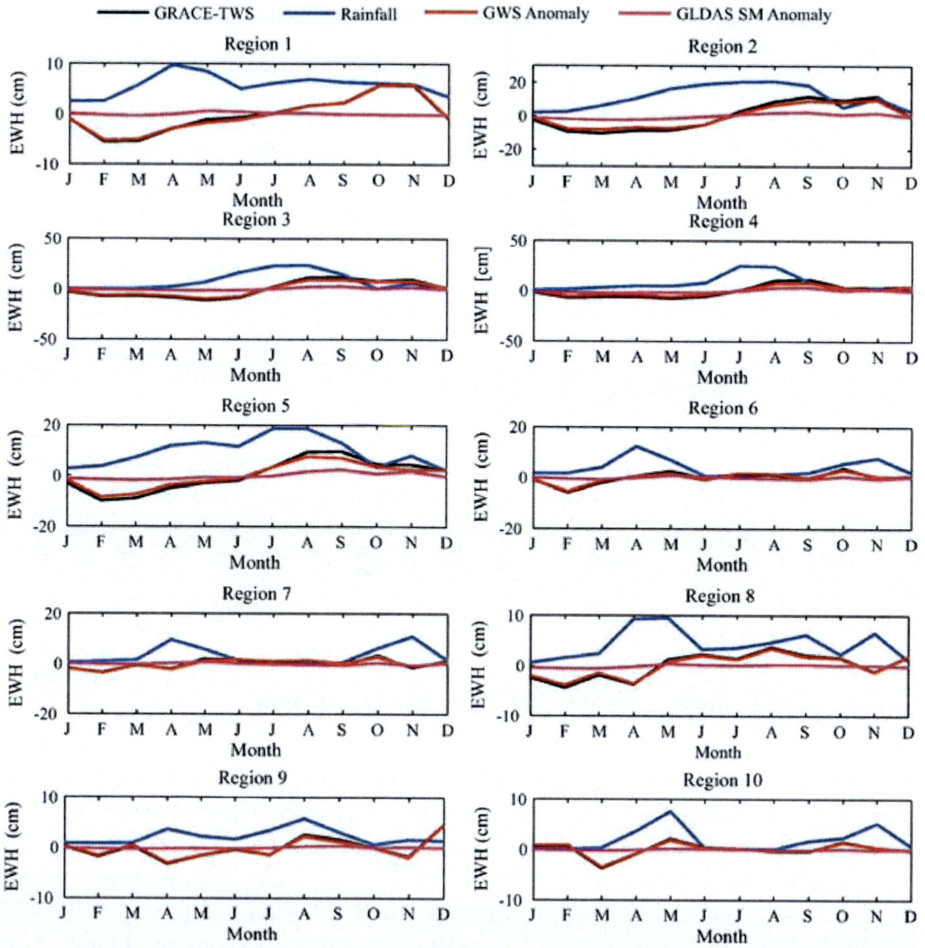

Fig. 5.8 Intra-annual variability of the TWS, SM, GW and rainfall EWH (equivalent water height) over Ethiopia. *Source* [90]

little rainfall. Furthermore, Fig. 5.9 indicates two regions of comparatively extreme wetness and dryness, i.e., regions 2 and 9, respectively. The possible reason for region 2's wetness is the fact that (i) this region receives rainfall almost throughout the year, comprising the area which obtains the highest annual rainfall amount in Ethiopia (see Sect. 5.3) and (ii), the land cover in the region is highly dominated by rainforest. In region 9, the Danakil depression region, which shows much lower soil moisture content, is known to be one of the driest regions of Ethiopia, where there is no significant river flow and only short period bimodal rainfall, very scarce vegetation cover, and high temperatures (see Sect. 5.3). Ertale, the area where volcanic magma can be seen at Earth's surface is also in this region, where high temperatures, combined with little rainfall, lead to the retention time of water in the soil to be very limited.

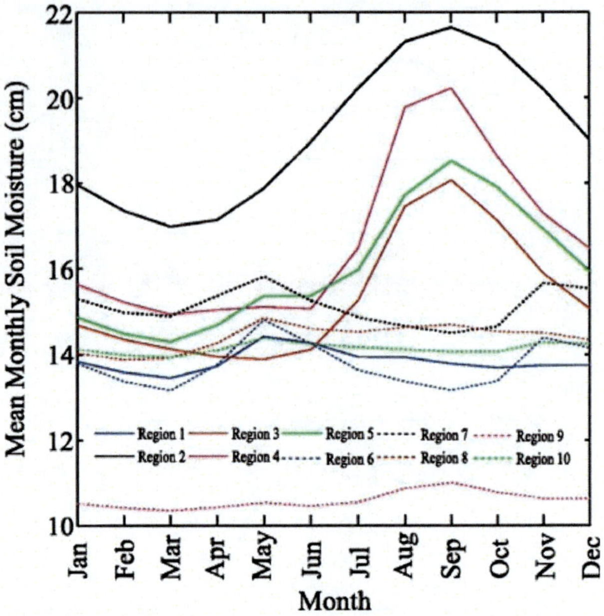

Fig. 5.9 Intra-annual variability of soil moisture over Ethiopia. *Source* [90]

5.5.5 *Correlation Between Different Data Sets*

A correlation analysis between available data sets was made using the mean monthly data and is presented in Fig. 5.10 to study the time lag between rainfall and TWS, soil moisture and TWS, and soil moisture and rainfall. The analysis shows almost no lag between soil moisture and TWS in all regions, except regions 1 and 9, where lags of 5 and 1 month, respectively are seen. For TWS and rainfall, lags of 0 (region 10), 1 month (regions 4 and 7), 2 months (region 3), 3 months (regions 2, 5 and 8), 4 months (region 9), and 6 months (regions 1 and 6) have been observed, with strong correlations in regions 2–5. The lags between TWS and rainfall obtained from Fig. 5.10 are tabulated in Table 5.2. Larger lag periods are indicative of water-storages properties such as large groundwater reservoirs, indirect recharge mechanisms, or a combination of both, while smaller values are more likely to be attributed to direct recharge mechanisms, smaller groundwater reservoir or again their combinations.

5.5.6 *TRMM-Rainfall and GRACE-TWS: Relationship*

From Figs. 5.7 and 5.8, one can see that there is little significant delineation between the plots of TWS and GW. It is, therefore, thought that the very large influence of TWS over Ethiopia comes from the groundwater storage. The relationship between rainfall and TWS is presented in Fig. 5.11. This is developed by accounting for the lag duration of TWS from rainfall. For example, in region 1, a lag of 6 months is

Table 5.2 Recharge mechanisms, recharge rates, hydrogeology, aquifer flow, and storage type characteristics over different regions of Ethiopia (*Source* [1, 18]). The Table also includes the slope m of the TDC derived from GRACE-TWS for each region and the lag between GRACE-TWS and TRMM-rainfall. *Source* [90]

Region	Recharge mechanisms	Recharge rate (mm/yr)	GRACE-mean Annual TWS(mm/yr)	Hydrogeologic setting	Aquifer flow and Storage type	TDC slope (m)	Lag (months)
1	Recharge from wadi beds through local wadi floods	50–150	−26.64	Unconsolidated sediment	Inter-granular	−0.15	6
2	Direct diffuse recharge through soil and root zone. Recharge from wetland during high flood	150–250	−1.08	Volcanic rocks, metamorphic rocks and intrusives	Fracture	−0.344	3
3	Wadi bed through local wadi floods, Selective recharge from heavy rainfall	50–150	−19.92	Volcanic rocks, metamorphic rocks and intrusives	Fracture	−0.276	2
4	Selective recharges from heavy rain falls, direct diffuse+floods from highlands	50–150	8.64	Metamorphic rocks and intrusives	Fracture	−0.236	1
5	Selective recharges from heavy rain falls, Mountain block and mountain font	50–250	30.12	Metamorphic rocks and intrusives	Fracture	−0.221	3

(continued)

Table 5.2 (continued)

Region	Recharge mechanisms	Recharge rate (mm/yr)	GRACE-mean Annual TWS(mm/yr)	Hydrogeologic setting	Aquifer flow and Storage type	TDC slope (m)	Lag (months)
6	Selective recharges from heavy rainfalls wadi beds through local wadi floods	50–150	17.04	Metamorphic rocks and intrusives, consolidated	Fracture	−0.146	6
7	Selective recharges from heavy rainfalls wadi beds through local wadi floods	<50	6.00	Consolidated sediments	Inter-granular and Fracture	−0.136	1
8	Selective recharges from heavy rainfalls wadi beds through local wadi floods	<50–150	8.64	Consolidated and unconsolidated sediments	Inter-granular and Fracture	−0.138	3
9	Flood Water from the highlands	<50	−4.80	Metamorphic rocks and intrusives	Fracture	−0.116	4
10	Recharge from wadi beds through local wadi floods	<50	12.48	Consolidated sedi-ments,unconsolidated Sediments	Fracture (Karst)	−0.138	0

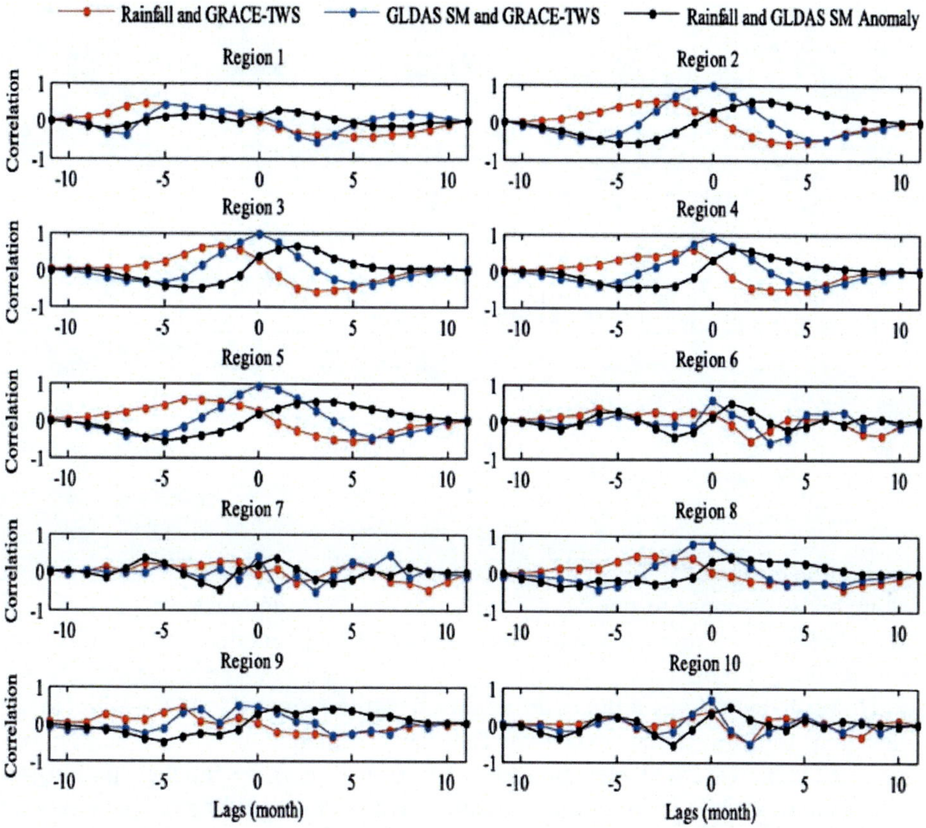

Fig. 5.10 Correlation between different data sets (TWS, soil moisture, and rainfall) at 95% confidence level. *Source* [90]

used to assess how much of the rainfall observed in January contributed to the TWS in July. In all regions, except 2, 4 and 5, coefficients of determinant R^2 of less than 0.5 were obtained. For these regions, where high values of R^2 are obtained, i.e., the western (region 2), northern (region 4) and central (region 5), two deductions could be made, i.e., (i) either the regions are source of water (e.g., regions 2, 4 and 5), or (ii), the reliance of the groundwater storage variability on the rainfall pattern is minimal.

5.5.7 Topographic Impact on TWS

As one can see from the results presented in Figs. 5.3 (top) and 5.4, the influence of topography on the TWS may be inferred. From these figures, the highland dominated regions of Ethiopia, i.e., the western highlands (Fig. 5.2, regions 2–5) show greater variability in TWS than the lowland-dominated regions. In addition, from the lag

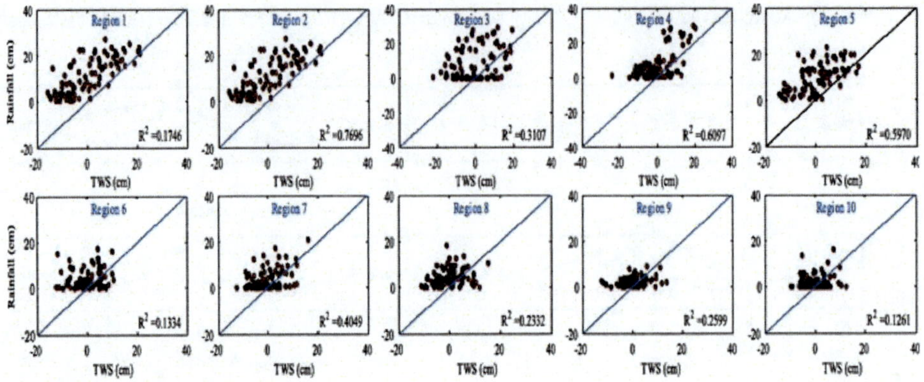

Fig. 5.11 Relationship between TRMM rainfall in January and GRACE TWS in July for each region at a lag of six months. This lag is chosen as an example to show the impact of the January rainfall on TWS in July. *Source* [90]

correlation presented in Fig. 5.10, one can see the influence of the regime's hydrogeology on its response to rainfall. For instance, the karst dominated regions (i.e., 10 in Table 5.2) has zero lag while the unconsolidated sediment dominated region 1 has a lag of around 6 months.

However, hydrological fluxes are expected to differ greatly from the outlet to high elevations for some of the areas in the Ethiopian highlands. Whereas precipitation is expected to increase with elevation (depending on aspect), potential evapotranspiration decreases with elevation due to reduced temperatures at higher elevations (cf., [13]). Based on water balance estimates, runoff is therefore expected to be greater at higher elevation than lower ones for high gradient basins facing the prevailing storm movement direction. The total amount of runoff from the basin will therefore depend on the hypsometric relationship, i.e., the relative proportions of drainage area at different elevations. Since evapotranspiration will vary with the type of vegetation, changes in land use will also have a potential impact on the amount and type of evaporation (from either surface or groundwater sources). In addition, the conversion of relatively undisturbed (e.g., forested) areas to more intensive agriculture will greatly increase surface runoff at the expense of groundwater recharge due to reduced infiltration capacity [97–99]. It is therefore critical to assess the role of elevation in hydrological fluxes coupled with seasonal variability and changes under potential climate scenarios.

5.5.8 Possible Human Influence on the Observed TWS

Ethiopia is a highly populated region that has seen a number of dams constructed and high levels of water withdrawal from groundwater and surface water reservoirs, which would impact upon the total water storage, e.g., [12, 83]. Large scale land-use and land-cover changes in Ethiopia due to agricultural expansion and heavy grazing (see Fig. 2.3d) have been reported, e.g., [14, 22]. In addition, Hurni and Nyssen et al. [41, 57] have reported that land degradation is a serious problem to Ethiopia. Ethiopia is an Agrarian country, where more than 85% of its population depend on irrigated agriculture, with production in certain regions being highly dependent on the available water resources (i.e., surface water and groundwater). The total area under irrigation in Ethiopia is about of 290,729 ha (~0.3% of the country's area), with about 2611 ha depending on groundwater irrigation [76]. The irrigation water demand in Ethiopia is estimated to be 40 km^3/year [27], with an estimated 2.6–6.5 km^3/year of groundwater potential [10]. The principal grain crops are wheat, barley (primarily cool-weather crops) and corn, sorghum, and millet (warm weather grain crops).

Since the GRACE products can resolve large areas to the tune of hundreds of km (i.e., 450 km $\times$ 450 km), the impact of anthropogenic activities at hill slope scales, such as those discussed in Descheemaeker et al. [22], on its signals will not be detectable. However, if such impacts occur over a wider scale that is of the same order as the resolution of GRACE, then they may be detectable.

TSD in percent (Eq. (5.4)) are computed from the GRACE-derived TWS anomalies and presented in Fig. 5.12. Depending upon their TSD patterns, Ethiopia can be divided into three different zones. In the first zone, we have regions 2–5, while regions 1 and 6 are in zone 2, with regions 7–10 in the third zone. For the three zones, TSD decreased from the beginning of 2003 until around mid-2006 for zones 1 and 2, and January 2005 for the third zone. Then, for the first and third zones, it started to rise in 2006, while for the second group this was from around the beginning of 2007. The decrease in TSD starting from Jan-2003 indicates increasing dryness, which may be due to the propagation of the 2002 drought in Ethiopia [3, 95]. In regions where significant TSD has been seen from 2003 to 2006 (e.g., Zone 1 in Fig. 5.12), the main crops are teff, sorghum, maize, wheat and millet. Despite the decrease in TSD in the regions, there has been an increase in crop production area and yield [26], indicating on the one hand that such crops can be sustained under such water deficit conditions. However, on the other hand, this could indicate that the greater portion of this deficit in TWS could be due to groundwater storage depletion. This, however, remains an open question for future investigations.

In regions 1–3, and 9, over the study period, very significant water loss was identified (see Fig. 5.5, bottom), especially when compared to the recharge amount per year, as stated in Table 5.2. This might be attributed in part to human exploitation and partly due to climate impacts discussed below.

From the developed TDC using 96 months of GRACE derived TWS data (Fig. 5.6), the lower end of the curve might be due to extreme water depletion cases, while the upper end results from high rainfall condition with long return periods. This is again an open question subject to future investigations. Nonetheless, to see the real variation of TWS, the slope of the TDC in the region where one can assume linearity is computed. If this slope is expressed in percentage, it varies between −14 and −35%. There are several factors that might affect the slope of the TDC, e.g., soil moisture, hydrogeological setting, aquifer flow, storage types, storage potential,

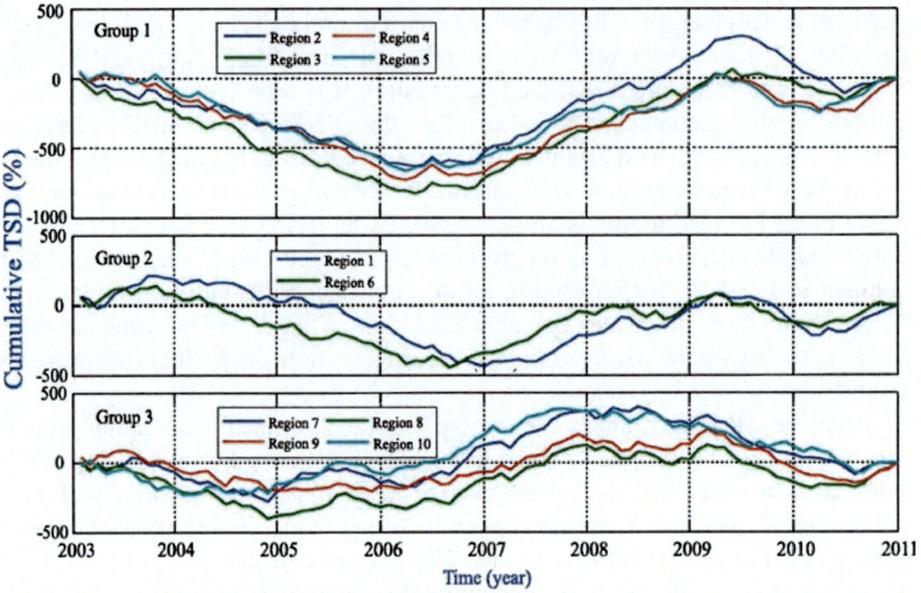

Fig. 5.12 Total Storage Deficit (TSD) derived for the main 10 regions defined over Ethiopia subdivided into different zones. The values are in percent (see Eq. (5.4))

recharge rates and recharge mechanisms. With the characteristics listed in Table 5.2, and their TDC slope, one notes that (i) region 2 has the smallest storage potential compared to the others, (ii) regions 3–5 have intermediate storage potentials, and (iii), all the other regions have relatively large storage potentials. From Fig. 5.11, although most regions have correlation coefficients of less than 0.5, Fig. 5.4 on the other hand show that certain regions gained TWS. This could indicate that a certain proportion of the TWS in the regions is contributed by subsurface inter flow and dominant indirect recharges.

5.5.9 *Climate Impact on the Observed TWS*

The extreme depletion in TWS and GW in certain regions during February 2003 (see PC1 of TWS in Fig. 5.4) could probably be due to the reported droughts in the year 2002 [3, 95]. This is further supported when they are considered together with the lag determined from TWS and rainfall correlation (Fig. 5.10). Rainfall over Ethiopia is known to be seasonal, having dry and rainy seasons [73]. The effect of this seasonality seems to be reflected in the TWS anomalies, having two distinct seasons on average, the water lose (winter, 21.3 mm/month and spring, 36.6 mm/month) and water gain (summer, 14.4 mm/month and autumn, 41.0 mm/month) seasons. PCA results showed that the annual peaks of rainfall are also very important and are highly correlated to the annual peaks of TWS, thus indicating that rainfall is the main driving factor for the rise in TWS levels in most parts of Ethiopia. Although climatic factors such as drought could be a contributor towards the fall in TWS levels, as pointed out earlier, human influences may also contribute.

The dominant hydrogeologic regimes, recharge mechanisms, recharge rates, aquifer flow and storage types for each region are presented, e.g., in [1, 18]. Here, the results obtained in relation to the properties of the regions described in the Table 5.2 are discussed. For the same aquifer flow and storage type, one can notice that there are different lag values, see, e.g., regions 2 and 3 in Table 5.2, which could be attributed to the nature of the recharge mechanism in each region. Furthermore, regions that are dominated by the inter-granular aquifer flow and storage type are seen to have a longer lag periods (regions 1, and 6), while karst dominated regions show a direct response to rainfall (region 10).

In the north-west (region 3) a lag of 2 months is noted (Table 5.2). This shows that it takes 2 months from when it rains to the time when the impact on TWS is visible, hence making the transfer of water from one season to another impossible. This implies that for most parts of Ethiopia, the rainfall in the previous season might have an impact on the TWS of the current season at most, but cannot have further effect later. In regions 2, 5 and 8, the cross correlation between rainfall with TWS indicated a lag of 3 months (Table 5.2). Figure 5.11 also shows that there is a high correlation between rainfall and TWS in regions 2 and 5, indicating that the influence of rainfall on TWS in these regions is higher than subsurface inflow from adjacent regions. TWS in region 9 recorded a lag of 4 months from the rainfall (Table 5.2), although this region has low rainfall and also displays a poor correlation between rainfall and TWS anomaly (Fig. 5.11).

The mean annual rainfall for regions 1 and 6 have two small peaks, with a lag of 6 months obtained from the TWS/Rainfall cross correlation, implying that the autumn water gain is mainly from the rainfall received in winter. With the characteristics described in Table 5.2, these regions have the property of being able to transfer water mass gained from rainfall in a certain year to the next one. According to Kebede [42], 75% of Ethiopia's aquifers receive localized and indirect recharge. This implies that only 25% of Ethiopia's aquifers should exhibit a good correlation between TWS and rainfall. The R^2 results obtained from correlation analysis in Fig. 5.11 supports this assertion.

In the northern (region 4) and south eastern (region 7) regions, a lag of one month between TWS and rainfall was obtained from the cross correlation, implying that the aquifer responds to the rainfall after about one month. From the slopes of the TDC, region 4 has a lower storage potential than region 7, while the rainfall in region 4 is a bit higher than in region 7. One may therefore conclude that the chance that water can be transferred from a rainy to non-rainy season in such regions is very low. In this case, the rise or fall in TWS in certain season is mostly subject to the rainfall for that particular season.

5.6 Concluding Remarks

Time series of terrestrial water storage anomalies (TWS), soil moisture, and rainfall over Ethiopia were derived from GRACE, GLDAS, and TRMM data sets and products, respectively, and analyzed with the aim of better understanding the relationship between rainfall and groundwater variations over Ethiopia's aquifers. The inter-annual and seasonal variability of water storage, the relationship between water storage changes with rainfall and soil moisture, and the aquifers' characteristics and TWS anomalies have been derived over a period of 8 years from 2003 to 2011. The chapter indicates that the western part and the north-eastern lowland of Ethiopia lost water during that period. From seasonal analysis, the seasons when water is gained are found to be summer and autumn, while the loss seasons are spring and winter for most parts of Ethiopia. Soil moisture and rainfall variations showed the dominant annual water variability in the western, north western, northern and central regions, and the dominant seasonal variability in the western and eastern regions. Change in soil moisture was seen to have less influence over the total water storage, while groundwater storage has a dominant influence. From a correlation analysis between TWS and rainfall, lags of 2–3 months for most regions were observed, indicating the capacity of a large portion of the Ethiopian groundwater storage (aquifers) to transfer water mass gained in certain periods to be less than 3 months. If validated at a finer spatial resolution, this could be vital information for water managers to know, that if an increase in TWS is required to back up water availability in dry seasons, one has to focus on increasing soil moisture and surface water storage rather than groundwater storages. The information could also be vital in land-use planning issues.

References

1. Abiye TA (2010) An overview of the transboundary aquifers in East Africa. J Afr Earth Sci 58:684–691. https://doi.org/10.1016/j.jafrearsci.2009.10.003. [Africa and the International Year of Planet Earth Promoting Earth sciences-based decision making in Africa 22nd Colloquium of African Geology]
2. Alemayehu T (2006) Groundwater occurrence in Ethiopia. Addis Ababa, Ethiopia: Addis Ababa University, 111 p. http://www.eah.org.et/docs/Ethiopian%20groundwater-Tamiru.pdf
3. Anderson S, Choularton, R (2004) Retrospective analysis: 2002/3 crisis in Ethiopia: early warning and response. Technical report the regional economic development services office for East and Southern Africa (REDSO)
4. Awange J, Forootan E, Kuhn M, Kusche J, Heck B (2014) Water storage changes and climate variability within the Nile Basin between 2002–2011. Adv Water Resour 1–15. https://doi.org/10.1016/j.advwatres.2014.06.010
5. Awange J, Forootan E, Kusche J, Kiema J, Omondi P, Heck B, Fleming K, Ohanya S, Gonçalves R (2013) Understanding the decline of water storage across the Ramser-Lake Naivasha using satellite-based methods. Adv Water Res 60:7–23. https://doi.org/10.1016/j.advwatres.2013.07.002
6. Awange JL (2012) Environmental monitoring. In: Environmental monitoring using GNSS environmental science and engineering, pp 1–12. Springer, Berlin. https://doi.org/10.1007/978-3-540-88256-5_1

7. Awange JL, Anyah R, Agola N, Forootan E, Omondi P (2013) Potential impacts of climate and environmental change on the stored water of lake Victoria Basin and economic implications. Water Resour Res 49:8160–8173. https://doi.org/10.1002/2013WR014350
8. Awange JL, Kyalo Kiema JB (2013) Environmental monitoring and management, environmental science and engineering. Springer, Berlin. https://doi.org/10.1007/978-3-642-34085-7_1
9. Awange JL, Sharifi MA, Ogonda G, Wickert J, Grafarend EW, Omulo MA (2008) The falling Lake Victoria water level: GRACE, TRIMM and CHAMP satellite analysis of the Lake Basin. Water Resour Manag 22:775–796. https://doi.org/10.1007/s11269-007-9191-y
10. Awulachew S, Yilma A, Loulseged M, Loiskandl W, Ayana M, Alamirew T (2007) Water resources and irrigation development in Ethiopia. In: Working paper 123 international water management institute Colombo, Sri Lanka
11. Ayenew T, Demlie M, Wohnlich S (2008) Hydrogeological framework and occurrence of groundwater in the Ethiopian aquifers. J Afr Earth Sci 52:97–113. https://doi.org/10.1016/j.jafrearsci.2008.06.006
12. Berhane G, Martens K, Farrah NA, Walraevens K (2013) Water leakage investigation of micro-dam reservoirs in Mesozoic sedimentary sequences in Northern Ethiopia. J Afr Earth Sci 79:98–110. https://doi.org/10.1016/j.jafrearsci.2012.10.004
13. Berhanu B, Melesse AM, Seleshi Y (2013) GIS-based hydrological zones and soil geodatabase of Ethiopia. CATENA 104:21–31. https://doi.org/10.1016/j.catena.2012.12.007
14. Bewket W, Abebe S (2013) Land-use and land-cover change and its environmental implications in a tropical highland watershed. Ethiopia. Int J Environ Stud 70:126–139. https://doi.org/10.1080/00207233.2012.755765
15. Bewket W, Conway D (2007) A note on the temporal and spatial variability of rainfall in the drought-prone Amhara region of Ethiopia. Int J Climatol 27:1467–1477. https://doi.org/10.1002/joc.1481
16. Beyene EG, Meissner B (2010) Spatio-temporal analyses of correlation between NOAA satellite RFE and weather stations' rainfall record in Ethiopia. Int J Appl Earth Obs Geoinf 12(Suppl. 1):S69–75. https://doi.org/10.1016/j.jag.2009.09.006 [Supplement Issue on Remote Sensing for Africa - A Special Collection from the African Association for Remote Sensing of the Environment (AARSE)]
17. Bonsor HC, Mansour MM, MacDonald AM, Hughes AG, Hipkin RG, Bedada T (2010) Interpretation of grace data of the Nile Basin using a groundwater recharge model. Hydrol Earth Syst Sci Discuss 7:4501–4533. https://doi.org/10.5194/hessd-7-4501-2010
18. Chernet T (1993) Hydrogeology of Ethiopia and water resources development. Ethiopia: Technical Report Ethiopian Institute of Geological Surveys, Ministry of Mines and Energy Addis Ababa
19. Cheung WH, Senay GB, Singh A (2008) Trends and spatial distribution of annual and seasonal rainfall in Ethiopia. Int J Climatol 28:1723–1734. https://doi.org/10.1002/joc.1623
20. Chukalla A, Haile A, Schultz B (2013) Optimum irrigation and pond operation to move away from exclusively rainfed agriculture: the Boru Dodota spate irrigation scheme. Ethiopia. Irrig Sci 31:1091–1102. https://doi.org/10.1007/s00271-012-0390-9
21. Conway D, Schipper ELF (2011) Adaptation to climate change in Africa: challenges and opportunities identified from Ethiopia. Global Environ Change 21:227–237. https://doi.org/10.1016/j.gloenvcha.2010.07.013
22. Descheemaeker K, Raes D, Nyssen J, Poesen J, Haile M, Deckers J (2009) Changes in water flows and water productivity upon vegetation regeneration on degraded hillslopes in northern Ethiopia: a water balance modelling exercise. Rangeland J 31:237–249. https://doi.org/10.1071/RJ09010
23. Dinku T, Ceccato P, Grover-Kopec E, Lemma M, Connor SJ, Ropelewski CF (2007) Validation of satellite rainfall products over East Africa's complex topography. Int J Remote Sens 28:1503–1526. https://doi.org/10.1080/01431160600954688
24. Dinku T, Chidzambwa S, Ceccato P, Connor SJ, Ropelewski CF (2008) Validation of high-resolution satellite rainfall products over complex terrain. Int J Remote Sens 29:4097–4110. https://doi.org/10.1080/01431160701772526

25. Dinku T, Pietro C, Keith C, Stephen JC (2010) Evaluating detection skills of satellite rainfall estimates over desert locust recession regions. J Appl Meteorol Climatol 49:1322–1332. https://doi.org/10.1175/2010JAMC2281.1
26. Eberhardt M (2008) Cereal production in Ethiopia: a brief historical overview of land under cultivation and yields. http://users.ox.ac.uk/econstd/Brief_MEberhardt.pdf
27. Egziabher TG (2000) Regional development planning in Ethiopia: past experience, current initiatives and future prospects. East Africa Soc Sci Res Rev 16:65–94
28. Engida AN, Esteves M (2011) Characterization and disaggregation of daily rainfall in the Upper Blue Nile Basin in Ethiopia. J Hydrol 399:226–234. https://doi.org/10.1016/j.jhydrol.2011.01.001
29. Famiglietti JS, Rodell M (2013) Water in the balance. Science 340:1300–1301. https://doi.org/10.1126/science.1236460
30. Fang H, Beaudoing HK, Rodell M, Teng WL, Vollmer BE (2009) Global land data assimilation system (GLDAS) products, services and application from NASA hydrology data and information services center (HDISC). In: ASPRS, Annual Conference. Baltimore, Maryland, p 2009
31. FAO. Irrigation in Africa in figures: AQUASTAT Survey – 2005. Karen Frenken, FAO Water Reports number 29 Food and Agriculture Organization of the United Nations Rome. 2005
32. Ferguson A, Harbott B (1982) Geophysical, physical and chemical aspects of Lake Turkana. In: Hopson E (ed) Lake Turkana: a report of the findings of the Lake Turkana project 1972–1975. Overseas development administration, London, UK
33. Fleming K, Awange JL (2013) Comparing the version 7 TRMM 3B43 monthly precipitation product with the TRMM 3B43 version 6/6A and Bureau of Meteorology datasets for Australia. Aust Meteorol Oceanogr J 63:421–426. http://www.bom.gov.au/amm/docs/2013/fleming.pdf
34. Forootan E, Awange J, Kusche J, Heck B, Eicker A (2012) Independent patterns of water mass anomalies over Australia from satellite data and models. Remote Sens Environ 124:427–43. https://doi.org/10.1016/j.rse.2012.05.023
35. Forootan E, Kusche J (2012) Separation of global time-variable gravity signals into maximally independent components. J Geodesy 86:477–547. https://doi.org/10.1007/s00190-011-0532-5.10.1007/s00190-011-0532-5
36. Forootan E, Rietbroek R, Kusche J, Sharifi M, Awange J, Schmidt M, Omondi P, Famiglietti J (2014) Separation of large scale water storage patterns over Iran using GRACE, altimetry and hydrological data. Remote Sens Environ 140:580–595. https://doi.org/10.1016/j.rse.2013.09.025
37. Furi W, Razack M, Abiye TA, Kebede S, Legesse D (2012) Hydrochemical characterization of complex volcanic aquifers in a continental rifted zone: the Middle Awash Basin, Ethiopia. Hydrogeol J 20:385–400. https://doi.org/10.1007/s10040-011-0807-1
38. Helsel D, Hirsch R (2002) Statistical Methods in Water Resources Techniques of Water Resources Investigations volume chapter A3 of Book 4. U.S. Geological Survey
39. Huffman GJ, Bolvin DT (2012) TRMM and other data precipitation data set documentation. (visited on 2012-03-23). ftp://meso-a.gsfc.nasa.gov/pub/trmmdocs/3B42_3B43_doc.pdf
40. Huffman GJ, Bolvin DT, Nelkin EJ, Wolff DB, Adler RF, Gu G, Hong Y, Bowman KP, Stocker EF (2007) The TRMM multisatellite precipitation analysis (TMPA): quasiglobal, multiyear, combined-sensor precipitation estimates at fine scales. J Hydrometeorol 8:38–55. https://doi.org/10.1175/JHM560.1
41. Hurni H (1988) Agroecological Belts of Ethiopia: explanatory notes on three maps at a scale of 1:1,000,000. Research Report Centre for Development and Environment University of Bern, Switzerland in association with The Ministry of Agriculture, Ethiopia Bern. Soil Conservation Research Programme Ethiopia
42. Kebede S (2013) Groundwater in Ethiopia: features, numbers and opportunities. Springer, Berlin [Springer Hydrogeology]
43. Kebede S, Travi Y, Alemayehu T, Ayenew T (2005) Groundwater recharge, circulation and geochemical evolution in the source region of the Blue Nile River. Ethiopia. Appl Geochem 20:1658–1676. https://doi.org/10.1016/j.apgeochem.2005.04.016

44. Kebede S, Travi Y, Asrat A, Alemayehu T, Ayenew T, Tessema Z (2008) Groundwater origin and flow along selected transects in Ethiopian Rift volcanic aquifers. Hydrogeol J 16:55–73. https://doi.org/10.1007/s10040-007-0210-0
45. Kummerow C, William B, Toshiaki K, James S, Simpson J (1998) The tropical rainfall measuring mission (TRMM) sensor package. J Atmos Oceanic Tech 15:809–817. https://doi.org/10.1175/1520-0426(1998)015<0809:TTRMMT>2.0.CO;2
46. Kusche J (2007) Approximate decorrelation and non-isotropic smoothing of timevariable GRACE-type gravity field models. J Geodesy 81:733–749 (2007) https://doi.org/10.1007/s00190-007-0143-3
47. Kusche J, Schmidt R, Petrovic S, Rietbroek R (2009) Decorrelated GRACE time-variable gravity solutions by GFZ, and their validation using a hydrological model. J Geodesy 83:903–913. https://doi.org/10.1007/s00190-009-0308-3
48. Liu Z, Ostrenga D, Teng W, Kempler S (2012) Tropical rainfall measuring mission (TRMM) precipitation data and services for research and applications. Bull Am Meteorol Soc 93:1317–1325. https://doi.org/10.1175/BAMS-D-11-00152.1
49. Longuevergne L, Scanlon BR, Wilson CR (2010) GRACE Hydrological estimates for small basins: evaluating processing approaches on the high plains aquifer, USA. Water Resour Res 46. https://doi.org/10.1029/2009WR008564 [n/ a–n/a]
50. Marshall M, Funk C, Michaelsen J (2012) Examining evapotranspiration trends in Africa. Clim Dyn 38:1849–1865. https://doi.org/10.1007/s00382-012-1299-y
51. Melesse A, Abtew W, Dessalegne T, Wang X (2010) Low and high flow analyses and wavelet application for characterization of the Blue Nile River system. Hydrol Process 24:241–252. https://doi.org/10.1002/hyp.7312
52. Melesse AM (ed) (2011) Nile River Basin: hydrology, climate and water use. Springer, Netherlands
53. Mengistu D, Bewket W, Lal R (2013) Recent spatiotemporal temperature and rainfall variability and trends over the upper Blue Nile river basin, Ethiopia. Int J Climatol. n/a–n/a. https://doi.org/10.1002/joc.3837
54. MoWE. Ministry of Water & Energy of Federal Democratic Republic of Ethiopia (2012) (visited on 2012-11-05). http://www.mowr.gov.et/index.php?pagenum=2.1&pagehgt=5495px
55. North GR, Bell TL, Cahalan RF, Moeng FJ (1982) Sampling errors in the estimation of empirical orthogonal functions. Mon Weather Rev 110:699–706. https://doi.org/10.1175/1520-0493(1982)110<0699:SEITEO>2.0.CO;2
56. Nyssen J, Clymans W, Descheemaeker K, Poesen J, Vandecasteele I, Vanmaercke M, Zenebe A, Van Camp M, Haile M, Haregeweyn N, Moeyersons J, Martens K, Gebreyohannes T, Deckers J, Walraevens K (2010) Impact of soil and water conservation measures on catchment hydrological response – a case in north Ethiopia. Hydrol Process 24:1880–1895. https://doi.org/10.1002/hyp.7628
57. Nyssen J, Poesen J, Moeyersons J, Deckers J, Haile M, Lang A (2004) Human impact on the environment in the Ethiopian and Eritrean highlands - a state of the art. Earth-Sci Rev 64:273–320. https://doi.org/10.1016/S0012-8252(03)00078-3
58. Omondi P, Awange J, Ogallo L, Ininda J, Forootan E (2013) The influence of low frequency sea surface temperature modes on delineated decadal rainfall zones in Eastern Africa region. Adv Water Resour 54:161–180. https://doi.org/10.1016/j.advwatres.2013.01.001
59. Omondi P, Awange J, Ogallo L, Okoola R, Forootan E (2012) Decadal rainfall variability modes in observed rainfall records over East Africa and their relations to historical sea surface temperature changes. J Hydrol 464–465:140–156. https://doi.org/10.1016/j.jhydrol.2012.07.003
60. Omondi PA, Awange JL, Forootan E, Ogallo LA, Barakiza R, Girmaw GB, Fesseha I, Kululetera V, Kilembe C, Mbati MM, Kilavi M, King'uyu SM, Omeny PA, Njogu A, Badr EM, Musa TA, Muchiri P, Bamanya D, Komutunga E (2014) Changes in temperature and precipitation extremes over the greater horn of Africa region from 1961 to 2010. Int J Climatol 34:1262–1277. https://doi.org/10.1002/joc.3763

61. Preisendorfer R (1988) Principal component analysis in meteorology and oceanography. Elsevier, Amsterdam
62. Reager JT, Famiglietti JS (2009) Global terrestrial water storage capacity and flood potential using grace. Geophys Res Lett 36. https://doi.org/10.1029/2009GL040826, [n/a–n/a]
63. Rientjes THM, Haile AT, Kebede E, Mannaerts CMM, Habib E, Steenhuis TS (2011) Changes in land cover, rainfall and stream flow in Upper Gilgel Abbay catchment, Blue Nile Basin - Ethiopia. Hydrol Earth Syst Sci 15:1979–1989. https://doi.org/10.5194/hess-15-1979-2011
64. Rodell M, Chen J, Kato H, Famiglietti JS, Nigro J, Wilson CR (2007) Estimating groundwater storage changes in the Mississippi River Basin (USA) using GRACE. Hydrogeol J 15:159–166. https://doi.org/10.1007/s10040-006-0103-7
65. Rodell M, Famiglietti JS (2001) An analysis of terrestrial water storage variations in illinois with implications for the gravity recovery and climate experiment (grace). Water Resour Res 37:1327–39. https://doi.org/10.1029/2000WR900306
66. Rodell M, Houser PR, Jambor U, Gottschalck J, Mitchell K, Meng C-J et al (2004) The global land data assimilation system. Bull Amer Meteorol Soc 85:381–394. https://doi.org/10.1175/BAMS-85-3-381
67. Rodell M, Velicogna I, Famiglietti JS (2009) Satellite-based estimates of groundwater depletion in India. Nature 460:999–1002. https://doi.org/10.1038/nature08238
68. Romilly TG, Gebremichael M. Evaluation of satellite rainfall estimates over Ethiopian river basins. Hydrol Earth Syst Sci 15:1505–1514. https://doi.org/10.5194/hess-15-1505-2011
69. Sawicz K, Wagener T, Sivapalan M, Troch PA, Carrillo G (2011) Catchment classification: empirical analysis of hydrologic similarity based on catchment function in the eastern USA. Hydrol Earth Syst Sci 15:2895–2911. https://doi.org/10.5194/hess-15-2895-2011
70. Schmidt R, Flechtner F, Meyer U, Neumayer K-H, Dahle C, König R, Kusche J (2008) Hydrological signals observed by the GRACE satellites. Surv Geophys 29:319–334. https://doi.org/10.1007/s10712-008-9033-3
71. Segele ZT, Lamb PJ, Leslie LM (2009) Seasonal-to-interannual variability of Ethiopia/ Horn of Africa monsoon. Part I: associations of wavelet-filtered large-scale atmospheric circulation and global sea surface temperature. J Climate 22:3396–3421. https://doi.org/10.1175/2008JCLI2859.1
72. Seleshi Y, Camberlin P (2006) Recent changes in dry spell and extreme rainfall events in Ethiopia. Theor Appl Climatol 83:181–191. https://doi.org/10.1007/s00704-005-0134-3
73. Seleshi Y, Zanke U (2004) Recent changes in rainfall and rainy days in Ethiopia. Int J Climatol 24:973–983
74. Senay GB, Asante K, Artan G (2009) Water balance dynamics in the Nile Basin. Hydrol Process 23:3675–3681. https://doi.org/10.1002/hyp.7364
75. Shang H, Yan J, Gebremichael M, Ayalew SM (2011) Trend analysis of extreme precipitation in the Northwestern Highlands of Ethiopia with a case study of Debre Markos. Hydrol Earth Syst Sci 15:1937–1944. https://doi.org/10.5194/hess-15-1937-2011
76. Siebert S, Henrich V, Frenken K, Burke J (2013) Update of the Digital Global Map of Irrigation Areas (GMIA) to Version 5. Documentation Institute of Crop Science and Resource Conservation Rheinische Friedrich Wilhelms Universitä Bonn and FAO Bonn, Germany and Rome, Italy
77. Sutcliffe J, Parks Y (1999) The hydrology of the Nile. Technical report international association of hydrological sciences Oxfordshire, UK. Special Publication No. 5
78. Swenson S, Famiglietti J, Basara J, Wahr J (2008) Estimating profile soil moisture and groundwater variations using GRACE and Oklahoma Mesonet soil moisture data. Water Resour Res 44. https://doi.org/10.1029/2007WR006057 [n/a–n/a]
79. Swenson S, Wahr J (2006) Post-processing removal of correlated errors in GRACE data. Geophys Res Lett 33:L08402. https://doi.org/10.1029/2005GL025285
80. Syed TH, Famiglietti JS, Rodell M, Chen J, Wilson CR (2008) Analysis of terrestrial water storage changes from GRACE and GLDAS. Water Resour Res 44. https://doi.org/10.1029/2006WR005779 [n/a–n/a]

81. Tapley BD, Bettadpur S, Ries JC, Thompson PF, Watkins MM (2004) GRACE measurements of mass variability in the Earth System. Science 305:503–5. https://doi.org/10.1126/science.1099192
82. Taye MT, Willems P (2012) Temporal variability of hydroclimatic extremes in the Blue Nile Basin. Water Resour Res 48. https://doi.org/10.1029/2011WR011466 [n/a–n/a]
83. Tesfagiorgis K, Gebreyohannes T, Smedt F, Moeyersons J, Hagos M, Nyssen J, Deckers J (2011) Evaluation of groundwater resources in the Geba Basin. Ethiopia. Bull Eng Geology Environ 70:461–466. https://doi.org/10.1007/s10064-010-0338-3
84. Wahr J, Molenaar M, Bryan F (1998) Time variability of the Earth's gravity field: hydrological and oceanic effects and their possible detection using GRACE. J Geophys Res: Solid Earth 103:30205–30229. https://doi.org/10.1029/98JB02844
85. Werth S, Güntner A, Schmidt R, Kusche J (2009) Evaluation of GRACE filter tools from a hydrological perspective. Geophys J Int 179:1499–1515. https://doi.org/10.1111/j.1365-246X.2009.04355.x
86. Yadav M, Wagener T, Gupta H (2007) Regionalization of constraints on expected watershed response behavior for improved predictions in ungauged basins. Adv Water Resour 30:1756–1774. https://doi.org/10.1016/j.advwatres.2007.01.005
87. Yirdaw S, Snelgrove K, Agboma C (2008) GRACE satellite observations of terrestrial moisture changes for drought characterization in the Canadian prairie. J Hydrol 356:84–92. https://doi.org/10.1016/j.jhydrol.2008.04.004
88. Yitbarek A, Razack M, Ayenew T, Zemedagegnehu E, Azagegn T (2012) Hydrogeological and hydrochemical framework of Upper Awash River basin, Ethiopia: with special emphasis on inter-basins groundwater transfer between Blue Nile and Awash Rivers. J Afri Earth Sci 65:46–60. https://doi.org/10.1016/j.jafrearsci.2012.01.002
89. Nile Basin Water Resources Atlas (2017). http://atlas.nilebasin.org/
90. Awange JL, Gebremichael M, Forootan E, Wakbulcho G, Anyah R, Ferreira CG, Alemayehu T (2014) Characterization of Ethiopian mega hydrogeological regimes using GRACE, TRMM and GLDAS datasets. Adv Water Resour 74:64–78. https://doi.org/10.1016/j.advwatres.2014.07.012
91. Awange JL, Forootan E, Kuhn M, Kusche J, Heck B (2014) Water storage changes and climate variability within the NRB between 2002 and 2011. Adv Water Resour 73:64–78. https://doi.org/10.1016/j.advwatres.2014.06.010
92. Awange J, Forootan E, Fleming K, Odhiambo GO (2015) Dominant patterns of water storage changes in the Nile Basin during 2003–2013. In: Remote sensing of the terrestrial water cycle (ed.) Lakshmi V, pp 367–382. Wiley, New Jersey. https://doi.org/10.1002/9781118872086.ch22
93. Khaki M, Awange J, Forootan E, Kuhn M (2018) Understanding the association between climate variability and the Nile's water level fluctuations and water storage changes during 1992–2016. Sci Total Environ 645:1509–1521. https://doi.org/10.1016/j.scitotenv.2018.07.212
94. Khaki M, Awange J, Forootan E, Kuhn M (2018) Understanding the association between climate variability and the Nile's water level fluctuations and water storage changes during 1992–2016. Sci Total Environ 645:1509–1521. https://doi.org/10.1016/j.scitotenv.2018.07.212
95. Mpelasoka F, Awange JL, Zerihun A (2018) Influence of coupled ocean-atmosphere phenomena on the Greater Horn of Africa droughts and their implications. Sci Total Environ 610–611:691–702. https://doi.org/10.1016/j.scitotenv.2017.08.109
96. Awange JL, Palancz B, Völgyesi L (2020) Hybrid imaging and visualization. Employing machine learning with mathematica - python. Springer Nature International, Berlin. ISBN 978-3-030-26152-8. https://doi.org/10.1007/978-3-030-26153-5
97. Awange JL (2018) GNSS environmental sensing. Revolutionizing environmental monitoring, Springer, Berlin
98. Awange JL, Kiema JBK (2019) Environmental geoinformatics. Exploiting the Big Data, Extreme Hydro-Climatic and Food Security Challenges. Springer Nature, Heidelberg
99. Awange JL, Kiema JBK (2013) Environmental geoinformatics. Monitoring and management. Springer, Berlin

Part III
The Nile: Weighed from Space

Chapter 6
Water Storage Changes: Dominant Patterns

6.1 Summary

Water resources within the Nile basin are under intense pressure from both human use and the changing climate. With the increasing human population and growing demand for both water and electricity, the various countries within the basin are likely to further exploit the waters within their sub-basins, thus negatively impacting upon the entire Nile system. Remote sensing methodologies discussed in Chap. 3 allow the water storage (WS) changes to be assessed, thereby providing essential information for water resource management. This chapter presents the work of Awange et al. [6] who analysed water storage changes for the period 2003–2013 over the Nile basin derived from monthly Gravity Recovery And Climate Experiment (GRACE), Tropical Rainfall Measuring Mission (TRMM), and NDVI (Normalized Digital Vegetation Index) products. TRMM analysis showed no significant trend in the rainfall over the Nile basin. Over the period 2003–2013, the Principal Component Analysis (PCA) results indicate a general decrease in GRACE derived water storage in the northern part of the basin at a rate of 3 mm/yr. In contrast, Ethiopian Highlands in the east exhibited a mass gain, with an average linear rate of 2.8 mm/yr. A lag of one to two months was derived between TRMM-rainfall and GRACE-WS time series. This might represent the large-scale relationship between accumulated rainfall patterns and storage changes over the Nile River Basin (NRB). NDVI results shows a downward trend in vegetation cover during low rainfall.

6.2 Nile River Basin's Water Storage Changes

Arguably the world's longest river, the Nile traverses some 6,500 km from south to north through Tanzania, Uganda, Kenya, Rwanda, Burundi, Democratic Republic of Congo (DRC), Eritrea, Ethiopia, South Sudan, Sudan and Egypt [60, 65]. As already

J. Awange, *The Nile Waters*,
https://doi.org/10.1007/978-3-030-64756-8_6

stated in Chap. 2, the Nile River Basin (NRB), with an area of 3,400,000 km^2, supports the livelihoods of over 300 million people (e.g., [7]).

Yet, even though its basin is huge in area, the volume of the Nile's runoff is relatively low (e.g., [65]), thus making its meager waters very precious, particularly to the downstream countries such as Sudan and Egypt, which are almost completely dependent upon the Nile for their very survival. One significant factor that has been reported to threaten the river's water resources involves direct anthropogenic influences (see e.g., [37]). This is largely due to increased human population (e.g., Fig. 2.2), which puts increased pressure on domestic water needs, the supply of hydroelectric power (e.g., Fig. 2.9), and increased agricultural activities, all of which are coupled to the need to sustain economic growth. Moreover, not only are the demands on the basin's water increasing, but the available water supplies appear to be decreasing, with environmental degradation of the upper Blue Nile catchment having increased throughout the 1980s [74]. For instance, Ethiopia, with its extreme hydrological variability and seasonality, is currently constructing the continent's largest hydroelectric dam (e.g., Fig. 2.9b) within the Blue Nile sub-basin [16]. Moreover, there is the threat that emanates from natural factors such as the changing climate, which has been the subject of numerous studies (e.g., [77, and the references therein]). Combined and individually, direct human influences (population growth, unsustainable water usage and development) and climatic change (droughts, floods, and desertification, both natural and anthropogenic) threaten the Nile's ability to supply crucially needed water to the people of the basin.

Therefore, for the proper management of the Nile basin's water resources, accurate monitoring of its spatial and temporal changes is required to allow informed policy decisions. Most studies dealing with the Nile basin to date, however, have dealt mainly with modelling the impacts of climate change (e.g., [14, 18, 21, 22]), with very little being reported on how to monitor the spatial and temporal variations in the basin's overall stored water (surface, groundwater and soil moisture) in a holistic manner. The reason for such few studies (e.g., [2, 15, 19, 20, 23, 61, 64]) has been partly attributed to its large size, as well as the lack of appropriate monitoring techniques that could cover such a vast spatial extent. For instance, the hydrological water balance involves the flow of surface water, the movement of deeper groundwater, and the coupling of the land, ocean and atmosphere through evaporation and precipitation. Monitoring these components requires an accuracy and completeness of geographical data coverage that challenges conventional ground-based measurement capabilities.

The acquisition of continuous measurements of land surface properties is important for the analyses of environmental change over any scale, with satellite-derived information becoming increasingly important when considering regional to global extents. For example, time series of satellite images have an important role in the monitoring of regional and global land cover change (e.g., [66]). Satellite sensors are well-suited to large-scale monitoring tasks, because they provide repeatable measurements at spatial scales that are appropriate for capturing the effects of changes in key properties of an ecosystem. Likewise, time series analysis is useful for identifying, modelling and forecasting the dynamics (trends, cycles, periodicity) of variables observed sequentially over time. One example of satellite monitoring is the Gravity

Recovery And Climate Experiment (GRACE; see Sect. 3.3.3), which was designed to monitor climate-related mass redistribution within the Earth system using measurements of changes in the terrestrial gravity field. After tidal, ocean, and atmospheric variations have been corrected for, the temporal variation in the gravity field is largely caused by changes in the distribution of water (e.g., [7, 9–11, 32, 57, 70, 71, 73, 83, 84]). Currently, river basins of the order of 400,000 km^2 in area can be successfully studied using the GRACE products [67, 69]. Given that the Nile basin has an area of more that 3,400,000 km^2, GRACE products can potentially be used to analyze changes in its water storage. Another satellite mission of interest is the Tropical Rainfall Measuring Mission (TRMM), which is observing precipitation over the Earth between latitudes 50°N and 50°S, hence providing data for regions poorly served by ground-based rain gauges, which includes much of the NRB. Similarly, among modern satellite observation-based products used to monitor terrestrial ecosystems, the Normalized Digital Vegetation Index (NDVI) is of primary importance thanks to its capability of providing synoptic information about the status of vegetation over wide areas with high acquisition frequency [47].

The main task of this chapter is to apply space-based satellite approaches to remotely sense the changes in the Nile basin's water storage over the last decade. To achieve this, use is made of GRACE products whose results are compared to precipitation, as observed by TRMM over the same period to assess whether changes in the NRB's waters are due to climatic change and/or human effects. This is achieved by constructing total water storage (TWS) time series for the NRB from the GRACE gravity data for the period between 2003 to 2013. The computation is done while taking care of the leakage caused by the basin's shape. GRACE spatio-temporal TWS changes reveal the general pattern of water changes within the basin, including all the tributaries in its catchments, lakes, soil moisture and groundwater. TRMM products have previously been applied to the region, e.g., by [4, 5, 48, 51], and have been used to study the impact of precipitation variations on TWS changes within the basin [78–81]. Furthermore, Moderate Resolution Imaging Spectroradiometer (MODIS) 500-meter 16-day maximum value Normalized Digital Vegetation Index (NDVI) composites (MOD13A1) were used to evaluate surface vegetation conditions, which are then related to changes in stored water within the basin (e.g., [50]). Note that [6] did not make use of satellite altimetry products, and the interested reader is referred to studies by [15, 23, 78, 79] for the use of such information.

In Sect. 6.3, the space products and the adopted methodology followed (data used and the analysis methods) are described. Section 6.4 presents and discusses the NRB's stored waters' dominant patterns. Finally, Sect. 6.5 summarizes the major findings. For the purpose of this chapter, the Nile basin in Fig. 2.1 is divided into two sectors at 15°N (see e.g., Fig. 6.1). The reason for this subdivision is that the Nile basin falls within equatorial and tropical sectors where rainfall seasons are both bimodal (March-April-May and October-November-December) and unimodal (June-July-August) and hence the southern and northern sectors, respectively.

6.3 Space Products and the Adopted Methods

6.3.1 Gravity Recovery And Climate Experiment (GRACE)

The Gravity Recovery And Climate Experiment (GRACE) discussed in Sect. 3.3.3 operates in what is termed as satellite-to-satellite tracking in the combined high-low and low-low mode. By this, it is meant that the satellites measure the separation between them using a K-band microwave ranging system, while at the same time employing GNSS (Global Navigation Satellite System) to determine their precise location ([59, 71]). Satellite-laser-ranging reflectors (SLR) are also used as an independent check (e.g., [70]) while the effects of non-gravitational forces acting on the satellite are removed using measurements from on-board accelerometers ([71]).

The main product of the GRACE mission is its time-variable gravity fields in the form of normalised spherical harmonic solutions, of which release five (RL05) monthly solutions, provided by the German Research Centre for Geosciences (GFZ) ([24]), covering 2003 to 2013, were used. GRACE level-02 products employed in this chapter contained correlated errors among the higher-order spherical harmonics, which is known as the striping pattern [43]. In order to remove these stripes, decorrelation filter denoted by DDK2 of [44] was applied to the GFZ-RL05 solutions. The filtered solutions can also be downloaded from http://icgem.gfz-potsdam.de/ICGEM/TimeSeries.html.

For computing monthly TWS change over the NRB, the following items were considered:

1. Residual gravity field solutions with respect to the temporal average of 2003 to 2013 were computed.
2. The residual coefficients were then convolved with a basin function, considering the basin boundary shown in Fig. 6.1 (see e.g., [28, 29]). The selected basin boundary had a convex shape, which itself reduced the leakage error, as outlined in [46].
3. To account for the leakage, a simple approach that assumes a uniform mass distribution with the value of 1 inside the NRB and 0 outside the basin ($S1 = 1$, is a uniform mass in the basin) was used, where the uniform mass was transformed into spherical harmonics. The obtained coefficients were then filtered with the same DDK2 filter as applied to the GRACE products. Afterward, the total surface mass of the basin was calculated from the filtered coefficients (S2, synthesized uniform mass in the basin). The ratio of S1/S2 reflects the effect of the truncation of the spherical harmonics as well as signal attenuation due to filtering. A more detailed discussion of the leakage problem can be found, for example, in [39]. An alternative statistical approach for reducing the leakage is addressed in [35].
4. The derived ratio was multiplied by the coefficients derived in item 2 and the results were transformed into $0.5^\circ \times 0.5^\circ$ TWS maps within the NRB, following [73]. An average uncertainty of $\sim$15 mm was computed based on the formal error of the GFZ-RL05 products.

6.3.2 *Tropical Rainfall Measuring Mission (TRMM)*

The TRMM was launched in 1997 and is a joint mission of NASA and the Japan Aerospace Exploration Agency (JAXA). TRMM orbits at an altitude of about 400 km with an inclination of 35° and orbital period of around 91 min, thus completing 16 revolutions per day. TRMM was designed to monitor and study tropical rainfall over the latitude range ±50° over inaccessible areas such as the oceans and poorly sampled terrain. The primary on-board instruments are the TRMM Microwave Imager (TMI), the Precipitation Radar (PR), and the Visible and Infrared Radiometer System (VIRS) ([41, 42]). TRMM products have been employed in a number of studies of precipitation in Africa, where they have been found to be adequate when compared with ground truth observations (e.g., [48]).

TRMM product employed in this chapter is referred to as the *TRMM and Other Data Precipitation Data Set* (denoted as 3B43). 3B43 provides monthly rainfall (average hourly rate) between latitudes 50°N/50°S over a 0.25° × 0.25° grid. Over time, the products 3B43 are updated as the processing techniques and methods for integrating the different data sets are improved upon. In this chapter, version, number 7 was used, which has been found in the case of Australia to be a significant improvement over the previous version 6 [31], which itself showed a high correlation with in-situ derived products [30], owing to such changes as the use of additional satellites and a superior means of incorporating rain gauge information from the Global Precipitation Climatological Centre [31, 38]. The use of TRMM satellite data in this chapter was informed by the works of [4, 5, 17, 25–27, 36] on the validation of satellite product in the region, which includes nearly half of Ethiopia (i.e., the Blue Nile portion). In particular, [25] reported good performance of TRMM-3B43 rainfall product. Correlations between time series of TRMM version 7 and its former version 6 products were compared over the NRB where a correlation of greater than 0.7 for more than 75% of the time series was obtained thus justifying its use for studying the precipitation pattern over the Nile basin.

6.3.3 *NDVI Data*

Vegetation monitoring by remotely sensed data has traditionally been carried out using vegetative indices, which are mathematical transformations designed to assess the spectral contributions of green plants to multispectral observations [45]. These indices are mainly derived from reflectance data from discrete red (R) and near-infrared (NIR) bands. Although there are several vegetation indices (see e.g., [3]), one of the most widely used is the Normalized Difference Vegetation Index (NDVI, see Sect. 3.2). NDVI values range from +1.0 to −1.0. It operates by contrasting intense chlorophyll pigment absorption in the red against the high reflectance of leaf mesophyll in the near infrared, hence NDVI = [NIR-R]/[NIR + R]. NDVI is therefore indicative of plant photosynthetic activity and has been found to be highly related to the green leaf area index and the fraction of photosynthetically active radiation absorbed by vegetation [75]. Areas of barren rock, sand, or snow usually show very low NDVI values (for example, 0.1 or less). Sparse vegetation such as shrubs and grasslands or crops may result in moderate NDVI values (approximately 0.2 to 0.5). High NDVI values (approximately 0.6 to 0.9) correspond to dense vegetation such as that found in tropical forests or crops at their peak growth stage. Because of these properties, NDVI can be utilized as an indicator of vegetation stress, particularly due to water shortage, which may result from reduced precipitation [47]. This is the case whenever water availability is the main limiting factor for vegetation processes and therefore controls leaf pigment content and integrity. NDVI values of vegetation usually offer an efficient and objective means for evaluating phenological characteristics and have long been used to monitor vegetation conditions and changes in vegetation cover [50, 76, 85].

In this chapter, time series analysis of the NDVI data for the NRB was carried out using MODIS data from NASA, in the form of 16-day composite MODIS surface-reflectance products (MOD13A1) for the years 2000 to 2009, provided by the NASA Distributed Active Archive Center in HDF-EOS format.[1] These data were derived from the Earth Observing System (EOS) Terra MODIS surface reflectances, which have been corrected for molecular scattering, ozone absorption, and aerosols. The index is produced globally over land at 16-day compositing intervals and enables consistent spatial and temporal comparisons of vegetation condition. Over Lake Victoria, they have been used e.g., by [3, 50, 82].

The MODIS re-projection tool[2] was used to merge the 20 tiles that cover the Nile basin. The data were then subject to the Nile basin study boundary, re-projected from a sinusoidal to UTM projection using a nearest-neighbour re-sampling routine, and entered into a 250 m × 250 m grid cell multilayer image stack. Separate data stacks were developed for both the original NDVI data and Quality Assessment Science Data Sets (QASDS). Details documenting the MODIS NDVI compositing process and Quality Assessment Science Data Sets (QASDS) can be found at NASA's MODIS web site.[3] The NDVI data stack was first filtered to eliminate anomalous high (hikes) and low (drops) values. These were effectively eliminated by removing data values that suddenly decreased or increased and then immediately returned to near the previous NDVI value. The threshold for the removal of pseudo hikes and drops was set at $\pm$ 0.15% (see, e.g., [40]) to achieve the best setting (determined qualitatively) to eliminate most anomalous points, while not inadvertently removing good data points, resulting in smoother temporal profiles. It should be pointed out that 0.15% was not by any means a standard value, but is somewhat subjective, and was mainly based on the fact that sudden spikes and drops in NDVI would, in most cases, refer to errors in the data as one would not expect that to occur even in cases where conditions change, i.e., the phenological effect takes time. What this simply means is that it takes a while (several weeks) before noticeable changes in vegetative

[1] http://reverb.echo.nasa.gov/reverb.

[2] http://edc.usgs.gov/programs/sddm/modisdist/.

[3] https://lpdaac.usgs.gov/products/modis_products_table/mod13a3.

cover occurs as a result of weather (mainly precipitation) changes. So, sudden spikes (drops and hikes) in NDVI would simply indicate error in the data, which must be corrected.

6.4 NRB's Stored Water Analysis of the Dominant Patterns

6.4.1 Analysis of TRMM Data

The statistical method of Principal Component Analysis (PCA) was applied to spatio-temporal rainfall grids in order to identify and extract their dominant variability [55]. PCA, was applied through an eigenvalue decomposition of the auto-covariance matrix of the data into sets (modes) of empirical orthogonal functions (EOFs) and principal components (PCs) corresponding to the spatial and temporal variability, respectively. Details of the PCA approach, when it is applied to time series of TWS and rainfall maps, can be found in e.g., [32–34]. All the PCA-derived orthogonal components shown here passed [49]'s significance test.

From the results of PCA applied to the TRMM products, only the first three components are presented. A point that should be made immediately is that the TRMM data over the region has been already subjected to a trend analysis in [52] who found no significant trend over the period of study and also over the longer period of 1998 to 2012. In this chapter, no trend assessment was performed but refer an interested reader to the study above. Examining the PCA results, we see from the EOFs (Fig. 6.1) that there is very little signal north of latitude 15°N. Considering the 1st mode, representing 69% of the total variability, which corresponds to the annual signal (EOF1 and PC1 in Fig. 6.1), we note the difference in sign between the eastern (Ethiopia) and western-most parts when compared to the southern extreme of the basin in the vicinity of Lake Victoria (see EOF1 in Fig. 6.1). This is similar to the 2nd mode, representing 7% of the total variability (EOF2 and PC2 in Fig. 6.1), which in turn corresponds to the semi-annual cycles. It is also worth noting that the peak of PC1 to be around the middle of the year, corresponding to the rain season in the Ethiopian Highlands. The 3rd mode representing 3% of the total variability appears to show variability in precipitation to be restricted to an area corresponding to a band extending from the Sudd Swamp region to the Ethiopian Highlands, with a change in sign from west to east (EOF3 and PC3 in Fig. 6.1), again with a variability approaching semi-annual. Correlation between PC2 and PC3 in Fig. 6.1 shows the maximum lag of 2 to 3 months between the peak of rainfall over Lake Victoria (see EOF2 and PC2 in Fig. 6.1) and the Ethiopian highlands (see EOF3 and PC3 in Fig. 6.1). From the PCA results, there seems to be very little change in the rate of precipitation over the NRB during the period studied.

In order to better illustrate the inter-annual variability of rainfall over the basin, the dominant annual and semi-annual cycles were removed from the original rainfall time series using a least squares adjustment procedure. The residuals were then subjected

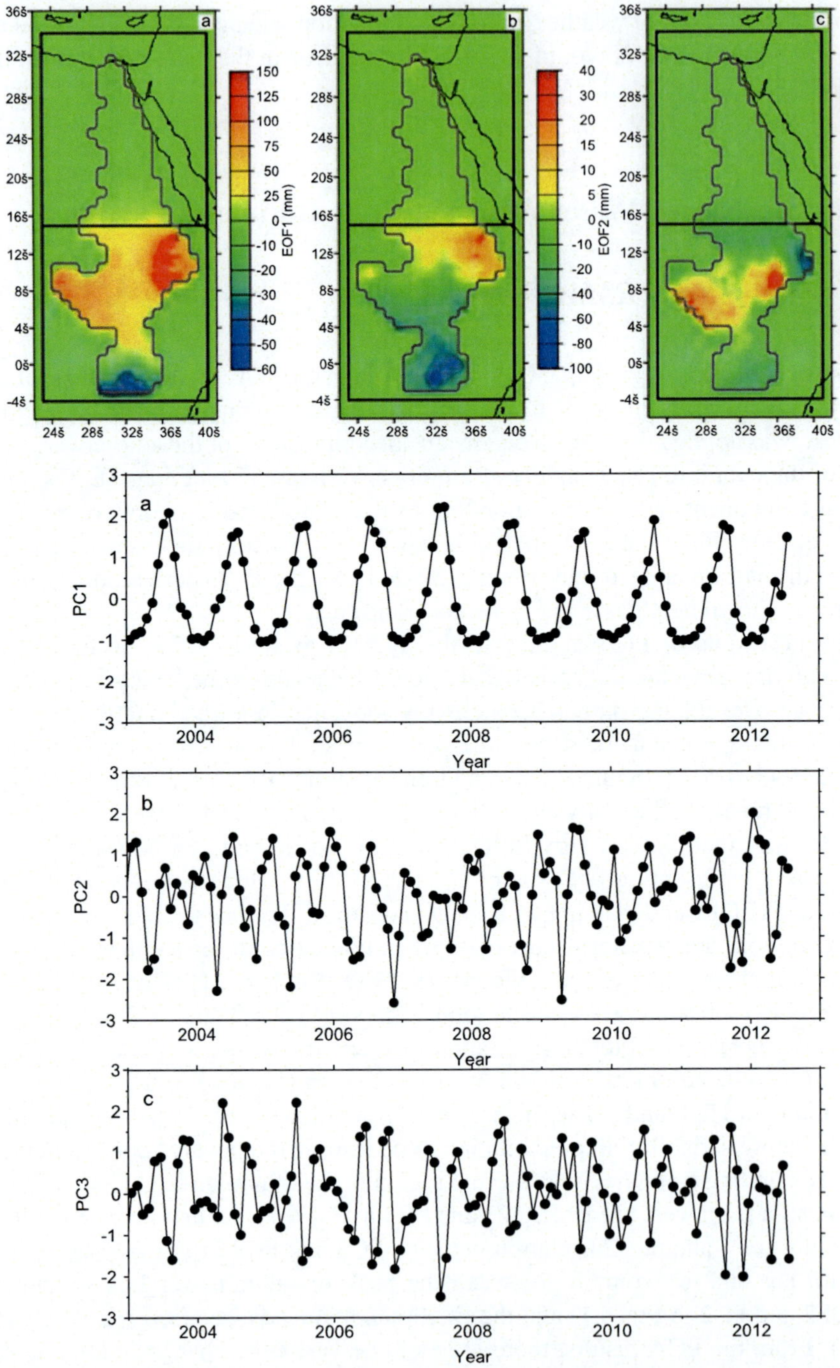

Fig. 6.1 PCA decomposition of the TRMM-derived rainfall variations over the Nile Basin. EOFs of rainfall changes over the Nile basin are shown on top and arranged with respect to the variance they represent from left to right. The corresponding temporal variations (PCs) are shown on bottom. PC1 and EOF1 presents 69% of the rainfall variance, PC2 and EOF2 presents 7% of the variance, and 5% of the variance is the contribution of PC3 and EOF3. *Source* [6]

to the PCA decomposition, which are shown in Fig. 6.2. The temporal components were subjected to the least squares spectral analysis [72, LSSA,] as described. e.g., in [63]. The reported dominant cycles passed a Fisher significant test (for details on Fisher test see e.g., [63]. PC1 in Fig. 6.2 contains two main dominant cyclic components with the period of 1.66 years and 4 months. Its corresponding spatial pattern (EOF1 in Fig. 6.2) indicates that these cycles are dominant over the southern part of the basin. The remaining amplitude of the cyclic component with the period of 4 months is capture by PC2 and EOF2 of Fig. 6.2. The EOF2 shows the dipole structure of the rainfall variability over the lake Victoria Basin and parts of South Sudan and Ethiopia. The computed spectrum of PC3 in Fig. 6.2 contained several peaks that barely passes the performed Fisher significant test and are therefore not interpreted further.

6.4.2 Analysis of the GRACE's Signals

The PCA method was also applied to the GRACE-TWS maps of the NRB, where the three dominant EOFs of the GRACE-TWS and their corresponding temporal components (PCs) are shown in Fig. 6.3. Similar to the rainfall results, the components are arranged with respect to the variance they represent. EOF1 representing 79% of the total variability shows a spatial north-south dipole structure, which as PC1 indicates, corresponds to the annual TWS changes of the NRB. EOF2 representing 9% of the total variability is concentrated over the Lake Victoria basin, with PC2 showing an average decline of −22 mm/yr between 2002 to middle of 2006 and a positive trend of 13 mm/yr between 2007 to 2013. Mode 3 (EOF3 and PC3) representing 5% of the total variability shows the northern (including Egypt) and western (including south Sudan) parts of the NRB are exhibiting decreasing mass between 2003 to 2012, with an average linear rate of −3 mm/yr. In contrast, the Ethiopian Highlands in the east, the water tower of the NRB, exhibited a mass gain, with an average linear rate of 2.8 mm/yr. Note that the patterns derived from PCA represent the dominant variability of TWS signal over the region. Hence the reason why it is preferred here over a simple computation of the trend and seasonal components. The first two dominant modes of GRACE-TWS, shown in Fig. 6.3, indicate the seasonal transport of water from the head water regions (e.g., Lake Victoria) and the reservoirs of the lower Nile Basin e.g., South Sudan.

Similar to the rainfall data, the dominant annual and semi-annual cycles were removed from TWS and the residuals were decomposed using the PCA (shown in Fig. 6.4). Using the LSSA, PC1 in Fig. 6.4 contains a dominant seasonal cycle with the period of ~3 to 4 months. A long-term trend, which is dominant over the Lake Victoria Basin is also noticed. PC2 contains a dominant cycle of 1.42 years period and several cycles with inter-annual periods. Justification of the reliability of the computed cycles in the second mode (PC2 and EOF2) and several seasonal cycles in the third mode (PC3 and EOF3) requires longer data sets. To assess the relationships between rainfall and TWS changes derived from GRACE within the NRB, first the

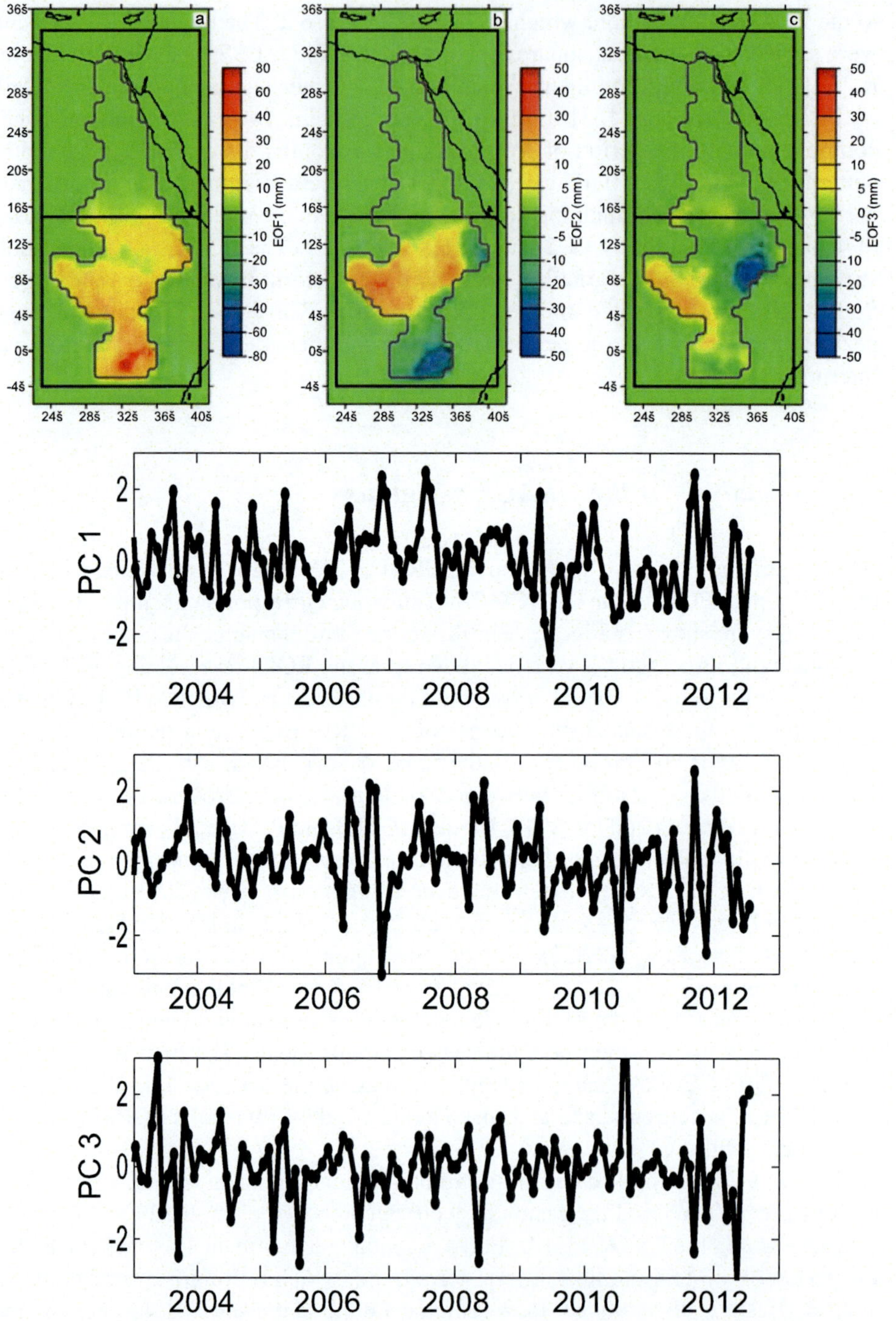

Fig. 6.2 PCA decomposition of the TRMM-derived rainfall variations over the NRB after removing the dominant annual and semi-annual cycles. EOFs and PCs are arranged similar to Fig. 6.1. *Source* [6]

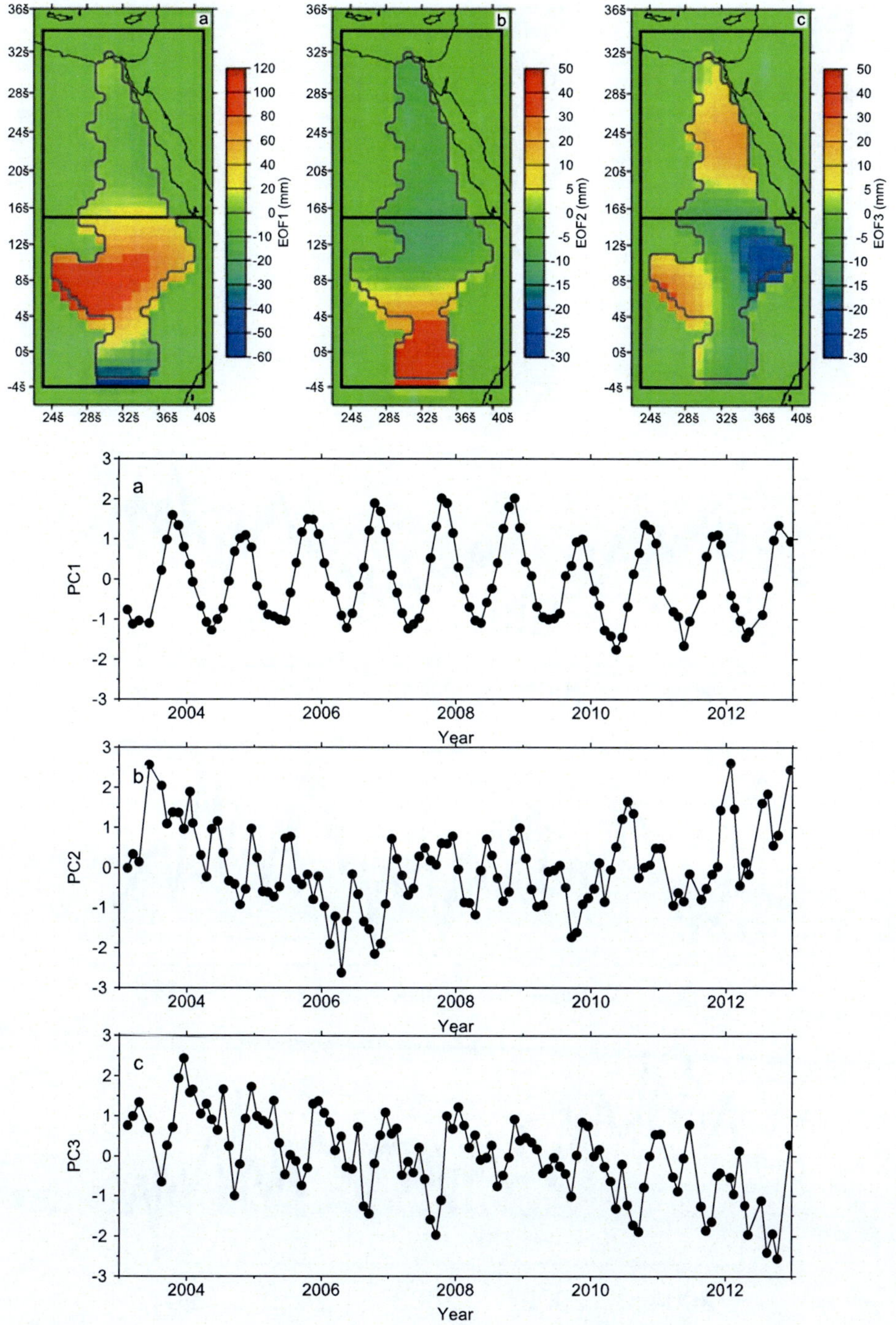

Fig. 6.3 An overview of the PCA decomposition of TWS changes from the GRACE products over the Nile basin. EOFs represent anomaly maps of the TWS variations. PCs are their unit-less normalized temporal evolution. PC1 and EOF1 presents 79% of the TWS variance, PC2 and EOF2 presents 9% of the variance, and 5% of the variance is the contribution of PC3 and EOF3. *Source* [6]

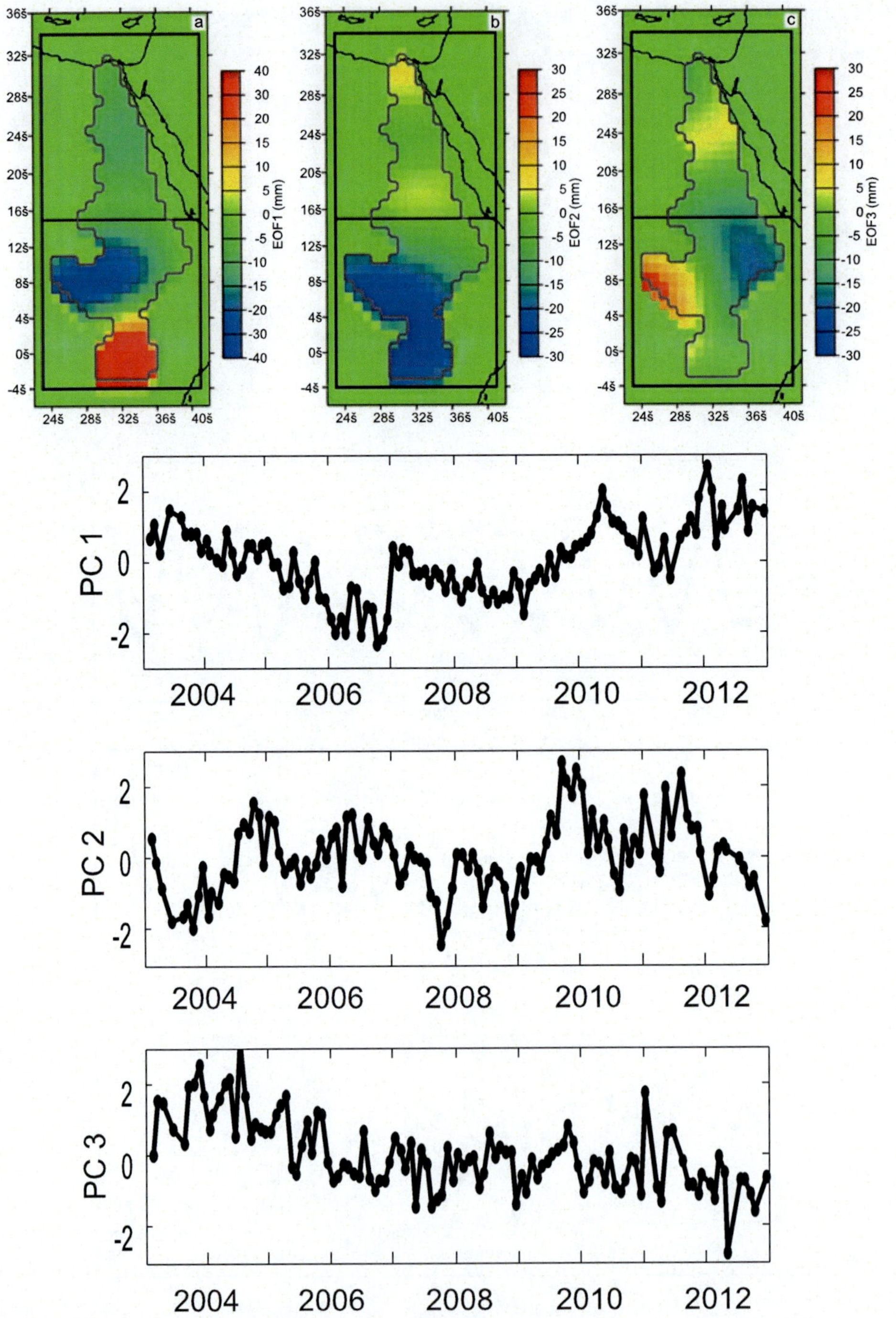

Fig. 6.4 PCA decomposition of TWS variations over the NRB after removing the dominant annual and semi-annual cycles. EOFs and PCs are arranged similar to Fig. 6.3. *Source* [6]

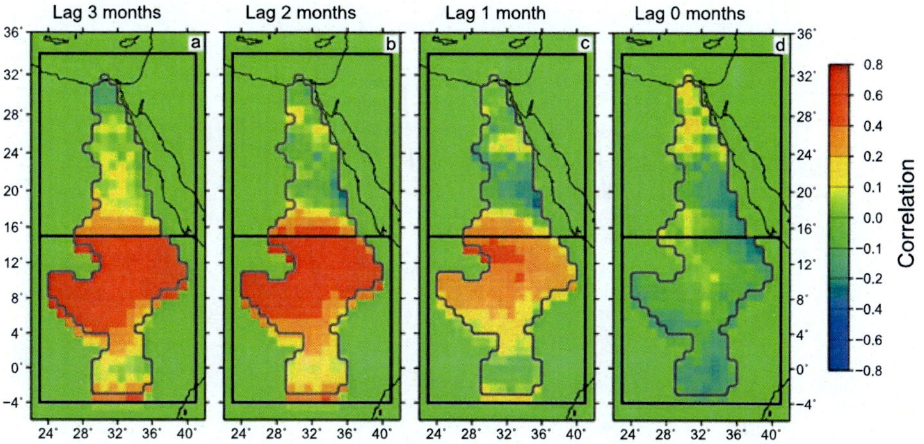

Fig. 6.5 Correlations between the GRACE and TRMM time series for different time lags over the NRB. The figure shows a lag of 2 and 3 months is the possible optimum correlation between rainfall and TWS changes of the region. *Source* [6]

main variability of TWS and rainfall were reconstructed using the three PCA modes shown above. Then, correlations between the TWS and rainfall time series, with a lag of 0, 1, 2 and 3 months at 95% level of confidence were computed. The results (Fig. 6.5) show the maximum correlation arises for the lag of 2 month. Correlations derived from 3 month lag are similar to that derived from 2 months lag. Note that before computing those correlations, TRMM grids were interpolated to $0.5° \times 0.5°$ grids, to be comparable with the GRACE grids.

6.4.3 Analysis of the NDVI Data

The NDVI trend analysis results are presented in Fig. 6.6. This analysis reveals a cyclic seasonal trend with peak values recorded during the rainfall seasons (Fig. 6.6). A downward trend is also noticed, albeit with some irregularity, with the NDVI monthly mean values for 2004 and 2006 being particularly lower than the rest of the years, coinciding with drought years [58, 62] when rainfall was significantly lower within the NRB. In these years, there was no month with NDVI values equal to or greater than 0.6, indicating that vegetative cover was more sparse. In other words, the years referred to were characterized by minimal/low precipitation hence the low vegetative cover/growth as shown by the low NDVI values.

The downward trend in the NDVI values coincided with the decreasing rainfall trend observed by the ground stations within the southern half of the Nile basin such as Bukoba. These findings corroborate those of [50], who found that for Lake Victoria, weather was a major factor governing the water level trends. A further study by [56] on rain use efficiency (the ratio of vegetation productivity to annual precipitation)

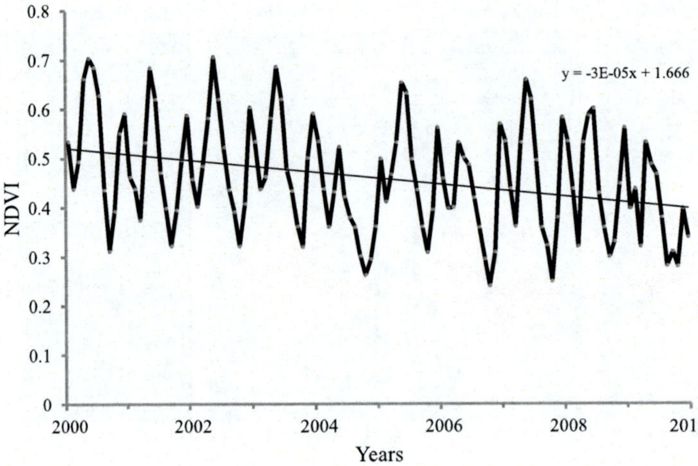

Fig. 6.6 Nile basin's monthly mean NDVI between the years 2000 and 2009. *Source* [6]

revealed similar patterns of NDVI trends with rainfall. However, the NDVI analysis in [6] covers only part of the period for which the rainfall analysis was carried out. The analysis reveals that climate has substantial control on NDVI through annual rainfall (see, e.g., [50]). That said, much more detailed statistical analysis involving other factors, such as land use, overgrazing, soil type, fire and floods, which might also influence vegetation at local scales, should be used to separate the rainfall impacts on NDVI from other factors. While high positive correlations between NDVI and rainfall portrays 'normal' land cover performance (e.g., [1]), low correlation might indicate areas impacted upon by human influence [45].

6.5 Concluding Remarks

This chapter has highlighted the following features of the state of the Nile River Basin (NRB) over the period 2003–2013:

1. Trend (not shown here) and PCA analysis do not show a significant overall decrease in rainfall magnitude over the NRB despite the high variability of precipitation, e.g., low rainfall during the 'short rains' of 2005 or the massive rainfall during the El-Niño of 2007 and 2012.
2. PCA results indicate a decrease in TWS in the northern part of the basin at a rate of 3 mm/year, while the southern part experienced a decrease at a rate of 22 mm/year between 2002 to 2006, and an increase thereafter from 2007 to 2013 at a rate of 13 mm/year. Over the whole period of study (2003–2013), the Ethiopian Highland gained stored water at a rate of 2.8 mm/year. A damp in the magnitude of the annual signals are found in 2005 and 2010 form the PCA analysis of TRMM rainfall and GRACE-TWS maps over the basin (e.g., PC1 in Figs. 6.1 and 6.3),

which indicates the importance of the rainfall magnitude in the total water storage changes over the basin.

3. A lag of one to two months is derived between rainfall and GRACE-TWS time series. This might represent the large-scale relationship between accumulated rainfall patterns and storage changes over the basin. From the maps of Fig. 6.5, this relationship is stronger over the southern part of the basin.
4. NDVI results show a downward trend in vegetation cover for the years characterized by low precipitation.

Considering the fact that rainfall conditions over the last decade have stayed relatively constant over the basin, it can be concluded that the 2002–2006 decline in the stored water within the southern part of the NRB, home to the Lake Victoria Basin, [3, 13] was attributed to anthropogenic factors consistent with previous findings of [8, 68]. The subsequent upsurge in the water levels from 2007 to 2013 could be due to the ENSO rainfalls, see e.g., [12, 52–54].

References

1. Aguilara C, Zinnert JC, Polo MJ, Young DR (2012) NDVI as an indicator for changes in water availability to woody vegetation. Ecol Ind 23:290–300. https://doi.org/10.1016/j.ecolind.2012.04.008
2. Ahmed M, Sultan M, Wahr J, Yan E, Milewski A, Sauck W, Becker R, Welton B (2011) Integration of GRACE (Gravity recovery and climate experiment) data with traditional data sets for a better understanding of the time-dependent water partitioning in African watersheds. Geology. https://doi.org/10.1130/G31812.1
3. Awange JL (2021) Lake Victoria monitored from space. Springer Nature International
4. Awange JL, Ferreira VG, Forootan E, Khandu Andam-Akorful SA, Agutu NO, He XF (2016) Uncertainties in remotely sensed precipitation data over Africa. Int J Climatol 36:303–323. https://doi.org/10.1002/joc.4346
5. Awange JL, Hu KX, Khaki M (2019) The newly merged satellite remotely sensed, gauge and reanalysis-based multi-source weighted-ensemble precipitation: evaluation over Australia and Africa (1981–2016). Sci Total Environ 670:448–465. https://doi.org/10.1016/j.scitotenv.2019.03.148
6. Awange J, Forootan E, Fleming K, and Odhiambo GO (2015) Dominant patterns of water storage changes in the nile basin during 2003–2013. In: Venkataraman Lakshmi (ed) Remote sensing of the terrestrial water cycle. John Wiley, New Jersey, pp 367–382. https://doi.org/10.1002/9781118872086.ch22
7. Awange JL (2012) Environmental monitoring using GNSS. Springer, Berlin, p 354
8. Awange JL, Sharifi M, Ogonda G, Wickert J, Grafarend E, Omulo M (2008) The falling Lake Victoria water levels: GRACE, TRIMM and CHAMP satellite analysis of the lake basin. Water Resourc Manag 22:775–796. https://doi.org/10.1007/s11269-007-9191-y
9. Awange J, Forootan E, Kusche J, Kiema JKB, Omondi P, Heck B, Fleming K, Ohanya S, Gonçalves RM (2013) Understanding the decline of water storage across the Ramser-lake Naivasha using satellite-based methods. Adv Water Resour 60:7–23. https://doi.org/10.1016/j.advwatres.2013.07.002
10. Awange JL, Kiema JBK (2012) Environmental geoinformatics. Monitoring and management. Springer, Berlin, p 354

11. Awange JL, Sharifi M, Ogonda G, Wickert J, Grafarend E, Omulo M (2008) The falling Lake Victoria water levels: GRACE, TRIMM and CHAMP satellite analysis of the lake basin. Water Resour Manag 22:775–796. https://doi.org/10.1007/s11269-007-9191-y
12. Awange JL, Aluoch J, Ogallo L, Omulo M, Omondi P (2007) An assessment of frequency and severity of drought in the Lake Victoria region (Kenya) and its impact on food security. Climate Res 33:135–142
13. Awange JL, Ong'ang'a O (2006) Lake Victoria: ecology resource and environment. Springer, Berlin, p 354
14. Baldassarre GD, Elshamy M, van Griensven A, Soliman E, Kigobe M, Ndomba P, Mutemi J, Mutua F, Moges S, Xuan Y, Solomatine D, Uhlenbrook S (2011) Future hydrology and climate in the River Nile basin: a review. Hydrol Sci J 56(2) https://doi.org/10.1080/02626667.2011.557378
15. Becker M, LLovel W, Cazenave A, Güntner A, Crétaux J-F (2010) Recent hydrological behavior of the East African great lakes region inferred from GRACE, satellite altimetry and rainfall observations. C R Geosci 342:223–233
16. Berhane G, Kristine M, Nawal A, Kristine W (2013) Water leakage investigation of micro-dam reservoirs in Mesozoic sedimentary sequences in Northern Ethiopia. J Afr Earth Sci 79:98–110. https://doi.org/10.1016/j.jafrearsci.2012.10.004
17. Beyene EG, Meissner B (2010) Spatio-temporal analyses of correlation between NOAA satellite RFE and weather stations' rainfall record in Ethiopia. Int J Appl Earth Obs Geoinf 125:569–575
18. Beyene T, Lettenmaier DP, Kabat P (2010) Hydrologic impacts of climate change on the Nile River Basin: implications of the 2007 IPCC scenarios. Clim Change 100:433–461. https://doi.org/10.1007/s10584-009-9693-0
19. Bonsor HC, Mansour MM, MacDonald AM, Hughes AG, Hipkin RG, Bedada T (2010) Interpretation of GRACE data of the Nile basin using a groundwater recharge model. Hydrol Earth Syst Sci Dis 7:4501–4533. https://doi.org/10.5194/hessd-7-4501-2010
20. Boy JP, Hinderer J, De Linage C (2012) Retrieval of large-scale hydrological signals in Africa from GRACE time-variable gravity fields. Pure appl Geophys 169(8):1373–1390
21. Conway D (2005) From headwater tributaries to international river: observing and adapting to climate variability and change in the Nile basin. Glob Environ Change 15:99–114. https://doi.org/10.1016/j.gloenvcha.2005.01.003
22. Conway D, Mould C, Bewket Woldeamlak (2004) Over one century of rainfall and temperature observations in Addis Ababa, Ethiopia. Int J Climatol 24:77–91
23. Crétaux JF, Jelinski W, Calmant S, Kouraec A, Vuglinski V, Bergé-Nguyen M, Gennero M-C, Nino F, Abarca Del Rio R, Cazenave A, Maisongrande P (2011) SOLS: A lake database to monitor in the Near real-time water level and storage variations from remote sensing data. Adv Space Res 47:1497–1507. https://doi.org/10.1016/j.asr.2011.01.004
24. Dahle C, Flechtner F, Gruber C, König R, Michalak G, Neumayer K-H (2012) GFZ GRACE Level-2 processing standards document for Level-2 product release 0005. Scientific technical report STR12/02-Data, GFZ, Potsdam, Germany
25. Dinku T, Ceccato P, Cressman K, Connor SJ (2010) Evaluating detection skills of satellite rainfall estimates over desert locust recession regions. Int J Appl Meteorol Climatol 49:1322–1332
26. Dinku T, Ceccato P, Grover-Kopec E, Lemma M, Connor SJ, Ropelewski CF (2007) Validation of satellite rainfall products over East Africa's complex topography. Int J Remote Sens 28(7):1503–1526. https://doi.org/10.1080/01431160600954688
27. Dinku T, Chidzambwa S, Ceccato P, Connor SJ, Ropelewski CF (2008) Validation of high-resolution satellite rainfall products over complex terrain. Int J Remote Sens 29(14):4097–4110. https://doi.org/10.1080/01431160701772526
28. Fenoglio-Marc L, Kusche J, Becker M (2006) Mass variation in the Mediterranean Sea from GRACE and its validation by altimetry, steric and hydrologic fields. Geophys Res Lett 33(19). https://doi.org/10.1029/2006GL026851

29. Fenoglio-Marc L, Rietbroek R, Grayek S, Becker M, Kusche J, Stanev E (2012) Water mass variation in the Mediterranean and Black Sea. J Geodyn 59–60:168–182. https://doi.org/10.1016/j.jog.2012.04.001
30. Fleming K, Awange JL, Kuhn M, Featherstone WE (2011) Evaluating the TRMM 3B43 monthly precipitation product using gridded rain-gauge data over Australia. Aust Meteorol Ocean J 61:171–184
31. Fleming K, Awange JL (2013) Comparing the version 7 TRMM 3B43 monthly precipitation product with the TRMM 3B43 version 6/6A and BoM datasets for Australia. Aust Meteorol Ocean J (in press)
32. Forootan E, Awange JL, Kusche J, Heck B, Eicker A (2012) Independent patterns of water mass anomalies over Australia from satellite data and models. Remote Sens Environ 124:427–443. https://doi.org/10.1016/j.rse.2012.05.023
33. Forootan E, Kusche J (2012) Separation of global time-variable gravity signals into maximally independent components. J Geodesy 86(7):477–497. https://doi.org/10.1007/s00190-011-0532-5
34. Forootan E, Kusche J (2013) Separation of deterministic signals, using independent component analysis (ICA). Stud Geophys Geod 57:17–26. https://doi.org/10.1007/s11200-012-0718-1
35. Forootan E, Rietbroek R, Kusche J, Sharifi MA, Awange J, Schmidt M, Omondi P, Famiglietti JS (2014) Separation of large scale water storage patterns over Iran using GRACE, altimetry and hydrological data. J Remote Sens Environ 140:580–595. https://doi.org/10.1016/j.rse.2013.09.025
36. Funk C, Verdin J (2003) Comparing satellite Rainfall Estimates and reanalysis precipitation fields with station data for Western Kenya. In: Proceedings of the international workshop on crop monitoring for food security in Africa, European Joint research Centre and UN FAO, Nairobi, Kenya
37. Hamouda MA, Nour El-Din MN, Moursy FI (2009) Vulnerability assessment of water resources systems in the Eastern Nile basin. Water Resour Manage 23:2697–2725. https://doi.org/10.1007/s11269-009-9404-7
38. Huffmann G, Bolvin D (2012) TRMM and other data precipitation data set documentation. http://precip.gsfc.nasa.gov/pub/trmmdoc/3B42_3B43_doc.pdf Accessed 25 Sep 2012
39. Klees R, Zapreeva EA, Winsemius HC, Savenije HHG (2007) The bias in GRACE estimates of continental water storage variations. Hydrol Earth Syst Sci Dis 11:1227–1241
40. Knight JF, Lunetta RL, Ediriwickrema J, Khorram S (2006) Regional scale land-cover characterization using MODIS-NDVI 250 m multi-temporal imagery: a phenology based approach. GIScience Remote Sens 43(1):1–23
41. Kummerow C, Barnes W, Kozu T, Shiue J, Simpson J (1998) The tropical rainfall measuring mission (TRMM) sensor package. J Atmos Ocean Technol 15(3):809–817. https://doi.org/10.1175/1520-0426(1998)015$<$0809:TTRMMT$>$2.0.CO;2
42. Kummerow C, Simpson J, Thiele O, Barnes W, Chang A, Stocker E, Adler R, Hou A, Kakar R, Wntz F, Aschroft P, Kozu T, Hing Y, Okamoto K, Iguchi T, Kuroiwa H, Im E, Haddad Z, Huffman G, Ferrier B, Olson W, Zipser E, Smith E, Wilheit T, North G, Krishnamurti T, Nakamura K (2000) The status of the tropical rainfall measuring mission (TRMM) after two years in orbit. J Appl Meteorol 39(12):1965–1982. https://doi.org/10.1175/1520-0450(2001)040$<$1965:TSOTTR$>$2.0.CO;2
43. Kusche J (2007) Approximate decorrelation and non-isotropic smoothing of time-variable GRACE-type gravity field models. J Geodesy 81:733–749. https://doi.org/10.1007/s00190-007-0143-3
44. Kusche J, Schmidt R, Petrovic S, Rietbroek R (2009) Decorrelated GRACE time-variable gravity solutions by GFZ, and their validation using a hydrological model. J Geodesy 83:903–913. https://doi.org/10.1007/s00190-009-0308-3
45. Li J, Lewis J, Rowland G, Tappan L, Tieszen L (2004) Evaluation of land performance in Senegal using multi-temporal NDVI and rainfall series. J Arid Environ 59(3):463–480
46. Longuevergne L, Scanlon BR, Wilson CR (2010) GRACE hydrological estimates for small basins: evaluating processing approaches on the high plains aquifer, USA. Water Resour Res 46:W11517. https://doi.org/10.1029/2009WR008564

47. Millington AC, Wllens J, Settle JJ, Saull RJ (1994) Explaining and monitoring land cover dynamics using multi-temporal analysis of NOAA-AVHRR imagery. In: Foody GM, Curran PJ (eds) Environmental remote sensing from regional to global scales. Wiley, Chichester, pp 223–232
48. Nicholson S, Some B, McCollum J, Nelkin E, Klotter D, Berte Y, Diallo B, Gaye I, Kpabeba G, Ndiaye O, Noukpozounkou J, Tanu A, Thiam A, Toure AA, Traore A (2003) Validation of TRMM and other rainfall estimates with a high-density gauge dataset for West Africa: part II: validation of TRMM rainfall products. J Appl Meteorol 42(10):1355–1368. https://doi.org/10.1175/1520-0450(2003)042$<$1355:VOT
49. North GR, Bell TL, Cahalan RF, Moeng FJ (1982) Sampling errors in the estimation of empirical orthogonal functions. Mon Weather Rev 110:699–706. https://doi.org/10.1175/1520-0493(1982)110$<$0699:SEITEO$>$2.0.CO;2
50. Omute P, Corner R, Awange JL (2012) The use of NDVI and its derivatives for monitoring Lake Victoria's water level and drought conditions. Water Resour Manage 26:1591–1613. https://doi.org/10.1007/s11269-011-9974-z
51. Owor M, Taylor RG, Tindimugaya C, Mwesigwa D (2009) Rainfall intensity and groundwater recharge: empirical evidence from the Upper Nile basin. Environ Res Lett 4:035009. https://doi.org/10.1088/1748-9326/4/3/035009
52. Omondi P, Awange J, Forootan E et al (2013) Changes in temperature and precipitation extremes over the greater horn of Africa region from 1961 to 2010. Int J Climatol. https://doi.org/10.1002/joc.3763
53. Omondi P, Awange J, Ogallo Ininda J, Forootan E (2013) The influence of low frequency sea surface temperature modes on delineated decadal rainfall zones in Eastern Africa region. Adv Water Resour. https://doi.org/10.1016/j.advwatres.2013.01.001
54. Omondi P, Awange J, Ogallo LA, Okoola RA, Forootan E (2012) Decadal rainfall variability modes in observed rainfall records over East Africa and their relations to historical sea surface temperature changes. J Hydrol 464–465:140–156. https://doi.org/10.1016/j.jhydrol.2012.07.003
55. Preisendorfer R (1988) Principal component analysis in meteorology and oceanography. Elsevier, Amsterdam
56. Rasmus F, Kjeld R (2011) Analysis of trends in the Sahelian 'rain-use efficiency' using GIMMS NDVI, RFE and GPCP rainfall data. Remote Sens Environ 115(2):438–451
57. Rodell M, Famiglietti JS (2001) An analysis of terrestrial water storage variations in illinois with implications for the gravity recovery and climate experiment (GRACE). Water Resourc Res 37:1327–1340
58. Rojasa O, Vrieling A, Rembold F (2011) Assessing drought probability for agricultural areas in Africa with coarse resolution remote sensing imagery. Remote Sens Environ 115(2):343–352
59. Rummel R, Balmino G, Johannessen J, Visser P, Woodworth P (2002) Dedicated gravity field missions - principles and aims. J Geodyn 33:3–20
60. Sahin M (1985) Hydrology of the Nile basin. Developments in water Science No. 21. Elsevier, New York, 575p
61. Senay GB, Asante K, Artan G (2009) Water balance dynamics in the Nile basin. Hydrol Process 23:3675–3681. https://doi.org/10.1002/hyp.7364
62. Sergio MV-S, Santiago B, Luis G, Lars E, Giuliani G, Weston D, Kenawyg A, López-Moreno JI, Nieto R, Tenalem A, Diawoy K, Ardö J, Pegram GS (2012) Challenges for drought mitigation in Africa: the potential use of geospatial data and drought information systems. Appl Geogr 34:471–486
63. Sharifi MA, Forootan E, Nikkhoo M, Awange J, Najafi-Alamdari M (2013) A point-wise least squares spectral analysis (LSSA) of the Caspian Sea level fluctuations, using TOPEX/Poseidon and Jason-1 observations. Adv Space Res 51(5):858–873. https://doi.org/10.1016/j.asr.2012.10.001
64. Sultan M, Ahmed M, Sturchio N, Yan YE, Milewski A, Becker R, Wahr J, Becker D, Chouinard K (2012) Assessment of the vulnerabilities of the Nubian sandstone fossil aquifer. Water encyclopedia (in press). Elsevier, North Africa

65. Sutcliffe JV, Parks YP (1999) An outline of the Nile basin. The hydrology of the Nile. IAHS Special Publication no. 5
66. Sulla-Menashe D, Friedl MA, Krankina ON, Baccini A, Woodcock CE, Sibley A, Sun G, Kharuk V, Elsakov V (2011) Hierarchical mapping of Northern Eurasian land cover using MODIS data. Remote Sens Environ 115:392–403
67. Swenson S, Wahr J, Milly P (2003) Estimated accuracies of regional water storage variations inferred from the gravity recovery and climate experiment (GRACE). Water Resour Res 39(8):1223. https://doi.org/10.1029/2002WR001808
68. Swenson S, Wahr J (2009) Monitoring the water balance of Lake Victoria, East Africa, from space. J Hydrol 370(1–4):163–176. https://doi.org/10.1016/j.jhydrol.2009.03.008
69. Swenson S, Wahr J (2006) Post-processing removal of correlated errors in GRACE data. Geophys Res Lett 33:L08402. https://doi.org/10.1029/2005GL025285
70. Tapley B, Bettadpur S, Ries J, Thompson P, Watkins M (2004) GRACE measurements of mass variability in the Earth system. Science 305:503–505. https://doi.org/10.1126/science.1099192
71. Tapley B, Bettadpur S, Watkins M, Reigber C (2004) The gravity recovery and climate experiment: mission overview and early results. Geophys Res Lett 31(L09607). https://doi.org/10.1029/2004GL019920
72. Vanicék P (1969) (1969) Approximate spectral analysis by least-squares fit. Successive spectral analysis. Astrophys Space Sci 4:387–391. https://doi.org/10.1007/BF00651344
73. Wahr J, Molenaar M, Bryan F (1998) Time variability of the Earth's gravity field: hydrological and oceanic effects and their possible detection using GRACE. J Geophys Res 103(B12):30205–30229. https://doi.org/10.1029/98JB02844
74. Whittington D, McClelland E (1992) Opportunities for regional and international cooperation in the Nile basin. Water Int 17(3):144–154. https://doi.org/10.1080/02508069208686134
75. Veroustraete F, Sabbe H, Eerens H (2002) Estimation of carbon mass fluxes over Europe using the C-Fix model and Eruflux data. Remote Sens Environ 83:376–399
76. Woodcock CE, Macomber SA, Pax-Lenney M, Cohen WB (2001) Monitoring large areas for forest change using landsat: generalization across space, time and landsat sensors. Remote Sens Environ 78:194–203
77. Yates DN, Strzcpek KM (1998) Modelling the Nile basin under climate change. J Hydrol Eng 3(2):98–108. https://doi.org/10.1061/(ASCE)1084-0699(1998)3:2(98)
78. Khaki M, Awange J, Forootan E, Kuhn M (2018) Understanding the association between climate variability and the Nile's water level fluctuations and water storage changes during 1992–2016. Sci Total Environ 645:1509–1521. https://doi.org/10.1016/j.scitotenv.2018.07.212
79. Khaki M, Awange J (2020) Altimetry-derived surface water data assimilation over the Nile Basin. Sci Total Environ. https://doi.org/10.1016/j.scitotenv.2020.139008
80. Agutu NO, Awange JL, Zerihun A, Ndehedehe CE, Kuhn M, Fukuda Y (2017) Assessing multi-satellite remote sensing, reanalysis, and land surface models' products in characterizing agricultural drought in East Africa. Remote Sens Environ 194:287–302. https://doi.org/10.1016/j.rse.2017.03.041
81. Agutu NO, Awange JL, Ndehedehe C, Kirimi F, Kuhn M (2019) GRACE-derived groundwater changes over greater horn of Africa: temporal variability and the potential for irrigated agriculture. Sci Total Environ. https://doi.org/10.1016/j.scitotenv.2019.07.273
82. Morgan B, Awange JL, Saleem A, Hu K (2020) Understanding vegetation variability and their "hotspots" within Lake Victoria Basin (LVB: 2003–2018), 122. https://doi.org/10.1016/j.apgeog.2020.102238
83. Awange JL (2018) GNSS environmental sensing. Revolutionizing environmental monitoring, Springer International Publishing AG
84. Awange JL, Kiema JBK (2019) Environmental geoinformatics. Exploiting the Big Data, Extreme Hydro-Climatic and Food Security Challenges. Springer Nature, Heidelberg
85. Saleem A, Awange JL, John B, Kuhn M, Hu K (2021) Impacts of extreme climate on Australia's green cover (2003–2018): A MODIS and mascon probe. Sci Total Environ. https://doi.org/10.1016/j.scitotenv.2020.142567

Chapter 7
Exploitation of the Nile's Waters: Space View

For the Lake Nasser region, a remarkable signal is noticeable indicating the possibility of the region not only losing water through evaporation, but also possibly through over extraction from wells in the Western Plateau (Nubian aquifer) for irrigation purposes. The situation is not helped much by the fact that there could be insufficient recharge from the south-north flow of the groundwater due to the presence of the Uweinat–Aswan uplift.
—*Awange et al.*, [78].

7.1 Summary

Understanding water storage changes within the Nile's main sub-basins and the related impacts of climate variability is an essential step in managing its water resources. The Gravity Recovery And Climate Experiment (GRACE; Sect. 3.3.3) satellite mission provides a unique opportunity to monitor changes in total water storage (TWS) of large river basins such as the Nile. Use of GRACE-TWS changes for monitoring the Nile is, however, difficult since stronger TWS signals over the Lake Victoria Basin (LVB), the mother of the Nile (see Chap. 4) and the Red Sea obscure those from smaller sub-basins making their analysis difficult to undertake. To mitigate this problem, [78] employed Independent Component Analysis (ICA) to extract statistically independent TWS patterns over the sub-basins from GRACE and the Global Land Data Assimilation System (GLDAS) model. Monthly precipitation from the Tropical Rainfall Measuring Mission (TRMM) over the entire Nile River Basin (NRB) were also analysed by ICA. Such extraction enabled an in-depth analysis of water storage changes within each sub-basin and provided a tool for assessing the influence of anthropogenic as well as climate variability caused by large scale ocean–atmosphere interactions such as the El Niño Southern Oscillation (ENSO) and the Indian Ocean Dipole (IOD). This chapter presents the results of [78], which indicate that LVB experienced effects of both anthropogenic and climate

J. Awange, *The Nile Waters*,
https://doi.org/10.1007/978-3-030-64756-8_7

variability (i.e., a correlation of 0.56 between TWS changes and ENSO at 95% confidence level) during the 2002–2011 period, with a sharp drop in rainfall between November and December 2010, the lowest during the entire 2002–2011 period, and coinciding with the drought that affected the Greater Horn of Africa [79]. Ethiopian Highlands (EH) generally exhibited a declining trend in the annual rainfall over the 2002–2011 period, which worsened during 2007–2010, possibly contributing to the 2011 drought over GHA. A correlation of 0.54 was found between ENSO and TWS changes over EH indicating ENSO's dominant influence. TWS changes over Bar-el-Ghazal experienced mixed increase–decrease, with ENSO being the dominant climate variability in the region during the 2002–2011 period. A remarkable signal is noticed over the Lake Nasser region indicating the possibility of the region losing water not only through evaporation, but also possibly through over extraction from wells in the Western Plateau (Nubian aquifer).

7.2 Challenges Facing Nile's Stored Water

The Nile River Basin (NRB, Fig. 2.1) is one of the largest basins in the world, with an area of about 3,400,000 km^2 (almost one-tenth of Africa), see Chap. 2. It traverses about 6,500 km from the White Nile in the south to the Mediterranean Sea in the north as it winds its way across the boundaries of eleven countries supporting livelihoods of over 300 million people [2, 46, 56]. Because of this huge size of the NRB, climate variability and change that is manifested locally or regionally may have regional and even international consequences through its effects on Nile river flows [19].

The Nile's water resources have come under threat from both anthropogenic and natural factors (see, e.g., [33]). Indeed, that hydrological regimes respond to climate change and anthropogenic influences have been reported, e.g., in [13, 22, 38]. For the NRB, anthropogenic influences are attributed to increased human population that has put pressure on domestic water needs and hydroelectric power supply (see e.g., Fig. 2.9), all coupled with the need to sustain economic growth, e.g., [2, 3, 6, 7]. Furthermore, not only are the demands on water increasing, but also the available water supplies appear to be decreasing, with environmental degradation of the upper Blue Nile catchment having increased throughout the 1980s [76]. For example, due to the large and increased population pressure, insufficient agricultural production, a low number of developed energy sources, and drought episodes; Ethiopia, which contributes about 85% of the Nile's annual flow [66], but has almost 94% of its population depending on wood fuel, has constructed a major hydropower (Fig. 2.9) and has irrigation development [12, 15, 72, 73]. In addition to irrigation and hydroelectric power, land degradation and changes in land cover in Ethiopia where forest lands are being converted to agricultural land are having impact on the Nile flow, see e.g., [52, 63]. Such measures are likely to impact on the downstream countries of Egypt and Sudan whose populations have been increasing (see Fig. 2.3), thus posing a challenge to water allocation, see [2, 3, 7, 76].

Natural factors have been the subject of numerous studies as shown in the works of Beyene et al. [13], Conway [19] and Yates [77, and the references therein] who investigated the influence of the changing climate on the Nile waters; Ghoubachi [32] and Sefelnasr [62] who considered groundwater movement. For the Nile river discharge, for example, the influence of climate variability and change has been shown e.g., by Eltahir [24] and Amarasekera et al. [1] to account for 25% of natural variability in annual discharge. Knowledge of climate variability is essential not only for predicting its floods and droughts [24, 39], but also for understanding of global atmospheric dynamics since stream flow is an index of precipitation integrated over large areas [1].

To support the management of the NRB's water resources, e.g., supply, demand and sustainable use; Hanson et al. [34], it is essential to understand the changes in its stored water (surface, groundwater, and soil moisture) and their relation to climate variability such as the El Niño Southern Oscillation (ENSO) and the Indian Ocean Dipole (IOD). Due to its large spatial extent, however, changes in NRB's stored water cannot be monitored using traditional methods (e.g., piezometric-based). This calls for satellite based methods of observation. One such satellite is the Gravity Recovery And Climate Experiment (GRACE; see Sect. 3.3.3) whose use offers a unique chance to monitor changes in the total water storage (TWS, i.e., an integral of surface, groundwater, and soil moisture storage) within the NRB, see, e.g., [2, 3, 6, 10, 16, 45, 63, 69].

The main challenges in using GRACE-TWS changes over the NRB is that on the one hand, the derived hydrological signals are dominated by stronger signals, e.g., those around the lakes such as Lakes Victoria (the mother of the Nile) and Tana (see, e.g., [3]), as well as regions close to the Red Sea and the Mediterranean Sea [9]. Computing basin average water variations, therefore, is normally influenced by dominant signals such as that of Lake Victoria Basin (LVB), while the relatively weaker signals from the other regions are masked making analysis of changes in TWS over such sub-basins difficult. On the other hand, Rodell and Famiglietti [54] pointed to the limitation of GRACE products to study basins of less than 200,000 km^2. Since some of the Nile's sub-basins (e.g., Lake Nasser region) have areas less than 200,000 km^2, computing their TWS changes is limited by the spatial resolution of GRACE (e.g., 400,000 km^2 in [67, 70]). Previous analysis of the NRB based on the GRACE satellite data have thus been unable to separate the signals into their respective sub-basins for the purpose of providing an in-depth analysis of spatial variations, see e.g., [3].

To overcome these challenges, in the present chapter; (i) we apply a higher order statistical tool of Independent Component Analysis (ICA) (Spatial ICA in [29, 30]) over the NRB (between 10°S–35°N and 25°E–45°E) to separate GRACE-TWS changes between 2002 and 2011 into their spatially independent sources (i.e., sub-basins shown in Fig. 2.1). This is then followed, in (ii), by an evaluation of the impacts of global climate change on the TWS within these sub-basins using global climate forcing by IOD and ENSO.

The remainder of the chapter is organized as follows; in Sect. 7.3, the space products and the analysis methods employed are presented. Section 7.4 analyses Nile's stored water changes while Sect. 7.6 concludes the major findings.

7.3 Employed Products and Analysis Methods

This section provides highlights on the data and methodology adopted in this chapter.

7.3.1 Satellite, Model and Climate Products

The data used in this chapter consisted of remotely sensed GRACE-TWS changes from 2002 to 2011, Tropical Rainfall Measuring Mission (TRMM)-derived precipitation, and water storage data from Global Land Data Assimilation System (GLDAS) hydrological model over the same period. In addition, ENSO and IOD climate indices are used.

7.3.1.1 Gravity Recovery and Climate Experiment (GRACE)

GRACE, a joint US-German satellite project launched in March 2002, detects spatio-temporal variations of the Earth's gravity field [70]. There are a number of institutions delivering GRACE products, each applying their own processing methodologies and, often, different background models. In chapter, gravity field time series provided by the German GeoForschungsZentrum (GFZ), Potsdam [25] were examined. GFZ's release (RLO4) gravity field solutions are provided at monthly resolutions and consist of a set of fully normalized spherical harmonic coefficients of the geopotential, up to degree and order 120. These coefficients are contaminated by correlated errors, manifesting as stripes in the spatial domain. These striping and high-frequency effects potentially mask hydrological signals, making their detection extremely difficult, e.g., [5], [42]. In this regard, the GFZ solutions were smoothed using the DDK2 de-correlation filter [43]. Filtered solutions can also be downloaded from the official website of the International Center for Global Gravity Field models (http://icgem.gfz-potsdam.de/ICGEM/TimeSeries.html). Version RL04 and the DDK2 filter were chosen to be consistent with previous studies, e.g., those of Awange et al. [2, 3]. In addition, any estimates for the degree 1 coefficients were neglected due to its small contribution to water storage estimation of the basin.

In order to derive the TWS changes, residual gravity field solutions taken with respect to 9 years (assumed static) average were derived. The resulting residual geopotential coefficients were then transformed into monthly TWS changes using Wahr et al.'s [74] approach. Besides the total signals of the respective sub-basins, this chapter explored the possibility of analyzing the residual signals once the dominant signals of Lake Victoria, Lake Tana and the Red Sea were removed. To remove the lakes' signals from GRACE-TWS changes, the approach in [68] was followed. The surface height of each lake was assumed to be changing homogeneously. For each lake, a global grid was defined by a function f, where its surface was given a value 1 and 0 over the rest of the globe (i.e., equivalent to 1 mm Equivalent Water Thickness (EWT) over the lake and zero over the rest of the globe). The lake grid function f is then defined over a sphere (with radius R) as

$$f(\lambda, \theta) = \frac{1}{4\pi} \sum_{n=1}^{120} \sum_{m=0}^{n} \bar{P}_{nm}(\cos\theta)[C^f_{nm}\cos(m\lambda)] + S^f_{nm}\sin(m\lambda)], \tag{7.1}$$

where $\bar{P}_{nm}$ are the normalized associated Legendre functions, λ and θ represent the geocentric positional vector in spherical coordinates. Since the function f was known by definition, the spectral values of C^f_{nm} and S^f_{nm} in Eq. 7.1 were derived by integration [75]. The estimated spherical harmonic coefficients (C^f_{nm} and S^f_{nm}) were then filtered using the DDK2 [43]. Finally, each field was scaled using the sea surface heights (SSH) time-series (in mm) derived from Jason 1 and 2 satellite altimetry data [21]. The altimetry data were downloaded from the LEGOS website (http://www.legos.obs-mip.fr/en/soa/hydrologie/hydroweb/). This procedure transferred SSH changes of the desired lake to what GRACE could see as a whole. As an example, in Fig. 7.1, the satellite altimetry-derived EWT changes and GRACE-TWS time series corresponding to Lake Victoria over the period 2002 to 2011 are shown.

7.3.1.2 Tropical Rainfall Measuring Mission (TRMM)

TRMM, a joint Japanese/USA satellite mission (see [41]), launched in 1997 provides measurements of the spatial and temporal variation of the tropical rainfall in the latitude range 35° over inaccessible areas such as the oceans and unsampled terrains. TRMM data has been validated and employed in a number of studies of African precipitation, where they have been found to be adequate, see, e.g., [2, 3, 48, 80, 81]. For example, Awange et al. [2, 3, 6] applied TRMM to study the NRB while Naumann et al. [47] found TRMM data to be reliable enough to be used for drought monitoring over Africa. The use of TRMM satellite data for this chapter, therefore, is informed by the works of Awange et al. [2] and [3] and the references therein], and utilized the DDK2-filtered TRMM 3B43 records for the period 2002 to May 2011 for comparisons with GRACE-TWS changes.

7.3.1.3 Global Land Data Assimilation System (GLDAS)

GLDAS hydrological model integrates a huge quantity of observation based data and modeling concepts, see e.g., [55] including a global Land Information System (LIS) at a spatial resolutions (25° x 25°) [40]. The monthly GLDAS data set is distributed in two parts: (i) the energy budget (EB), and (ii,) the water budget (WB).

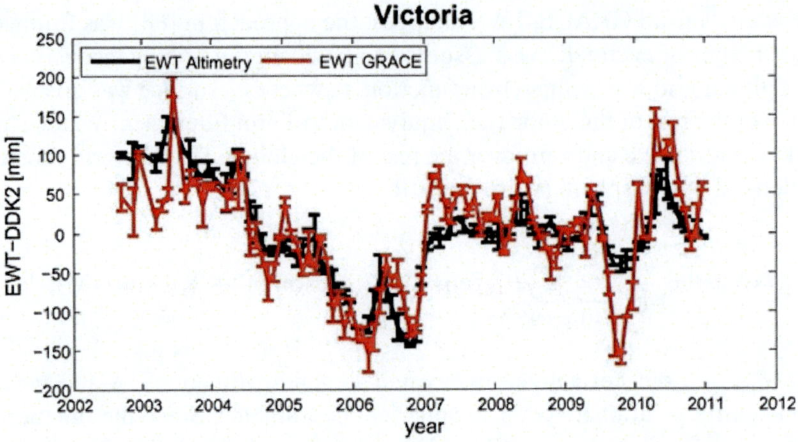

Fig. 7.1 Satellite altimetry-derived signals for Lake Victoria (after smoothing with the DDK2 filter [43] and conversion to equivalent water thickness (EWT)). Both time series are centered with respect to their temporal means. The altimetry derived EWT signals represents a smoother pattern compared to the GRACE signals, which might be due to the fact that it contains only water level changes of Lake Victoria and not water storage changes over the surrounding regions of the lake. Error-bars for the GRACE-derived TWS are obtained by propagating the errors of spherical harmonics, after DDK2 filtering, to TWS without considering the covariances. Computing the error-bars for altimetry-derived TWS is done by considering the accuracy provided in LEGOS data [21]. *Source* [78]

To assess the water storage variations within the NRB and their comparisons with GRACE-TWS changes, monthly WB data (including soil moisture, snow and water canopy storages) covering the period from August 2002 to October 2010.2 are used. The GLDAS data was also filtered using the same DDK2 filter used for smoothing GRACE-TWS changes to have similar spectral structure.

7.3.1.4 El Niiño and Southern Oscillation (ENSO)

El Niño/Southern Oscillation (ENSO) is a large scale climate variability that influences the global atmospheric circulation and its impacts on regional climates depend on the region and season, and the strength and spatial distribution of the phenomenon [17, 18]. ENSO has a strong influence on trade winds and climate of the East African region, and has been observed to weaken the Hadley circulation [18]. It has significant influence on rainfall over Eastern Africa [2, 3, 36, 39, 82]. The warm/cold (El Niño/La Niña) phase of ENSO is associated with enhanced/depressed seasonal rainfall over various parts of the region especially during the October-December (OND) rainfall season [36, 49–51, 71].

The Southern Oscillation (SO) is measured using Southern Oscillation Index (SOI) [35]. High SO phase is associated with high and low pressure over the western and eastern Pacific Ocean, respectively [35], resulting in enhanced westerly and cooling over the western parts of the ocean. The low phase has the opposite effect of enhancing rainfall over the western Indian Ocean and East Africa [35]. The low-level circulation pattern associated with the above-normal rainfall over the region is dominated by easterly inflow from the Indian Ocean and westerly inflow from the Congo tropical rain forest into the positive rainfall region [8]. This chapter used monthly SOI provided by the Australian Bureau of Meteorology covering the period from 2002 to 2011.[1]

7.3.1.5 Indian Ocean Dipole (IOD)

The Indian Ocean Dipole (IOD) and the related Indian Ocean Zonal Mode (IOZM) is reflected, e.g., in the sea surface temperature (SST) data over the Indian ocean. IOD has been observed to have significant influence on rainfall over the East African sub-regions and other areas neighbouring the Indian Ocean see, e.g., Behera et al. [11], Black et al. [14], Clark et al. [17], Taylor et al. [71] and Awange et al. [2]. IOD is caused by air-sea interactions in the tropical Indian Ocean leading to the warming/cooling of western/eastern tropical Indian Ocean during the positive/negative phases resulting into the reversal of SST gradients and changes in the zonal wind currents [58, 59].

The warm/cool western/eastern Indian Ocean is associated with enhanced/deficient OND seasonal rainfall over the sub-region [11, 14, 17, 57–59] resulting from the anomalous changes in Walker circulation that lead to anomalous moisture transport and convergence [11]. The influence of the IOD on rainfall over the sub-region is much higher than that associated with ENSO [11], and is strongest during the OND rainfall season. However, the co-occurred ENSO and IOD events are stronger than those that occur independently [2, 58, 59, 64]. The Dipole Mode Index (DMI) is the gradient of the IOD event, which is usually used as a measure of the IOD influence of climate variability. In this chapter, we used DMI time series provided by the Japan Agency for Marine-Earth Science and Technology (JAMSTEC) covering the period from 2002 to September 2010.[2]

7.3.2 *Statistical Analysis: Independent Component Analysis and Correlation*

7.3.2.1 Extracting Independent Patterns

At regional scale, temporal changes in total water storage (TWS), measured by GRACE is related to precipitation P through the water budget equation

[1] http://www.bom.gov.au/climate/enso/.

[2] http://www.jamstec.go.jp).

$$\frac{d(TWS(t))}{dt} = P(t) - E(t) - R(t), \tag{7.2}$$

e.g., [10, 55], where P is precipitation, E for evapotranspiration and R is river discharge measured in time (t), see details in Chap. 3. In order to evaluate TWS changes obtained by GRACE, this chapter explores the relationship between GRACE-TWS changes and TRMM precipitation product P over the NRB. Precipitation is used here based on the fact that previous studies have shown its capability to detect and monitor land hydrological changes obtained from time-variable gravity observations (see, e.g., [4, 26, 53]). When comparing TRMM-rainfall with GRACE-TWS changes, however, it should be kept in mind that they cannot be compared on a one-to-one basis since GRACE observes the total mass changes that arises from all components of the hydrological cycle, and not just precipitation. Nonetheless, precipitation plays an important role in the terrestrial water balance, being the main replenishment source of large-scale TWS changes. Analysis of the contribution of E and R over the basin are outside the scope of the present chapter and will be treated in future contributions.

In order to assess the water storage changes within the NRB, the Spatial Independent Component Analysis (Spatial ICA) approach (see, [27, 30]) was used to extract the spatially independent signals within the NRB boundary, selected to extend from 25°E $< \lambda <$ 45°E and 5°S $< \varphi <$ 35°N, including areas that do not actually belong to the Basin, such as the Red Sea (Fig. 2.1). For instance, assuming that time series of TWS fields, after removing their temporal mean, are stored in a matrix $\mathbf{X}_{TWS} = \mathbf{X}_{TWS}(t, s)$, where t is the time, and s stands for spatial coordinate (grid points). Applying Spatial ICA, $\mathbf{X}_{TWS}$ can be decomposed into spatial and temporal components as

$$\mathbf{X}_{TWS} = \mathbf{A}_j \mathbf{S}_j, \tag{7.3}$$

where $\mathbf{A}_j(t \times j)$ stores the j dominant unit-less temporally evolution of TWS changes in its columns, and the rows of $\mathbf{S}_j(j \times s)$ represent their corresponding statistically independent spatial maps. Each temporal pattern (a column of $\mathbf{A}_j$) along with its corresponding spatial pattern (a row of $\mathbf{S}_j$) provides an independent mode of variability [29, 30]. The use of the Spatial ICA method, which is simply called ICA from now on, is advantageous in two respects: first, it localizes the signals and isolates those outside the basin; and second, it provides statistically independent components, which enhance the interpretation of the results.

To understand the advantages of using ICA in localizing TWS changes over the NRB, let us use a simple illustration of Fig. 7.2 based on simulated data. Assuming that the water storage changes of the NRB consists of changes in lake water storage, soil moisture, and sea storage. We demonstrate the power of the ICA [30] in a two-step approach (see, also [27] for a simulation example on Australia). In the first step, we simulate the combined lake water storage, soil moisture, and sea storage over the NRB and in the second step, we apply the ICA approach to localize them into their respective separate components. To achieve the first step, altimetry data from the LEGOS website [21] are used to introduce the Lake water storage changes, which are then converted to EWT using the approach discussed in Sect. 3.3.3. The same approach is used to covert the level variations of the Red Sea[3] into EWT changes, while the soil moisture are introduced from the GLDAS model. From the first step, Fig. 7.2 (top-A) shows the introduced signals of lakes, Fig. 7.2 (top-B) corresponds to soil moisture changes, and Fig. 7.2 (top-C) represents the introduced signals over the sea. Combined, Fig. 7.2 (top-A, B, and C) leads to Fig. 7.2 (top-D) that contains a summation of all the signals. As can be seen from Fig. 7.2 (top-D), the composition of the mixed signals make it hard to interpret thereby necessitating the need for ICA method to separate them and make them interpretable. Now, applying the ICA method to the combined introduced patterns of Fig. (7.2, top-D) in order to separate the signals to their respective sources, the results of Fig. (7.2, bottom) show that the introduced water storage changes over the Red Sea and Lake Nasser having been separated as the first independent mode (IC1). Soil moisture pattern has been separated as the second independent mode (IC2), and finally, water storage changes of Lake Victoria are localized in the third independent mode (IC3). The successful performance of ICA in localizing the simulated water storage changes motivated its application to real GRACE-TWS data over the NRB. For more discussion on the ICA-method, its performance on decomposing GRACE-derived TWS fields, as well as its comparison with the second order statistical method of Principal Component Analysis (PCA), we refer to Forootan and Kusche [30] and Forootan et al. [27, 28]. The main independent patterns of GRACE/GLDAS-derived TWS changes are therefore compared to those of TRMM rainfall (considered as the main input of TWS changes over the NRB).

7.3.2.2 Correlation Analysis

In order to study the relations between the patterns of TWS changes within different Nile sub-basins and climate teleconnections (i.e., ENSO and IOD), correlation analysis between ICs of GRACE-TWS changes and those of ENSO and IOD indices was performed. To this end, first, the independent components (ICs) of GRACE-TWS changes and indices were smoothed using a 12-month moving average filter and interpolated to a regular monthly time steps covering the period from October 2002 to May 2011. The temporal filter was applied to decrease the effects of strong inter-annual variability on the computed correlations (see e.g., [27, 31]). The correlation analysis are then performed at 95% level of confidence over the complete period in order to study the long-term impacts of climate variability on TWS changes of the Nile's sub-basins.

[3] http://coastwatch.pfeg.noaa.gov.

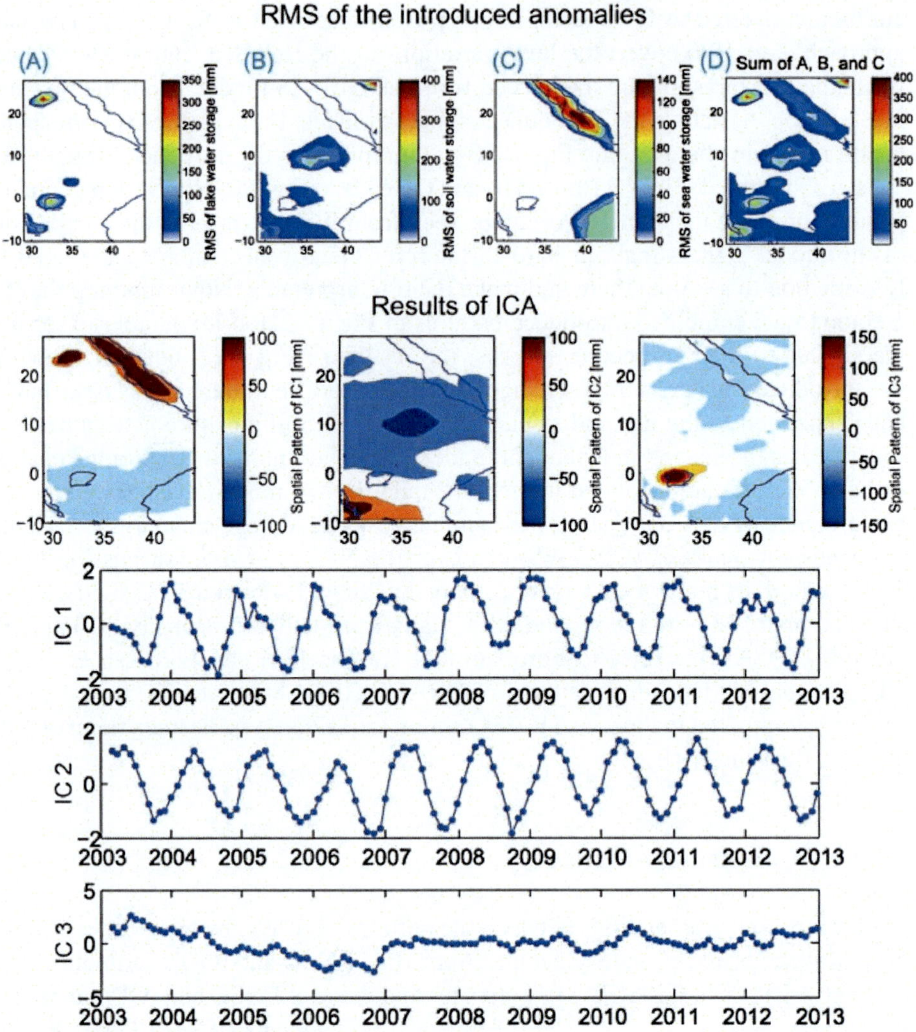

Fig. 7.2 (top) Root mean squares of the introduced water storage changes over the major lakes of the NRB, soil moisture changes of the basin and non-tidal water storage changes of the Red Sea; (Bottom) The first three independent modes, derived from ICA. The results show that the strong signal of Lake Victoria (IC3) can successfully be separated from those of the Red Sea and Lake Nasser (IC1) and those of soil moisture (IC2). (For interpretation of the references to color in this figure legend, the reader is referred to the web version of this book. *Source* [78])

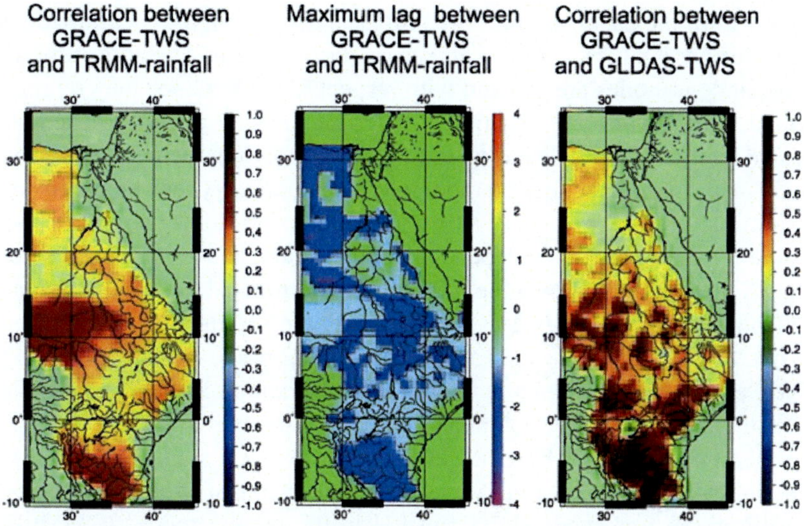

Fig. 7.3 (left) Correlation between GRACE and TRMM; (center) phase lag between GRACE and TRMM in months; and (right) correlation between GRACE and GLDAS. The results show high rainfall impacts on TWS changes over the tropical regions, Bar-el-Ghazal (BEG) and parts of Ethiopian Highlands (EH). *Source* [78]

7.4 Comparison of the Stored Water Products

Before undertaking a detailed analysis of the water storage changes within each sub-basin, overall comparison of the similarities and differences between GRACE, TRMM, and GLADS was performed by applying correlation analyses between data sets at 95% confidence level, where significant bounds based on asymptotic normal distribution with the variance of $1/(n-3)$ were set [44]. Those correlations that were smaller than the estimated bounds were set to zero. Figure 7.3 presents the spatial variation of the correlations, indicating the high rainfall impact on TWS changes over the tropical regions, Ber-el-Ghazal (BEG), and parts of Ethiopian Highlands (EH). Lower correlations in other regions might reflect a higher influence of unaccounted evaporation and runoff.

Implementing ICA on GRACE-TWS changes produced 4 significant spatially independent components over the NRB corresponding to 92% of the cumulative total variance of TWS changes (see Fig. 7.4). In comparison to GRACE, GLDAS and TRMM data sets each produced 3 significant modes of ICs, which corresponded to 86% and 91% of total variance, respectively (see Figs. 7.5 and 7.6). The variances suggest that the presented independent modes extract the dominant large scale variability of TWS (particularly from GRACE compared to GLDAS) and rainfall changes within the basin and thus are representative enough to be individually subjected to further analysis and interpretations. For all the ICA results presented here, the independent modes are ordered with respect to the variance they represent. The presented spatial maps are statistically independent with respect to each other and are scaled by the standard deviation of their corresponding temporal ICs to represent anomaly maps in mm, while the temporal ICs are divided by their standard deviations to be unitless.

Figure 7.4 shows the first independent mode (spatial and temporal patterns of IC1) of GRACE to be localized over the BEG region (dominated by an annual signals), the second independent mode (IC2) is localized over the EH, the third independent mode (IC3) over LVB, and finally the fourth independent mode (IC4) is found to be over the Red Sea and the northern parts of the basin. In comparison, the spatial pattern of Fig. 7.5 of GLDAS captures a similar pattern over the BEG, and the anomalies over the northeast part of the LVB. It, however, shows a weaker signal over EH. Figure 7.6 shows the ICA results of rainfall variability over the basin, where the rainfall signal over LVB in the first mode (IC1) and EH in the second mode (IC2). It does not show any rainfall related fluctuations over the Red Sea. Given the high percentage of variability captured, i.e., 92%, and its capability to capture signals from various regions, the proposed use of GRACE-based TWS changes of the Nile sub-basins are adequate for further analysis. The TWS from GLDAS and the TRMM-derived rainfall changes will thus not be treated further.

With the removal of Lake Victoria's signal using the procedure discussed above, the TWS from GRACE data did not change for BEG but is concentrated more over the catchments for LVB and EH (Fig. 7.7). For the Nasser region, the removal of the Red Sea and the Lake Nasser signal allowed the possible impact of ground water extraction in the Western Plateau (Nubian aquifer; e.g., [65]) to be more visible (e.g., Fig. 7.8). The results of correlations between TWS changes in different sub-basins and the climate variability of ENSO and IOD are summarized in Table 7.1.

7.5 Sub-Basins Analysis

7.5.1 Lake Victoria Basin (LVB); the White Nile

Figure 7.4 (spatial and temporal patterns of the third mode, IC3) presents the results of changes in water storage over LVB. From the spatial maps, GRACE-TWS changes indicate dominant anomalies over the western part of the Lake (Fig. 7.4, IC3). Looking at the spatial pattern of IC3 (Fig. 7.4) together with its associated temporal pattern, the most significant drop in water level within the basin occurred between October 2003 and March 2006, consistent with the findings of Awange et al. [6], Swenson and Wahr [69] and Becker et al. [10]. The rate of fall in water computed for LVB within this period, for the data within the NRB, was 84.5 mm/year. From March 2006 to May 2007, however, the TWS within LVB increased at a rate of 145.2 mm/year due

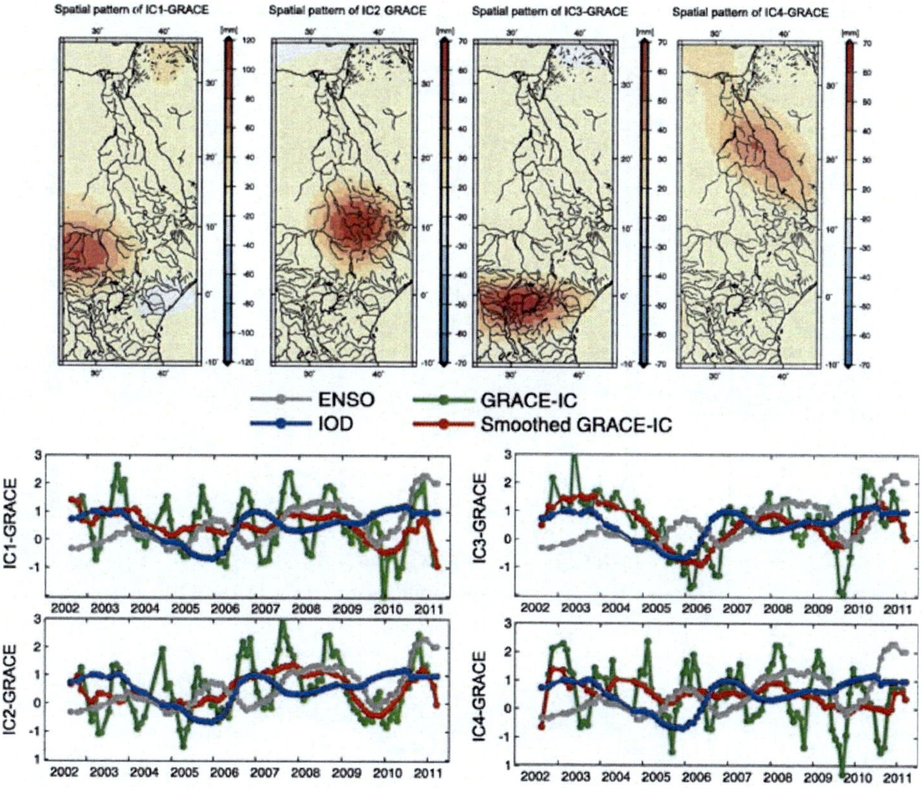

Fig. 7.4 ICA decomposition of GRACE TWS signal within the NRB using the proposed ICA method of [29, 30]. The upper panels are the spatial pattern of each IC. Note that the NRB's TWS signal is separated (localized) by the ICA method into 4 significant spatially independent components (BEG (left), EH (second from left), LVB (third from left) and Red Sea (right)). Together, these four signals computed over the NRB account for 92% of cumulative total variance of TWS changes. In the lower panel, their corresponding time series (ICs) overlaid on smoothed time series of the loading, as well as ENSO and IOD indices. (For interpretation of the references to color in this figure legend, the reader is referred to the web version of this book). *Source* [78]

to the ENSO related rainfall (i.e., Fig. 7.6, IC1 indicates an increase in the amplitude of rainfall between 2006 and 2007). Becker et al. [10] who used the Global Precipitation Climatology Project (GPCP) observed an increase in precipitation from the end of 2005 to the beginning of 2007. For the 2002–2011 period, the correlations between GRACE-TWS changes and IOD was 0.48 and 0.56 with ENSO suggesting that both IOD and ENSO influence TWS changes in LVB.

The evidence that rainfall is influenced by the ENSO is supported by the fact that a positive correlation of 0.72 is obtained between GRACE-TWS changes and ENSO over the period 2007–2011 (see Table 7.1) consistent with Becker et al. [10] who obtained a positive correlation value of 0.8 between TWS of GRACE and those of GPCP for the period 2005–2008. The period between May 2007 and August 2009

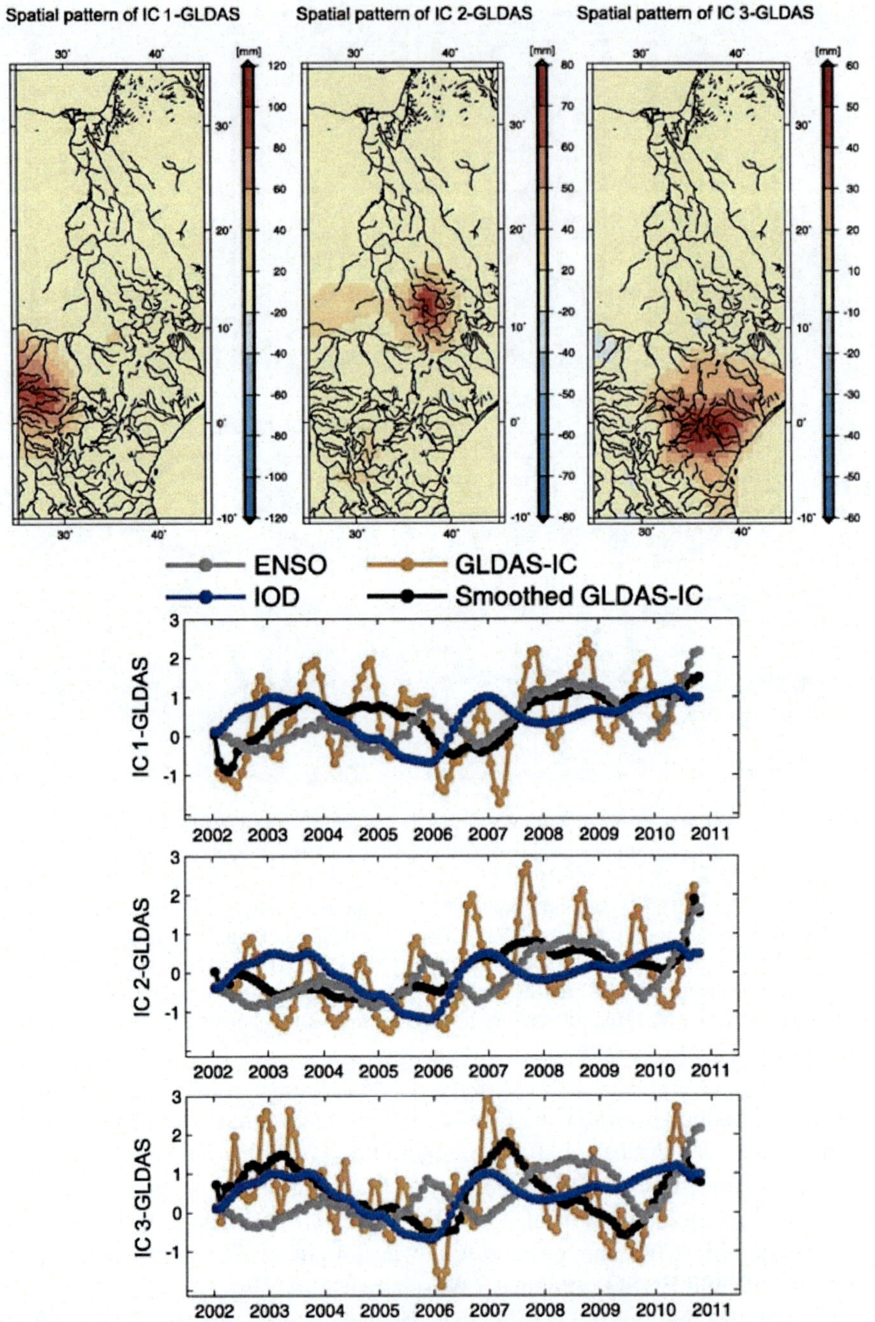

Fig. 7.5 ICA decomposition of GLDAS water storage data within the NRB accounting for 86% of cumulative total variance of GLDAS-TWS changes. The linear rates of TWS changes corresponding to each sub-basin are also reported in the figure. IC2 and IC3 of GLDAS are comparable to those of GRACE, although IC3 of GLDAS is concentrated over the eastern part of the LVB. However, no concentration over the northern part of the basin (e.g., IC4 of GRACE) is found from GLDAS data. *Source* [78]

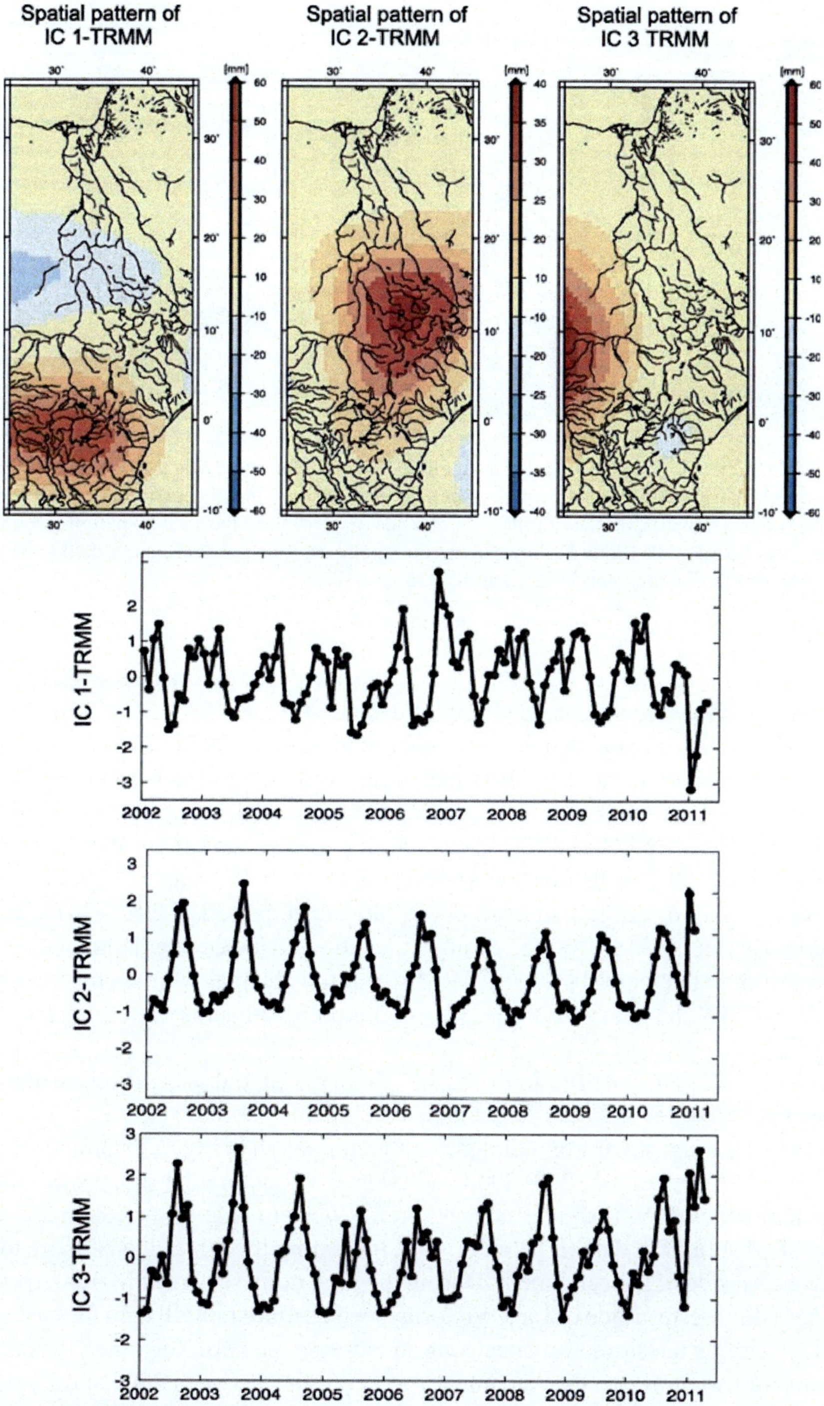

Fig. 7.6 ICA decomposition of TRMM derived rainfall data within the NRB accounting for 91% of cumulative total variance of rainfall changes. IC1 is localized over LVB, IC2 over the EH and IC3 over the BEG region. *Source* [78]

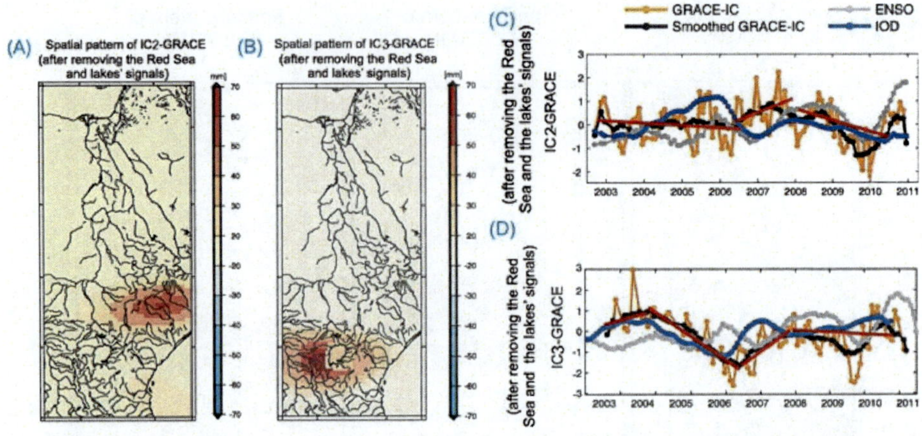

Fig. 7.7 Dominant independent patterns of GRACE TWS after removing the contribution of Lakes Tana and Victoria surface water signals. Spatial patterns show the concentration of the remaining signals (after removal) over the Ethiopian Highlands (panel A) and the western part of Lake Victoria Basin (panel B). Panels C and D shows their corresponding time series (ICs) overlaid on smoothed time series of the loading and ENSO and IOD. *Source* [78]

saw a drop in the TWS of LVB at a rate of 25.8 mm/year before it rose again at a rate of 49.8 mm/year due to increased rainfall from 2009 (see IC1 of Fig. 7.6).

As for the rainfall, the first independent mode (see, IC1 in Fig. 7.6 in the appendix) indicate a reduction between 2004 and about mid 2006, which coincides with the drought that was experienced in the region (see, e.g., [10]). During the period 2006–2007, there was ENSO and IOD induced increase in rainfall but the decline commenced after 2007 with the lowest level reached at the start of 2011. This was the lowest level of rainfall attained during the entire period of 2002–2011, and was mainly due to a failed season and drought that affected the Greater Horn of Africa that started around that period [49, 79]. TRMM-rainfall changes were found to supersede GRACE-TWS changes with a phase lag of 1 month with a maximum lag correlation of 0.52.

In Fig. 7.7b, the contribution of Lake Victoria's signal is removed as discussed in Sect. 7.3 in order to study the residual catchment's stored water signals. With the removal of the signals, the remaining catchment's signal in Fig. 7.7b indicate a similar behaviour for the period 2003–2004 and 2006–2007 before the signal was removed. The dominant GRACE signals appears on the western side of the catchment in line with TRMM rainfall data (IC1 in Fig. 7.6), indicating the rainfall to be more towards the western side of the catchment. During the long rainy season of March-April-May (MAM), the western side of Lake Victoria receives more rainfall than the eastern part thus recharging the stored water causing an increase (cf. [2]). Regarding the influence of climate variability on the catchment's stored water between 2002 and 2011, there is a decrease in correlations between ENSO (0.48) and GRACE-TWS changes after the removal of the Lake's signals (see Table 7.1). This could be attributed to the influence of climate variability on the rainfall that falls on the western side of Lake

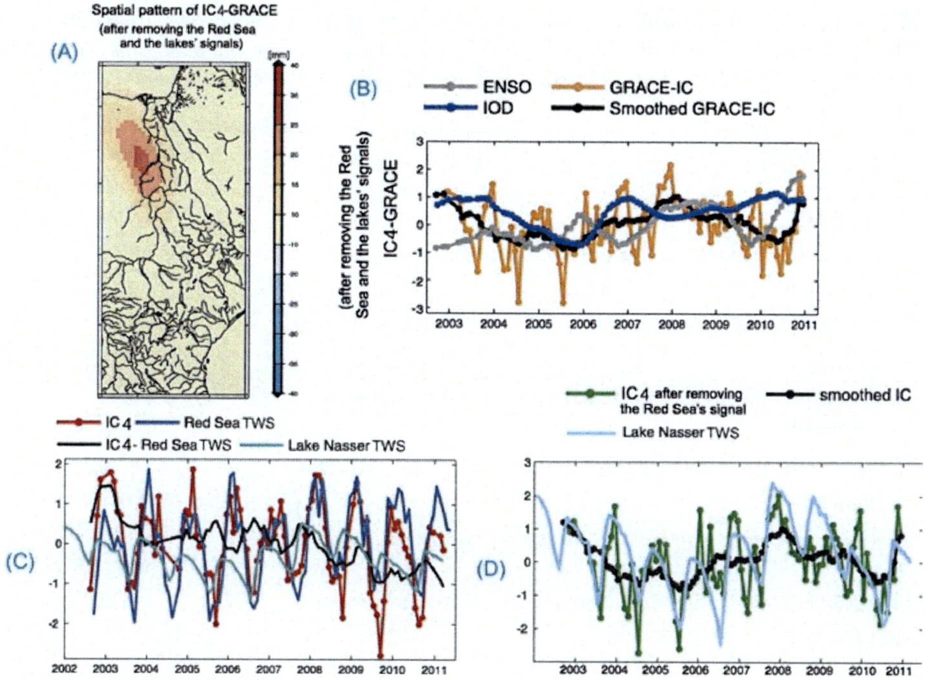

Fig. 7.8 Dominant independent pattern of GRACE-TWS changes for the Nasser region. **a** Spatial pattern of IC4 derived from GRACE-TWS changes after correction for the water storage changes of the Red Sea; **b** the corresponding temporal evolution of (A) and its comparison with the ENSO and IOD indices; **c** comparison of IC4 in **b** with the signal of the Red Sea and Lake Nasser; and **d** an overview of IC4 after removing the signal of the Red Sea and its comparison with the signal of Lake Nasser. We should mention here that the Lake Nasser TWS of **c** and **d** represent the same quantity. To enhance the visual comparison, that of **d**, is however, vertically shifted to the mean of IC4. (For interpretation of the references to colour in this figure caption, the reader is referred to the web version of this article. *Source* [78])

Victoria, which provides most of the source of the stored water. In interpreting the results of the removed Lake's signals, however, it should be pointed out that remnant residual Lake's signals (e.g., Fig. 7.7b) also contribute to the correlation, hence any conclusion needs to be taken with care.

In general, for LVB, over the study period of 2002 to 2011, and within the boundary of Fig. 2.1, the ICA analysis of GRACE-TWS changes showed that the changes in LVB's TWS experienced periods of significant decrease (e.g., 2002–2006) and significant increase (e.g., 2006–2007). Both anthropogenic and climate variability could have played significant contributions to these occurrences, i.e., with the 2003–2006 decrease hugely associated with the expansion of the Owen Falls/Nalubaale dam, e.g., [6, 69] while the 2005–2007 increase is associated with climate variability (ENSO and IOD) as seen by comparing indices with the ICs of GRACE-TWS changes. For the catchment's stored water, a similar deduction could be made, i.e.,

Table 7.1 Correlation coefficient between independent patterns of TWS derived from GRACE and climate variability of IOD (in bold) and ENSO (in normal text) over the Nile subbasins. Correlations are computed at 95% level of confidence. The insignificant values are marked. *Source* [78]

Basin	Correlation
Barh-El-Ghazal	**+0.18** (insignificant)
IC1 GRACE	+0.71
Ethiopian Highlands (with the red Sea and the lake's signals)	**+0.31**
IC2 GRACE	+0.54
Ethiopian Highlands (without the red Sea and the lake's signals)	**+0.47**
IC2 GRACE	+0.59
Lake Victoria Basin (with the red Sea and the lake's signals)	**+0.48**
IC3 GRACE	+0.56
Lake Victoria Basin (without the red Sea and the lake's signals)	**+0.48**
IC3 GRACE	+0.46
The Nasser region (with the red Sea and the lake's signals)	**+0.13** (insignificant)
The Nasser region (without the red Sea and the lake's signals)	**+0.33**
IC4 GRACE	+0.30

the expansion of the Owen Falls/Nalubaale dam could have affected not only the Lake but also its catchment.

In contrast to the White Nile's water discharge from LVB, where the influence of ENSO climate variability has been shown to be weak (i.e., a weak negative association with ENSO; [1, 24]), the influence of climate variability on TWS changes within the basin is strong (see Table 7.1). LVB's general TWS trend showed a decline even after the 2007 ENSO that saw a rise in its TWS. This general decline impacts upon the entire NRB as evident from the water levels of Lake Nasser (e.g., as reported also in [10, 21]), which follows the pattern of Lake Victoria even though the Blue Nile contributes most of the Nile's waters serving Egypt and Sudan.

7.5.2 *The Ethiopian Highlands (EH); the Blue Nile*

Analysing GRACE-TWS changes over Ethiopian Highlands (EH) (Fig. 7.4, IC2) showed a decline at a rate of 18.4 mm/year between 2002 and 2006. This decline can be explained by the temporal curve of the rainfall (Fig. 7.6, IC2), which shows a decline in annual rainfall for the period 2002 to the end of 2005. A similar decline was noted by Di Baldassarre et al. [23] who studied the changes in land cover, rainfall, and stream flow of Gilgel Abbay catchment, the largest contributor to the inflow of Lake Tana (the source of the upper Blue Nile; e.g., [20]), and found that changes in streamflow records for the period 2001–2005 could have been attributed to changes in land cover, and changes in the annual and seasonal distribution of rainfall. Between 2006 and 2007, GRACE-TWS changes showed an increase in stored water at a rate of 44.3 mm/year. This could be attributed to the more than normal rainfall during the peak season of July-August as indicated by the amplitude of TRMM rainfall for 2006–2007 in Fig. 7.6, IC2. Jury [37] found high peaks of the 2006–2007 in Ethiopian floods, which occurred in the periods 23–28 July 2006 and 26–31 July 2007. The floods are attributed to rainfall associated with sea surface temperature anomalies over the western Indian Ocean [37].

ENSO is reported to have a reverse effect, i.e., deficient rainfall tends to occur during ENSO summers (e.g., [24, 39]). Thus the Blue Nile sub-basin experienced drought episodes during ENSO warmer phases and enhanced rains during the ENSO cooler phases. Since 2006–2007 was an ENSO year in East Africa, one would expect less rainfall within the Blue Nile. The increase in the rainfall seen in TRMM's temporal graph (Fig. 7.6, IC2), therefore, could be attributed to other factors such as the influence of IOD, given that a correlation of 0.61 was obtained between the TRMM rainfall and IOD as opposed to 0.43 between TRMM and ENSO (results not shown). Indeed, that IOD exerts influence over the same region is reported by Korecha and Barnston [39] to have occurred during the 1997 ENSO year. From 2007 to 2011 (Fig. 7.4, IC2), the temporal curves of GRACE indicate reductions in TWS at a rate of 12.8 mm/year. This is due to a general reduction in the total annual rainfall as seen from the TRMM in Fig. 7.6, IC2.

EH receives most of its rains in the period JJAS, which accounts for 50% to 80% of annual rainfall totals over the region, contributing to high agricultural productivity and major water reservoirs [e.g., [39]. The peak of the JJAS rainfall occurs between July and August while the dry season takes place between October and May (e.g., [52]). The temporal results of TRMM rainfall for EH in Fig. 7.6, IC2 show a general decline in the total annual rainfall over the 2002–2011 period, with the period 2007–2010 being mostly affected. This could be due to the reduced amount of the maximum rainfall recorded during the long high rainfall (JJAS) season, and could possibly explain the cause of the drought that faced the Greater Horn of Africa (GHA) over that period. Di Baldassarre et al. [23] points to a decreasing seasonality in some key watersheds of the upper Nile in Ethiopia such as the southern Blue Nile.

For the Blue Nile sub-basin, Amarasekera et al. [1] found a significant negative correlation between discharge and ENSO. For TWS, likewise, the positive correlations between ENSO and GRACE, i.e., (0.54 in Table 7.1), support the fact that ENSO is the dominant climate variability. However, the influence of IOD on TWS is also noticeable, i.e., 0.31 with respect to GRACE. With the removal of Lake Tana's signal, there is a shift in the GRACE signals (Fig. 7.7a) further towards the highlands.

The magnitude of the rate of change in stored water remains relatively the same (i.e., 44.3 mm/year) for the period April 2006–December 2007 even after removing the signals of Lake Tana. For the period August 2002–April 2006 and December 2007–March 2012, the rate of decline in stored water within the highland is of the same order of magnitude as before the removal of Lake Tana's signal (e.g., 18.4 mm/year and 12.8 mm/year respectively). This further confirms the well known fact that the contribution of the Blue Nile's waters comes mainly from these highlands. This fact is supported by the computed correlations to climate variability, which increases slightly for IOD (0.47) but remains the same for ENSO (0.59, e.g., Table 7.1). The slight increase in correlation with IOD following the removal of Lake Tana's signals signifies that the stored water within the lake Tana dominates the entire sub-basin.

7.5.3 Bahr-el-Ghazal Region (BEG)

Bahr-el-Ghazal (BEG) sub-basin consists of a number of rivers that originate from the Congo-Nile River divide and is made up of a large area of very low slope such that nearly all the basin runoff and precipitation is evaporated or leaked into swamps (i.e., about 96% is lost), with only about 0.5 km^3 leaving the basin annually [20]. From the estimated GRACE-TWS patterns in (Fig. 7.4, IC1), increased water storage for this region in the period between 2002 and 2003 at a rate of 64.4 mm/year was noticed. This could be explained by the increase in total annual rainfall between 2002 and 2004 (see, e.g., Fig. 7.6, IC3). Between 2003 and 2006, GRACE-TWS changes showed a reduction at a rate of 29.2 mm/year, again, inline with the noticeable reduction in total annual rainfall from TRMM (see Fig. 7.6, IC3). From 2006 to 2007, the stored water increased at a rate of 74.9 mm/year although the TRMM rainfall over this period does not show significant increase. This could imply reduced evaporation since the correlation between GRACE-TWS changes and the TRMM-rainfall over BEG gives a phase lag of 1-month at a maximum correlation of 0.53.

The ICA-derived TRMM-rainfall pattern in BEG region showed similarities to that of EH (Fig. 7.6, IC3) but with higher amplitudes during 2002–2004 period. The BEG sub-basin experienced a reduction in total annual rainfall from 2003 to 2007, i.e., an increase in 2008, then a decreased between 2009 and 2010 before starting to increase again (Fig. 7.6, IC3). The drop seen in Fig. (7.6, IC3) supports the deductions of Di Baldassarre et al. [23] stating that the three sub-basins; Bahr-el-Ghazal, Sobat and Central Sudan within the NRB recorded significant drops in annual precipitation.

For the period 2002–2011, an insignificant correlation (0.18) between GRACE-TWS changes and IOD, and a strong positive correlation (0.71) between GRACE-TWS changes and ENSO (Table 7.1) was found. These correlations suggest that ENSO is the dominant climate variability associated with TWS changes over the BEG sub-basin.

7.5.4 Lake Nasser Region

For this region, the ICA-derived GRACE-TWS anomalies were localized to the northwest-southeast direction of the Red Sea (Fig. 7.4, IC4). This signal could be largely attributed to the water storage changes within the Red Sea and Lake Nasser and thus needs to be removed. The use of altimetry data to remove the dominant GRACE signals from the Red Sea is supported by a correlation of 0.71 between GRACE-TWS changes and the altimetry data for the Red Sea. Altimetry observations provided by NOAA ERDDAP (the Environmental Research Division's Data Access Program program) 6 over the Red Sea were converted to EWT changes (see the conversion procedure in Sect. 7.3) and scaled by its variance to unit-variance (i.e., IC4 of GRACE, c.f., Fig. 7.4d). Some differences in amplitude were detected (compare the amplitudes of red and blue in Fig. 7.8c) mainly in 2004, 2005, and after 2008. The residual of GRACE-TWS changes minus Red Sea-EWT is shown in black. Lake Nasser's altimetry are also converted into EWT time series (same procedure as for the Red Sea), the results of which are shown in cyan. As can be seen, after 2008, the EWT changes corresponding to Lake Nasser is very similar to the residuals (the black line; Fig. 7.8c).

After the removal of the dominant signal of the Red Sea, the resulting signal (c.f. Fig. 7.8a) indicates a decline in stored water in the Western Plateau within the Nubian Aquifer covering Lake Nasser at a rate of 2.6 mm/year (cf. -3.5 mm/year for the period April 2002 to November 2010 in [65]). The loss of water in this region is attributed by Sultan et al. [65] to the fact that most of the water is extracted from the Nubian Aquifer and used for agricultural purposes that largely occur throughout the winter season, and also due to the fact that the Uweinat–Aswan uplift prevents recharge of ground water flowing from the South to the North. To strengthen this argument is the fact that expansions of some large irrigation schemes such as East Uweinat project has seen heavy utilization of groundwater. In the East Uweinat project, for example, the lands reclaimed amounted to 1200 ha in 1992 and 4200 ha in 2003, with the target of reclaiming a total of 75,000 ha by 2022, all of which will be irrigated using groundwater (see, e.g., [60, 61]).

No dominant independent pattern was observed from TRMM and GLDAS data over the region surrounding lake Nasser, indicating that precipitation and soil moisture changes during the period 2002–2011 was not the source of the observed TWS changes (see Fig. 7.8a). This supports the fact that the stored water within the Lake Nasser basin originates from the other Nile sub-basins. Weaker negative correlations were observed between GRACE and ENSO (−0.10) for the study period. During the same period, the correlation between GRACE and IOD was 0.13. Without the Red sea and Lake Tana's signals, correlations between IOD and ENSO on the one hand and GRACE-TWS changes on the other hand were respectively 0.33 and 0.30, thus following closely to the LVB pattern (i.e., 0.48 and 0.46 for IOD and ENSO, respectively, for the same).

7.6 Concluding Remarks

Using the ICA method (Spatial ICA in [27, 30]), independent water storage patterns of the Nile sub-basins were extracted and related to climate variability over the period 2002–2011. For the individual sub-basins, the chapter found that:

1. The stored water within LVB could have been influenced both by anthropogenic as well as climate variability (ENSO and IOD). The removal of the dominant Lake Victoria signal shows the western side of Lake Victoria having increased TWS, a possible consequence of the long rains of the MAM season that pounds the western side of the lake more.
2. Ethiopian Highlands (EH) experienced a general reduction in rainfall as seen from the TRMM data over the study period from 2002 to the end of 2010, with the period between 2005 and 2010 recording low rainfall during the long rainy season of JJAS and low total annual rainfall, a possible explanation for the 2011 drought that hit the Greater Horn of Africa (GHA), e.g., [79]. This reduction in rainfall impacts on TWS changes over the GHA as seen from GRACE outputs (see also [49]).
3. Bar-el-Ghazal region showed a mixed increase/decrease in the stored water associated with increase/decrease in rainfall at different times of the 2002–2011 period. The dominant climate variability in the region was found to be ENSO.
4. For the Lake Nasser region, it is clear that the dominant TWS changes of the Red Sea obscure the real impact of over extraction of water in the Nubian aquifer region for irrigation purposes [e.g., [65]. The situation is not helped much by the fact that there could be insufficient recharge from the south-north flow of the groundwater due to the presence of the Uweinat–Aswan uplift.
5. In general, for almost all the sub-basins, the dominant climate variability on GRACE-TWS changes was the ENSO, with the exception of LVB and EH where the influence of IOD on GRACE-TWS changes was noticeable.

For the proper management of the NRB's water resources, therefore, focus should be on all the sub-basins; namely the Lake Victoria Basin, the Ethiopian Highlands, the Bahr-el-Ghazal river Basin, and the Lake Nasser region in order to ensure a sustainable use that does not impact on the overall NRB. The basin-wide analysis of GRACE signal does not provide this possibility as the dominant signals over LVB and Tana obscurr the catchment signal. The Spatial Independent Component Analysis (ICA) employed in this chapter offers a possible solution. Further, analysis of the stored water in these sub-basins should consider the influence of climate variability.

References

1. Amarasekera KN, Lee RF, Williams ER, Eltahir EAB (1997) ENSO and the natural variability in the flow of tropical rivers. J Hydrol 200:24–39. https://doi.org/10.1016/S0022-1694(96)03340-9
2. Awange JL, Anyah R, Agola N, Forootan E, Omondi P (2013) Potential impacts of climate and environmental change on the stored water of Lake Victoria Basin and economic implications. Water Resour Res 49:8160–73. https://doi.org/10.1002/2013WR014350
3. Awange JL, Forootan E, Fleming K, Omondi P, Odhiambo G. Dominant patterns of water storage changes in the NRB, during 2003–2013. AGU books: chapman conference on remote sensing of terrestrial waters in Hawaii, USA (2013)

4. Awange JL, Fleming KM, Kuhn M, Featherstone WE, Heck B, Anjasmara I (2011) On the suitability of the 4° × 4° GRACE mascon solutions for remote sensing Australian hydrology. Remote Sens Environ 115:864–75. https://doi.org/10.1016/j.rse.2010.11.014
5. Awange JL, Sharifi MA, Baur O, Keller W, Featherstone W, Kuhn M (2009) GRACE hydrological monitoring of Australia: current limitations and future prospects. J Spatial Sci 54(1):23–36. https://doi.org/10.1080/14498596.2009. 9635164
6. Awange JL, Sharifi MA, Ogonda G, Wickert J, Grafarend E, Omulo M (2008) The falling lake victoria water levels: GRACE, TRIMM and CHAMP satellite analysis of the lake Basin. Water Resour Manag 22:775–96. https://doi.org/10.1007/s11269-007-9191-y
7. Awange JL, Ong'ang'a O (2006) Lake victoria: ecology resource and environment. Springer, Berlin, p 354
8. Anyah RO, Semazzi FHM (2006) NCAR-AGCM ensemble simulations of the variability of the Greater Horn of Africa climate. Theor Appl Climatol 86:39–62. https://doi.org/10.1007/s00704-005-0203-7 [special issue]
9. Aus der Beek T, Menzel L, Rietbroek R, Fenoglio-Marc L, Grayek S, Becker M, Kusche J, Stanev E. Modeling the water resources of the Black and Mediterranean Sea river basins and their impact on regional mass changes. J Geodyn 59–60:157–67 (2012). https://doi.org/10.1016/j.jog.2011.11.011
10. Becker M, LLovel W, Cazenave A, Güntner A, Crétaux J-F. Recent hydrological behavior of the East African great lakes region inferred from GRACE, satellite altimetry and rainfall observations. C R Geosci 342:223–33 (2010). https://doi.org/10.1016/j.crte.2009.12.010
11. Behera SK, Luo JJ, Masson S, Delecluse P, Gualdi S, Navarra A (2005) Paramount impact of the Indian ocean dipole on the East African short Rains: ACGCM study. J Climate 18:41–54. https://doi.org/10.1175/JCLI3541.1
12. Berhane G, Kristine M, Nawal A, Kristine W (2013) Water leakage investigation of micro-dam reservoirs in Mesozoic sedimentary se-quences in Northern Ethiopia. J Afr Earth Sci 79:98–110. https://doi.org/10.1016/j.jafrearsci.2012.10.004
13. Beyene T, Lettenmaier DP, Kabat P (2010) Hydrologic impacts of climate change on the Nile River Basin: implications of the 2007 IPCC scenarios. Clim Change 100:433–61. https://doi.org/10.1007/s10584-009-9693-0
14. Black E, Slingo J, Sperber KR (2003) An observational study of the relationship between excessively strong short rains in the coastal East Africa and Indian Ocean SST. Mon Weather Rev 131:74–94. https://doi.org/10.1175/1520-0493(2003)131$<$0074:AOSOTR$>$2.0.CO;2
15. Block PJ, Strzepek K, Rajagopalan B (2007) Integrated management of the Blue NRB in Ethiopia. Hydropower and irrigation modeling. International Food Policy Research Institute, IFPRI Discussion Paper 00700. http://citeseerx.ist.psu.edu/viewdoc/summary?doi=10.1.1.139.3725
16. Bonsor HC, Mansour MM, MacDonald AM, Hughes AG, Hipkin RG, Bedada T (2010) Interpretation of GRACE data of the NRB using a groundwater recharge model. Hydrol Earth Syst Sci Discuss 7:4501–33. https://doi.org/10.5194/hessd-7-4501-2010
17. Clark CO, Webster PJ, Cole JE (2003) Interdecadal variability of the relationship between the Indian Ocean zonal mode and East African coastal rainfall anomalies. J Climate 16:548–54. https://doi.org/10.1175/1520-0442(2003)016$<$0548:IVOTRB$>$2.0.CO;2
18. Colberg F, Reason CJC (2004) South Atlantic response to El-Nino-Southern oscillation induced climate variability in an ocean general circulation model. J Geophys Res 109(C12015):14. https://doi.org/10.1029/2004JC002301
19. Conway D (2005) From headwater tributaries to international river: observing and adapting to climate variability and change in the NRB. Global Environ Change 15:99–114. https://doi.org/10.1016/j.gloenvcha.2005.01.003
20. Conway D, Hulme M (1993) Recent fluctuations in precipitation and runoff over the Nile sub-basins and their impact on main Nile discharge. Clim Change 25:127–51. https://doi.org/10.1007/BF01661202

21. Crétaux JF, Jelinski W, Calmant S, Kouraec A, Vuglinski V, Bergé-Nguyen M, Gennero M-C, Nino F, Abarca Del Rio R, Cazenave A, Maisongrande P (2011) SOLS: A lake database to monitor in the near real-time water level and storage variations from remote sensing data. Adv Space Res 47:1497–507. https://doi.org/10.1016/j.asr.2011.01.004
22. Destouni G, Jaramillo F, Prieto C (2013) Hydroclimatic shifts driven by human water use for food and energy production. Nature Clim Change 3:213–7. https://doi.org/10.1038/nclimate1719
23. Di Baldassarre G, Elshamy M, van Griensven A, Soliman E, Kigobe M, Ndomba P, Mutemi J, Mutua F, Moges S, Xuan J-Q, Solomatine D, Uhlenbrook S (2011) Future hydrology and climate in the River NRB: a review. Hydrol Sci J 56(2):199–211. https://doi.org/10.1080/02626667.2011.557378
24. Eltahir EAB (1996) El Niño and the natural variability in the flow of the Nile River. Water Resour Res 32(1):131–7. https://doi.org/10.1029/95WR02968
25. Flechtner F (2007) GFZ Level-2 processing standards document for level-2 product release 0004. GRACE 327–743, Rev 1
26. Fleming KM, Awange JL, Kuhn M, Featherstone WE (2012) Evaluating the TRMM 3B43 monthly precipitation product using gridded rain-gauge data over Australia. Aust Meteorol Oceanogr J 63:421–6
27. Forootan E, Awange JL, Kusche J, Heck B, Eicker A (2012) Independent patterns of water mass anomalies over Australia from satellite data and models. Remote Sens Environ 124:427–43. https://doi.org/10.1016/j.rse.2012.05.023
28. Forootan E, Rietbroek R, Kusche J, Sharifi MA, Awange JL, Schmidt M, Omondi P, Famiglietti J (2014) Separation of large scale water storage patterns over Iran using GRACE, altimetry and hydrological data. Remote Sens Environ 140:580–95. https://doi.org/10.1016/j.rse.2013.09.025
29. Forootan E, Kusche J (2013) Separation of deterministic signals, using independent component analysis (ICA). Stud Geophys Geod 57:17–26. https://doi.org/10.1007/s11200-012-0718-1
30. Forootan E, Kusche J (2012) Separation of global time-variable gravity signals into maximally independent components. J Geod 86(7):477–97. https://doi.org/10.1007/s00190-011-0532-5
31. García-García D, Ummenhofer CC, Zlotnicki V (2011) Australian water mass variations from GRACE data linked to Indo-Pacific climate variability. Remote Sens Environ 115:2175–83. https://doi.org/10.1016/j.rse.2011.04.007
32. Ghoubachi SY (2010) Impact of Lake Nasser on the groundwater of the Nubia sandstone aquifer system in Tushka area, south western desert. Egypt. J King Saud Univ - Sci. https://doi.org/10.1016/j.jksus.2010.04.005
33. Hamouda MA, Nour El-Din MN, Moursy FI (2009) Vulnerability assessment of water resources systems in the Eastern NRB. Water Resour Manag 23:2697–725. https://doi.org/10.1007/s11269-009-9404-7
34. Hanson RT, Newhouse MW, Dettinger MD (2004) A methodology to assess relations between climatic variability and variations in hydrologic time series in the southwestern United States. J Hydrol 287(1–4):252–69. https://doi.org/10.1016/j.jhydrol.2003.10.006
35. Hastenrath S, Nicklis A, Greishar L (1993) Atmospheric-hydrospheric mechanisms of climate anomalies over the western equatorial Indian Ocean. J Geophys Res 98:20219–35. https://doi.org/10.1029/93JC02330
36. Indeje M, Semazzi FHM, Ogallo LJ (2000) ENSO signals in East African rainfall seasons. Int J Climatol 20:19–46. https://doi.org/10.1002/(SICI)1097-0088(200001)20:1$<$19::AID-JOC449$>$3.0.CO;2-0
37. Jury MR (2011) Meteorological scenario of Ethiopian floods in 2006–2007. Theor Appl Climatol 104:209–19. https://doi.org/10.1007/s00704-010-0337-0
38. Konar M, Todd MJ, Muneepeerakul R, Rinaldo A, Rodriguez-Iturbe I (2013) Hydrology as a driver of biodiversity: Controls on carrying capacity, niche formation, and dispersal. Adv Water Resour 51:317–25. https://doi.org/10.1016/j.advwatres.2012.02.009
39. Korecha D, Barnston A (2007) Predictability of June-September rainfall in Ethiopia. Mon Weather Rev 135:628–50. https://doi.org/10.1175/MWR3304.1

40. Kumar SV, Peters-Lidard CD, Tian Y, Houser PR, Geiger J, Olden S, Lighty L, Eastman JL, Doty B, Dirmeyer P, Adams J, Mitchell K, Wood EF, Sheffield J (2006) Land information system: an interoperable framework for high resolution land surface modeling, environmental modelling & amp. Software 21(10):1402–15. https://doi.org/10.1016/j.envsoft.2005.07.004
41. Kummerow C, Barnes W, Kozu T, Shiue J, Simpson J (1998) The Tropical Rainfall Measuring Mission (TRMM) sensor package. J Atmos Oceanic Technol 15(3):809–17. https://doi.org/10.1175/1520-0426(1998)015$<$0809:TTRMMT$>$2.0.CO;2
42. Kusche J (2007) Approximate decorrelation and non-isotropic smoothing of timevariable GRACE-type gravity field models. J Geod 81:733–49. https://doi.org/10.1007/s00190-007-0143-3
43. Kusche J, Schmidt R, Petrovic S, Rietbroek R (2009) Decorrelated GRACE time-variable gravity solutions by GFZ, and their validation using a hydrological model. J Geod 83:903–13. https://doi.org/10.1007/s00190-009-0308-3
44. Lehmann EL, D'Abrera HJM (1998) Nonparametrics: statistical methods based on ranks, rev edn. Prentice-Hall, Englewood Cliffs. p. 464. ISBN 978-0-387-35212-1
45. Longuevergne L, Wilson CR, Scanlon BR, Crétaux JF (2012) GRACE water storage estimates for the Middle East and other regions with significant reservoir and lake storage. Hydrol Earth Syst Sci Discuss 9:11131–59. https://doi.org/10.5194/hessd-9-11131-2012
46. Mohamed YA, Savenije HHG, Bastiaanssen WGM, van den Hurk BJJM (2006) New lessons on the Sudd hydrology learned from remote sensing and climate modeling. Hydrol Earth Syst Sci 10:507–18. https://doi.org/10.5194/hess-10-507-2006
47. Naumann G, Barbosa P, Carro H, Singleton A, Vogt J (2012) Monitoring drought conditions and their uncertainties in Africa using TRMM Data. J Appl Meteor Climatol 51:1867–74. https://doi.org/10.1175/JAMCD-12-0113.1
48. Nicholson S, Some B, McCollum J, Nelkin E, Klotter D, Berte Y, Diallo B, Gaye I, Kpabeba G, Ndiaye O, Noukpozounkou J, Tanu A, Thiam A, Toure AA, Traore A (2003) Validation of TRMM and other rainfall estimates with a high-density gauge dataset for West Africa: part II: validation of TRMM rainfall products. J Appl Meteorol 42(10):1355–68. https://doi.org/10.1175/1520-0450(2003)042$<$1355:VOT
49. Omondi P, Awange J, Forootan E (2013) Changes in temperature and precipitation extremes over the Greater Horn of Africa region from, et al (1961) to 2010. Int J Climatol https://doi.org/10.1002/joc.3763
50. Omondi P, Ogallo Awange J, Ininda J, Forootan E (2013) The influence of low frequency sea surface temperature modes on delineated decadal rainfall zones in Eastern Africa region. Adv Water Resour. https://doi.org/10.1016/j.advwatres.2013.01.001
51. Omondi P, Awange J, Ogallo LA, Okoola RA, Forootan E (2012) Decadal rainfall variability modes in observed rainfall records over East Africa and their relations to historical sea surface temperature changes. J Hydrol 464–465:140–56. https://doi.org/10.1016/j.jhydrol.2012.07.003
52. Rientjes THM, Haile AT, Kebede E, Mannaerts CMM, Habib E, Steenhuis TS (2011) Changes in land cover, rainfall and stream flow in Upper Gilgel Abbay catchment, Blue NRB - Ethiopia. Hydrol Earth Syst Sci 15:1979–89. https://doi.org/10.5194/hess-15-1979-2011
53. Rieser D, Kuhn M, Pail R, Anjasmara I, Awange JL (2010) Relation between GRACE-derived surface mass variations and precipitation over Australia. Aust J Earth Sci 57(7):887–900. https://doi.org/10.1080/08120099.2010.512645
54. Rodell M, Famiglietti JS (2001) An analysis of terrestrial water storage variations in Illinois with implications for the gravity recovery and climate experiment (GRACE). Water Resour Res 37:1327–40. https://doi.org/10.1029/2000WR900306
55. Rodell M, Houser PR, Jambor U, Gottschalck J, Mitchell K, Meng K, Arsenault CJ, Cosgrove B, Radakovich J, Bosilovich M, Entin JK, Walker JP, Lohmann D, Toll D (2004) The global land data assimilation system. Bull Am Meteorol Soc 85(3):381–94. https://doi.org/10.1175/BAMS-85-3-381
56. Sahin M (1985) Hydrology of the NRB. Developments in water science, vol 21. Elsevier, New York. p 575

57. Saji NH, Goswami BN, Vinayachandran PN, Yamagata T (1999) A dipole mode in the tropical Indian Ocean. Nature 401:360–3. https://doi.org/10.1038/43854
58. Saji NH, Yamagata T (2003) Possible impacts of Indian Ocean dipole mode events on global climate. Climate Res 25(2):151–69. https://doi.org/10.3354/cr025151
59. Saji NH, Yamagata T (2003) Structure of SST and surface wind variability during Indian ocean dipole mode events: COADS observations. J Climate 16(16):2735–51. https://doi.org/10.1175/1520-0442(2003)016$<$2735:SOSASW$>$2.0.CO;2
60. Salem OM, Pallas P (2002) The Nubian sandstone aquifer system, paper presented at Managing Shared Aquifer Resources in Africa, ISARM-AFRICA, Tripoli, Libya, Accessed 2–4 June 2002
61. Salem OM (2007) Management of shared groundwater basins in Libya. Afr Water J 1(1):109–20. ISBN 92-1-125089-7
62. Sefelnasr AM (20107) Development of groundwater flow model for water resources management in the development areas of the western desert, Egypt. Dissertation, Faculty of Natural Sciences III of the Martn Luther University Halle-Wittenberg
63. Senay GB, Asante K, Artan G (2009) Water balance dynamics in the Nile basin. Hydrol Proces 23:3675–81. https://doi.org/10.1002/hyp.7364
64. Song Q, Vecchi GA, Rosati AJ (2007) Indian ocean variability in the GFDL coupled climate model. J Climate 20:2895–916. https://doi.org/10.1175/JCLI4159.1
65. Sultan M, Ahmed M, Sturchio N, Yan YE, Milewski A, Becker R, et al (2012) Assessment of the vulnerabilities of the Nubian sandstone fossil aquifer, North Africa, Water Encyclopedia. In: Pielke, RA (ed). Climate vulnerability: understanding and addressing threats to essential resources. Elsevier Inc. Academic. pp 311–33. ISBN:9780123847034
66. Sutcliffe JV, Parks YP (1999) The hydrology of the Nile, IAHS special publication No. 5. Wallingford, Oxford Shine: IAHS press, Institute of Hydrology
67. Swenson S, Wahr J, Milly P (2003) Estimated accuracies of regional water storage variations inferred from the gravity recovery and climate experiment (GRACE). Water Resour Res 39(8):1223. https://doi.org/10.1029/2002WR001808
68. Swenson S, Wahr J (2002) Methods for inferring regional surface-mass anomalies from gravity recovery and climate experiment (GRACE) measurements of time-variable gravity. J Geophys Res Solid Earth 107(B9):2193. https://doi.org/10.1029/2001JB000576
69. Swenson S, Wahr J (2009) Monitoring the water balance of Lake Victoria, East Africa, from space. J Hydrol 370(1–4):163–76. https://doi.org/10.1016/j.jhydrol.2009.03.008
70. Tapley BD, Bettadpur S, Ries JC, Thompson PF, Watkins M (2004) GRACE measurements of mass variability in the Earth system. Science 305:503–5. https://doi.org/10.1126/science.1099192
71. Taylor RG, Todd MC, Kongola L, Maurice L, Nahozya E, Sanga H, MacDonald AM (2012) Evidence of the dependence of groundwater resources on extreme rainfall in East Africa. Nature Clim Change 3:374–8. https://doi.org/10.1038/nclimate1731
72. Tesemma ZK, Mohamed YA, Steenhuis TS (2010) Trends in rainfall and runoff in the Blue NRB: 1964–2003. Hydrol Process 24:3747–58. https://doi.org/10.1002/hyp.7893
73. Tesfagiorgis K, Gebreyohannes T, De Smedt F, Moeyersons J, Hagos M, Nyssen J, Deckers J (2011) Evaluation of groundwater resources in the Geba basin. Ethiopia. Bull Eng Geol Environ 70(3):461–6. https://doi.org/10.1007/s10064-010-0338-3
74. Wahr J, Molenaar M, Bryan F (1998) Time variability of the Earth's gravity field: hydrological and oceanic effects and their possible detection using GRACE. J Geophys Res 103(B12):30205–29. https://doi.org/10.1029/98JB02844
75. Wang H, Wu P, Wang Z (2006) An approach for spherical harmonic analysis of nonsmooth data. Comput Geosci 32(10):1654–68. https://doi.org/10.1016/j.cageo.2006.03.004
76. Whittington D, McClelland E (1992) Opportunities for regional and international cooperation in the NRB. Water Int 17(3):144–54. https://doi.org/10.1080/02508069208686134
77. Yates DN, Strzepek KM (1998) Modelling the Nile basin under climate change. J Hydrol Eng 3(2):98–108. https://doi.org/10.1061/(ASCE)1084-0699(1998) 3:2(98
78. Awange JL, Forootan E, Kuhn M, Kusche J, Heck B (2014a) Water storage changes and climate variability within the Nile river basin between 2002 and 2011. Adv. Water Res. 73:64–78. https://doi.org/10.1016/j.advwatres.2014.06.010

79. Mpelasoka F, Awange JL, Zerihun A (2018) Influence of coupled ocean-atmosphere phenomena on the Greater Horn of Africa droughts and their implications. Sci Total Environ 610–611:691–702. https://doi.org/10.1016/j.scitotenv.2017.08.109
80. Awange J, Ferreira V, Forootan E, Khandu Andam-Akorful, S., Agutu, N., He, X. (2016) Uncertainties in remotely sensed precipitation data over Africa. Int J Climatol 36(1):303–323. https://doi.org/10.1002/joc.4346
81. Awange JL, Hu KX, Khaki M (2019) The newly merged satellite remotely sensed, gauge and reanalysis-based Multi-Source Weighted-Ensemble Precipitation: Evaluation over Australia and Africa (1981–2016). Sci Total Environ 670:448–465. https://doi.org/10.1016/j.scitotenv.2019.03.148
82. Anyah RO, Forootan E, Awange JL, Khaki M (2018) Understanding linkages between global climate indices and terrestrial water storage changes over Africa using GRACE products. Sci Total Environ 635:1405–1416. https://doi.org/10.1016/j.scitotenv.2018.04.159

Chapter 8
Compartmental Changes: Climate Variability

A considerable long-term (2002-2016) negative groundwater trend is found in the Lower Nile (Egypt) signifying a potential depletion. The rate of decline is seen to increase rapidly from 2008 despite increase in precipitation and TWS time series, thus signifying the possibility of human usage, e.g., for irrigation purposes. Smaller soil moisture and groundwater variations exist over the Lower Nile and to a lesser degree over the Central Nile compared to the Upper Nile. While no considerable trend is observed for soil moisture variations, groundwater changes exhibit short-term and long-term trends over different regions. Negative trends are found between 2002 and 2006 over the Upper and Central Nile.
—Khaki et al., [65].

8.1 Summary

With the construction of the largest dam in Africa, the Grand Ethiopian Renaissance Dam (GERD) along the Blue Nile, the Nile is back in the news. This, combined with Bujagali dam on the White Nile are expected to bring ramification to the downstream countries. A comprehensive analysis of the Nile's waters (surface, soil moisture and groundwater) is, therefore, essential to inform its management. Owing to its shear size, however, obtaining in-situ data from "boots on the ground" is practically impossible, paving way to the use of satellite remotely sensed and models' products. The present chapter,[1] based on the work of [65] employs multi-mission satellites and surface models' products to provide, for the first time, a comprehensive analysis of the changes in Nile's stored waters' compartments; surface, soil moisture and groundwater, and their association to climate variability (El Niño Southern Oscillation (ENSO) and Indian Ocean Dipole (IOD)) over the period 1992–2016 (see Fig. 8.1). In this regard, remotely sensed altimetry data from TOPEX/Poseidon

[1] This is an invited Chapter from Dr. Mehdi Khaki, School of Engineering, University of Newcastle, Callaghan, New South Wales, Australia.

J. Awange, *The Nile Waters*,
https://doi.org/10.1007/978-3-030-64756-8_8

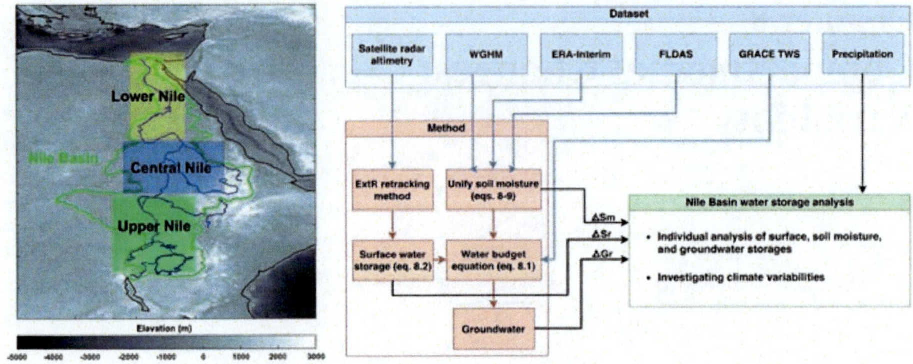

Fig. 8.1 Graphic abstract summarizing the chapters work. *Source* [65]

(T/P), Jason-1, and Jason-2 satellites along with the Gravity Recovery And Climate Experiment (GRACE) mission, and the Tropical Rainfall Measuring Mission Project (TRMM) rainfall products are applied to analyze the compartmental changes over the Nile River Basin (NRB). This is achieved through the creation of 62 virtual gauge stations distributed throughout the Nile River that generate water levels, which are used to compute surface water storage changes. Using GRACE total water storage (TWS), soil moisture data from multi-models based on the Triple Collocation Analysis (TCA) method, and altimetry derived surface water storage, Nile basin's groundwater variations are estimated. The impacts of climate variability on the compartmental changes are examined using TRMM precipitation and large-scale ocean-atmosphere ENSO and IOD indices. The results indicate a strong correlation between the river level variations and precipitation changes in the central part of the basin (0.77 on average) in comparison to the northern (0.64 on average) and southern parts (0.72 on average). Larger water storages and rainfall variations are observed in the Upper Nile in contrast to the Lower Nile. A negative groundwater trend is also found over the Lower Nile, which could be attributed to a significantly lower amount of rainfall in the last decade and extensive irrigation over the region.

8.2 Surface Fluctuations and Stored Water Changes

The Nile River, arguably the longest river in the world (6800 km), has a major impact on the livelihood of over 300 million people of 11 countries within the region [109]. This population is expected to double in the next twenty-five years [79] thereby putting extreme pressure on its water resources. Already now, this pressure is building up with the upper stream countries damming the Nile to exploit on its resources. On the White Nile, Uganda has constructed the Bujagali dam while along the Blue Nile, Ethiopia is constructing the continent's largest dam; the Grand Ethiopian Renaissance Dam (GERD) that is expected to take several years to fill. These human-induced impacts on the Nile, coupled with those of climate variability are expected to exacerbate tension with the low stream countries fearing the cut in the Nile's total volume. Its fresh water in the region that covers approximately 10% of the entire African continent is expected to continue sustaining the livelihood of the growing population thus making a large part of the African continent extremely vulnerable in aspects such as water supplies, agriculture, and industry [15, 77, 104]. An understanding of changes in its stored water (surface, soil moisture and groundwater), and their association to climate variability/change, is, therefore, crucial for environmental assessments and provides useful information for water resources management and climate impact studies, see e.g., [93]. Owing to its size however, the Nile predisposes itself to remotely sensed approaches with the vast spatial and temporal coverage as opposed to ground based in-situ networks.

Remote sensing has provided useful observations for studying water resources around the world, especially over areas with insufficient in situ measurements e.g., [7, 58, 59, 83, 108]. Over the Nile River Basin (NRB), [76] estimated reservoir discharges of Lake Nasser and Roseires Reservoir while flood monitoring using altimetry data was carried out by [23] over Lake Victoria. Ayana et al., [16] reviewed the application of satellite radar altimetry data in the water resource management in Ethiopia, while [102] introduced a post-processing approach to improve the accuracy of radar altimetry measurements over African lakes such as Lake Naivasha and Lake Victoria, see also [1, 10]. A number of hydroclimate variability studies over the basin using various satellite remotely sensed products have been documented, e.g., the Gravity Recovery And Climate Experiment (GRACE) for studying the Nile basin's total water storage changes e.g., [9, 12, 13, 53], satellite precipitation data for studying the basin's rainfall e.g., [11, 67], and a combination of both ground-based and remotely sensed observations for studying the lake's water balance e.g., [97, 107].

Despite the efforts above, a comprehensive long-term study of climate variability and its association with various water storages (TWS, groundwater, surface water storage, and soil moisture) separately, as well as water level fluctuations over the entire NRB is missing. For example, [13] studied water storage changes within the Nile's main sub-basins and the related impacts of climate variability by employing Independent Component Analysis ICA; see also [41–43, 66] to extract statistically independent TWS patterns over the sub-basins from GRACE and the Global Land Data Assimilation System (GLDAS) for the period 2002–2011. Their study, however, does not address variability of individual water storage compartments (surface, soil moisture and groundwater). Rather, they treated them as a combined entity and did not compute the fluctuations of the water level within the NRB. Fluctuations of surface water levels, which can be derived from satellite radar altimetry, are important as they can be related to seasonal variations of precipitation, evaporation, and anthropogenic use [50]. Surface water storages and their variations are also important to study the interactions between land and the atmosphere and oceans [84].

The present chapter addresses these missing gaps by exploiting multi-satellites and surface models' products to study changes in the various Nile River Basin's water storage compartments (surface, soil moisture, and groundwater), and relate them to climate variability. Specifically, the chapter aims at (i) analyzing the long-term (1992–2016) water level fluctuations through 62 virtual altimetry-derived tide gauges along the Nile River, (ii) deriving surface water storage from level variations in (i) above, and (iii), studying compartmental water storage changes separately; surface, soil moisture, and groundwater, and their association with El Niño Southern Oscillation (ENSO) and Indian Ocean Dipole (IOD) climate variability.

To provide the 62 virtual stations over the entire Nile River, TOPEX/Poseidon (T/P), Jason-1 and -2 satellite altimetry products (see Sect. 3.5) are applied to the Nile Basin divided into Lower Nile, Central Nile, and Upper Nile (see Fig. 8.2). The obtained water level fluctuations from these 62 virtual stations are improved using the Extremum Retracking (ExtR) algorithm [56, 61] and used to generate time series that are employed to derive surface water storage following the approach proposed by [46]. Furthermore, multiple models are used to estimate soil moisture variations, while their uncertainty is estimated using the Triple Collocation Analysis TCA [52] within the basin. These data together with TWS changes from GRACE are applied to estimate Nile River Basin's groundwater storage. The impact of climate variability (e.g., ENSO and IOD phenomena) and precipitation from the Tropical Rainfall Measuring Mission Project (TRMM) on surface water variations and water storage components are thereafter explored.

The reminder of the chapter is organized as follow; the Nile River Basin's climatic condition, datasets, and methods are presented in Sects. 8.3 and 8.4, respectively. The results are discussed in Sect. 8.5 and the study is concluded in Sect. 8.6.

8.3 NRB's Analysis Products and Approaches Adopted

8.3.1 Climatic Conditions

As already mentioned in Chap. 2, the climatic conditions in the NRB vary over different parts (see Fig. 2.3) and include different climate zones (e.g., Mediterranean climate), with an average temperature of about 30°C in summers and ranging between 5°C and 10°C during winters [40]. The arid region starts from Sudan and extends north to Egypt with average precipitation rates of 50 and 20 mm per year, respectively, representing almost rainless conditions during a given year [2, 40]. In contrast, the southern parts of the basin from the equatorial region of south-western Sudan to most of the Lake Victoria basin and the Ethiopian Highlands experience heavy rainfall of about 1520 mm per year [14, 29]. A modest increase in rainfall and stored water over Lake Victoria from 2007 to 2013 after the 2002–2006 decline [9] is captured by [11] while water loss in the north-eastern lowland of Ethiopia between 2003 and 2011, in contrast to the western parts, has been observed by [12, 13]. Becker et al., [18] studied the 2003–2008 water level changes in major lakes of East Africa and concluded that for Lake Victoria and Lake Turkana, changes were mainly due to individual lake's storages.

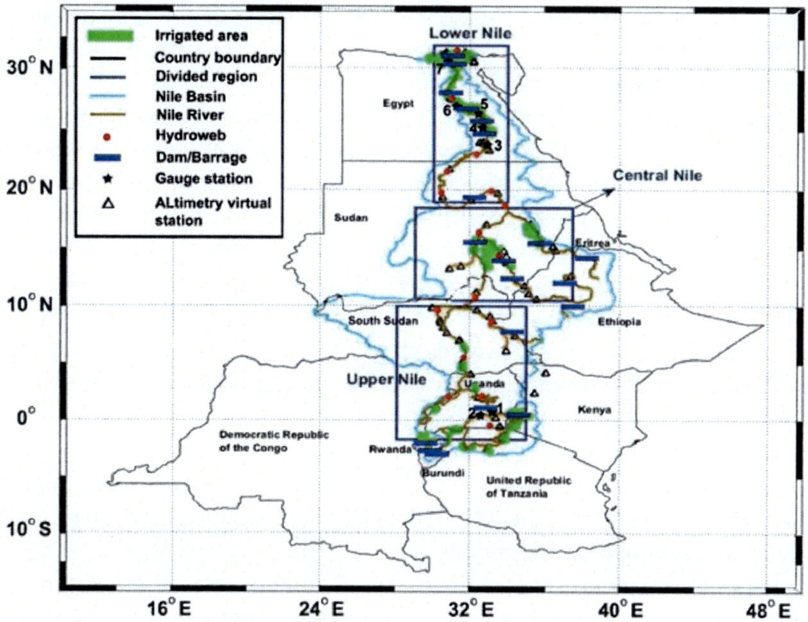

Fig. 8.2 Location of the 62 virtual altimetry stations (black triangles) and the three study regions (Green-Upper Nile (Kenya, Uganda, Tanzania, Rwanda, Burundi, Ethiopia and South Sudan); Blue-Central Nile (Sudan, Ethiopia and Eritirea); and Yellow-Lower Nile (Egypt)) in the Nile River Basin. The black stars show the positions of the gauge stations for measuring water levels while the red circles represent the locations of Hydroweb [30] virtual stations. *Source* [65]

The difference between climatic conditions and water availability along the Nile has become increasingly important especially for the northern areas facing increased water scarcity, i.e., Sudan and Egypt see, e.g., [32, 39, 98]. A number of studies have investigated the interactions between different areas along the NRB and various issues, e.g., sediments [4] and residents income inequality [5]. To better study the entire basin in regard to fluctuations and the impacts of climate, the present chapter divides the entire NRB into three different regions; the Upper Nile, Central Nile, and Lower Nile (Fig. 8.2) approximated according to the provenance of the water as suggested in [3].

8.3.2 Satellite Radar Altimetry

Satellite radar altimetry discussed in Sect. 3.5 is an effective tool for monitoring surface water level fluctuations and has been employed for a wide range of applications e.g., [18, 48, 54, 57, 68, 89]. Application of altimetry [89] for monitoring inland water lakes, see e.g., [22] and rivers e.g., [20, 24, 101, 105] is growing because of its vast coverage contrary to ground-based measurements [28]. In this Chapter, we use TOPEX/Poseidon (T/P), Jason-1, and Jason-2 data of the Sensor Geographic Data Records (SGDR), which contains 20-Hz waveform data. This includes 360 cycles of T/P covering 1992–2002, 260 cycles of Jason-1 from 2002 to 2008, and 277 cycles of Jason-2 covering 2008–2016. The temporal resolution of these observations is ∼9.915 days and their ground cross-track resolution is ∼280 km [19]. T/P and Jason-1 data are both derived from the Physical Oceanography Distributed Active Archive Center (PO.DAAC) and Jason-2 data is provided by AVISO (see, e.g., Table 8.1). Here, we apply geophysical corrections are applied, including solid earth tide, pole tide, and dry tropospheric [22]. Importantly, the waveform retracking, which refers to the re-analysis of the waveforms, a time-series of returned power in the satellite antenna [34, 51], is required to improve the accuracy of measured ranges [27]. In order to retrack satellite radar altimetry data, a developed Extrema Retracking (ExtR) algorithm proposed by [56] is applied. The motivation to select the ExtR is due to its processing speed and its promising results while various types of waveforms are observed, see e.g., [57, 62]. ExtR has already been compared to the Off Center of Gravity OCOG [103], the NASA β-Parameter Retracking [70], and Threshold Retracking [35] techniques.

The datasets are then used to build virtual time series for 62 different points (see black triangles in Fig. 8.2) located on the satellite ground tracks and distributed throughout the NRB. At each virtual point, several points belonging to the same satellite cycle are considered and the median value of the retracked altimetry-based water level is computed in order to address the hooking effects [45]. This effect is derived from off-nadir measurements when a satellite locks over a water body before or after passing above it [25, 92]. Afterward, the water level variations time series derived from T/P, Jason-1, and Jason-2 (covering the period from 1992 to 2016) are merged and the combined time series converted into a monthly scale. Note that the maximum number of stations available from satellite radar altimetry over the area is used to better represent the water level variations over the NRB. Nevertheless, when comparing with other datasets (e.g., precipitation), provided at $1° \times 1°$ spatial resolution, altimetry-derived data from stations located in a grid point (i.e., $1° \times 1°$) are averaged. The average of water level data in the grid point is then compared with other data from the same point. Details of altimetry data sources and pass numbers used in this study are presented in Table 8.1.

8.3.3 GRACE

Monthly GRACE level 2 (L2) potential coefficients products up to degree and order (d/o) 90 from the ITSG-Grace2014 gravity field model [72] are obtained and used to generate total water storage (TWS) within the NRB. Following [96], degree 1 coefficients[2] are replaced to account for the movement of the Earth's centre of mass.

[2]http://grace.jpl.nasa.gov/data/get-data/geocenter/.

Degree 2 and order 0 (C_{20}) coefficients[3] are not well determined, and therefore, they are replaced by those from [31]. Colored/correlated noise and leakage errors are reduced using the Kernel Fourier Integration (KeFIn) filter [63] in a two-step post-processing scheme. The first step accounts for the measurement noise and the aliasing of unmodelled high-frequency mass variations, and the second step reduces the leakage errors, see also [64].

8.3.4 Land Hydrological Models

Soil moisture data is provided from three sources; the Famine Early Warning Systems Network (FEWS NET) Land Data Assimilation System, FLDAS-NOAH; [73], the WaterGAP Global Hydrology Model WGHM; see more details in [37, 38], and ERA-Interim [36]. Soil moisture outputs from these models are acquired on a monthly scale and rescaled into $1^\circ \times 1^\circ$ spatial grid and merged into a single soil moisture product (using the Triple Collocation Analysis (TCA), see Sect. 8.4.3) to study the soil moisture variations within the NRB, as well as extracting groundwater from GRACE's TWS and surface water storage estimates (see Sect. 8.4.1).

8.3.5 Precipitation

The Tropical Rainfall Measuring Mission (TRMM-3B43; version 7) products [100] covering the period 1998 to 2016 are used to estimate how much water enters the NRB. The data is available on a 0.25° degree resolution, which is averaged to generate $1^\circ \times 1^\circ$ grids before being extended to 1992 (same starting period as the altimetry data) using the Global Precipitation Climatology Center (GPCC) reanalysis version 7.0 [90].

8.3.5.1 ENSO and IOD

Two major climate variability indices associated with the dominant SST variability; El Niño Southern Oscillation (ENSO) [17] and Indian Ocean Dipole (IOD) [87] are used to assess the association of climate variability and NRB's stored water changes. ENSO, provided by the NOAA National Centers for Environmental Information (NCEI) between 1992 and 2016, is the largest inter-annual climate variability phenomenon in the Tropical Pacific, which affects the climate of many regions of the Earth [44, 99]. El Niño refers to the positive phase of ENSO that brings warm water towards the east of the Americas causing a climate shift over the Pacific. The opposite phase La Niña causes less than normal precipitation variability [78] in the western

[3]http://grace.jpl.nasa.gov/data/get-data/oblateness/.

Pacific, and to the north of Australia. An ocean-atmosphere phenomenon measure that indicates changes in sea surface temperature in the Indian Ocean is IOD (Indian Ocean Dipole). Its data is acquired from NASA's Global Change Master Directory (GCMD). A positive IOD (often associated with El Niño) causes cooler waters (and droughts) near Australia and Southeast Asia and brings warmer than normal water and heavy rains in East Africa and India. A negative IOD, on the other hand, brings the opposite conditions, e.g., larger precipitation in the eastern Indian Ocean, and cooler conditions in the west. These indices are the result of interactions between the oceans and atmosphere on each corresponding area and their impact can be seen directly on rainfall that occurs around the world [80]. Here, the interest is to understand their influences on the Nile River fluctuations, thereby, further indicating the impact of climate variability. A summary of the datasets used in the present study are presented in Table 8.1.

8.4 Extrema Retracking (ExtR) and the Validation

In order to retrack satellite radar altimetry data over the NRB, the Extrema Retracking (ExtR) post-processing technique of [56] is employed. It is applied to the altimetry-derived waveforms to retrack datasets, what is vital for inland applications of satellite radar altimetry. The algorithm operates in three steps (1) a moving average filter is applied to reduce the random noise of the waveforms, (2) extremum points of the filtered waveforms are identified, and (3), the leading edges among all detected extremum points are explored. Range corrections are applied using the offsets between the positions of the leading edges and their on-board values. To assess the performance of the ExtR filter, its results are evaluated against those of in-situ (see Fig. 8.2) height variations. To this end, use is made of (i) monthly water level measurements from two gauge stations (Jinja 1992–1995 and Entebbe 1992–2009) obtained from the Ministry of Energy & Mineral Development (Kampala, Uganda), (ii) in-situ data obtained from [55]. These are; old Aswan 1996–2009, Esna Barrage 1996–2009, Naga Hammadi Barrage 1996–2007, and Assiut Barrage 1996–2009, and (iii), Nubaria (1997–2007) in-situ data obtained from [88]. The Root-Mean-Square Errors (RMSE) and the correlation coefficient between the variations of altimetry-derived height time series (with and without the application of the ExtR) at the closest virtual stations to the gauge locations and in-situ time series measurements are presented in Table 8.2.

The Results indicate that applying retracking method increases the correlation coefficients between altimetry results and the gauge levels (0.33 on average) and improves the RMSE by 37.56% (on average). Due to a limited number of validating gauge stations in the area, water level time series from the Hydroweb project by LEGOS (Laboratoire d'Etude en Geophysique et Oceanographie Spatiale) [30] and DAHITI (Database for Hydrological Time Series of Inland Waters) [91] were further used. Figure 8.3 shows a sample time series over Lake Victoria within the Upper Nile derived from the ExtR filter compared to the Hydroweb and DAHITI time series. It

Table 8.1 A summary of the datasets used in this chapter. *Source* [65]

Description	Source	Data resolution		Detail	Data access
		Spatial	Temporal		
Altimetry-derived level height	T/P, Jason-1	~280 km	~9.915	Pass numbers 18, 31, 44, 57, 83, 94, 107, 120, 133, 159, 170, 196, 209, 222, 235	http://podaac.jpl.nasa.gov
	Jason-2	~280 km	~9.915	Pass numbers similar to T/P, Jason-1	http://avisoftp.cnes.fr/
Precipitation	GPCC	1°	Monthly	Global precipitation climatology center (GPCC) reanalysis version 7.0	https://www.esrl.noaa.gov/psd/data/gridded/data.gpcc.html#detail
	TRMM-3B43	0.25°	Monthly	Tropical Rainfall Measuring Mission Project (TRMM) version 7.0	https://disc.gsfc.nasa.gov/datacollection/TRMM_3B43_7.html
Terrestrial water storage (TWS)	GRACE	~300 km	Monthly		https://www.tugraz.at/institute/ifg/downloads/gravity-field-models/itsg-grace2014/
Soil moisture	WGHM	0.5°	Monthly		https://www.uni-frankfurt.de/45218093/Global_Water_Modeling
Soil moisture	ERA-Interim	0.5°	Monthly		https://www.ecmwf.int/en/forecasts/datasets/reanalysis-datasets/era-interim
Soil moisture	FLDAS	0.5°	Monthly	Famine Early Warning Systems Network (FEWS NET) Land Data Assimilation System (first soil moisture layer)	https://disc.sci.gsfc.nasa.gov/uui/datasets/FLDAS_VIC025_C_SA_M_V001/summary?keywords=FLDAS
ENSO		~280 km	Monthly		https://www.ncdc.noaa.gov/teleconnections/enso/
IOD		~280 km	Monthly		http://gcmd.nasa.gov/records/GCMD_Indian_Ocean_Dipole.html
Water level measurements		–	–	Ministry of Energy & Mineral Development Kampala, Uganda	http://www.energyandminerals.go.ug/

Table 8.2 A comparison between satellite altimetry values derived from the ExtR retracking method and those from in-situ water level measurements. The improvements in the RMSE are calculated using the in-situ measurements in comparison to the raw altimetry data. *Source* [65]

	Raw altimetry data		ExtR retracking		
Stations	RMSE (cm)	Correlation coefficient	RMSE (cm)	Correlation coefficient	Improvement (%)
(1) Jinja	48.49	0.52	31.45	0.78	35.14
(2) Entebbe	61.14	0.66	36.04	0.95	41.05
(3) Old Aswan	36.77	0.54	22.44	0.88	38.97
(4) Esna	27.27	0.52	15.36	0.91	43.67
(5) Naga Hammadi	38.73	0.61	28.12	0.82	27.40
(6) Assiut	42.62	0.57	25.89	0.93	39.26
(7) Nubaria	25.87	0.62	16.19	0.92	37.44
(8) Hydroweb	56.13	0.59	29.28	0.94	47.83

can be seen from the figure that the ExtR time series are close to the retracked time series of DAHITI (i.e., 0.94 average correlation coefficient) and to a lesser degree to Hydroweb (i.e., 0.92 on average). Overall, the correlation coefficients from both Hydroweb and DAHITI are high (i.e., >0.90) and are statistically significant at 95% confidence level thus indicating a good performance of ExtR. More virtual stations are provided by Hydroweb along the Nile River (see Fig. 8.2), which are used for comparison with the ExtR results. The average estimated RMSE and correlation coefficients are presented in Table 8.2. The ExtR results are 37.44% (on average at 95% confidence level) more correlated to Hydroweb data. Based on these in-situ validations, the ExtR algorithm is further justified and thus employed in this chapter to retrack satellite radar altimetry data.

8.4.1 Water Storage Changes

Assuming the contribution of ice/snow and vegetation to be negligible over the NRB, changes in TWS (ΔTWS) can be sufficiently taken to be the summation of changes in surface water (ΔSr), soil moisture (ΔSm), and groundwater (ΔGr) storages see, e.g., [60, 64] as,

$$\Delta TWS = \Delta Sr + \Delta Sm + \Delta Gr. \tag{8.1}$$

GRACE products provide the total water storage changes ΔTWS while ΔSr are calculated from water level measurements as discussed in Sect. 8.4.2. The changes in soil moisture ΔSm are estimated from multi-models' outputs as discussed in Sect. 8.4.3. Using these estimates of ΔSr and ΔSm in Eq. 8.1, the estimates of groundwater ΔGr within the NRB are derived.

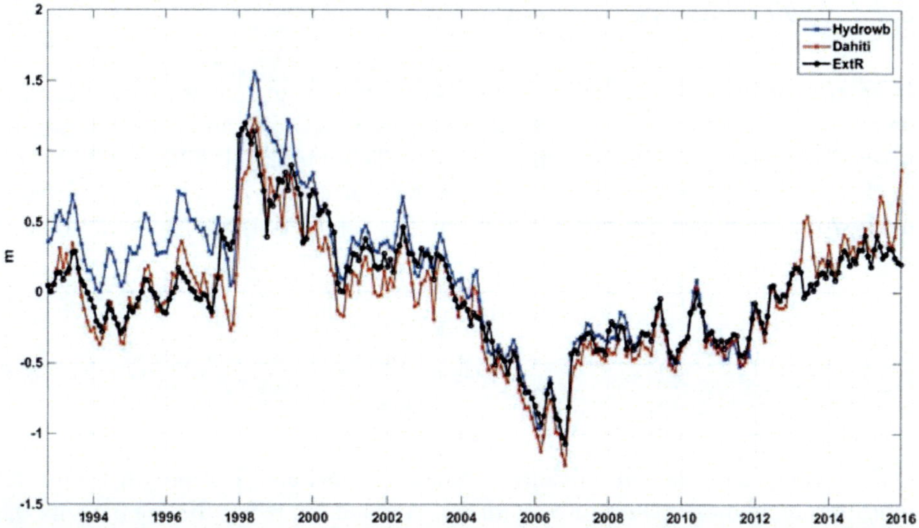

Fig. 8.3 A comparison between height time series of Lake Victoria obtained from the ExtR retracking method (black), DAHITI (red), and HYDROWEB (blue). The results of ExtR retracking method are highly correlated with those of HYDROWEB and DAHITI (i.e., >0.90; significant at 95% confidence level). This justifies the usage of the ExtR retracking method to obtain the surface heights of the 62 virtual stations along the NRB. *Source* [65]

8.4.2 Surface Water Storage Changes

To calculate changes in surface water storage from water level data, the approach proposed in [46] is used. The process begins by generating water level maps at monthly scales using altimetry-derived in-situ and Hydroweb time series across the NRB. These maps are constructed at $1^\circ \times 1^\circ$ (similar to those of GRACE TWS) using point-wise water level time series and a bilinear interpolation scheme to estimate water levels at each grid point. Afterwards, the surface water volume changes (ΔSr) between two consecutive months i and $i-1$ within the basin S, corresponding to the difference of surface water level maps, is estimated by see, e.g., [46, 47],

$$\Delta Sr(i, i-1) = R_e^2 \delta\lambda\delta\theta \sum_{j \in S} P_j \delta h_j(\lambda, \theta, i, i-1) sin(\theta_j), \tag{8.2}$$

where $\delta\lambda$ and $\delta\theta$ are the sampling grid steps in longitude (λ) and latitude (θ) directions, respectively. R_e is the mean radius of the Earth ($\sim 6378\,km$), δh represents the difference of surface water levels, and $R_e^2 sin(\theta_j)\delta\lambda\delta\theta$ corresponds to the surface of grid element j. The percentage of inundation P_j is acquired from Multisatellite Inundation Data Set approach [85, 86].

8.4.3 Soil Moisture Changes

In order to achieve more reliable estimates of soil moisture changes over the NRB, data from three different sources (FLDAS-NOAH, WGHM, and ERA-Interim) are merged using the Triple Collocation Analysis (TCA) [14, 52] following [94]. TCA is chosen since in the absence of ground reference data, it offers an alternative method for estimating random error variances [52]. TCA is applied here to merge soil moisture outputs;

$$S_1 = \alpha_1 S_t + e_1, \tag{8.3}$$

$$S_2 = \alpha_2 S_t + e_2, \tag{8.4}$$

$$S_3 = \alpha_3 S_t + e_3, \tag{8.5}$$

with S_t being the true soil moisture variation, S_1, S_2, and S_3 represents three soil moisture anomalies related to S_t with α_1, α_2, and α_3 being the coefficients that correspond to the errors of e_1, e_2, and e_3, respectively. The objective is to estimate error variances associated with e_1, e_2, and e_3 to be used in the weighting process. On the one hand, TCA solves this by considering the errors of the products to be independent of each other while on the other hand, it arbitrarily assumes any of the products as a reference. For more details regarding TCA implementation, see [94, 106]. By selecting any of the products as the reference, no impact is imposed on the merged time series [52]. Once the error variances are calculated, they are used in Eqs. 8.6–8.8 to estimate weights of each merged product through

$$w_1 = \frac{\sigma {S_2}^2 \sigma {S_3}^2}{\sigma {S_1}^2 \sigma {S_2}^2 + \sigma {S_1}^2 \sigma {S_3}^2 + \sigma {S_2}^2 \sigma {S_3}^2}, \tag{8.6}$$

$$w_2 = \frac{\sigma {S_1}^2 \sigma {S_3}^2}{\sigma {S_1}^2 \sigma {S_2}^2 + \sigma {S_1}^2 \sigma {S_3}^2 + \sigma {S_2}^2 \sigma {S_3}^2}, \tag{8.7}$$

$$w_3 = \frac{\sigma {S_1}^2 \sigma {S_2}^2}{\sigma {S_1}^2 \sigma {S_2}^2 + \sigma {S_1}^2 \sigma {S_3}^2 + \sigma {S_2}^2 \sigma {S_3}^2}, \tag{8.8}$$

where $\sigma {S_1}^2$, $\sigma {S_2}^2$, and $\sigma {S_3}^2$ are error variances of S_1, S_2, and S_3, respectively, with the corresponding weights of w_1, w_2, and w_3. The final merged soil moisture estimate (Sm) is obtained by,

$$Sm = w_1 S_1 + w_2 S_2 + w_3 S_3. \tag{8.9}$$

A schematic illustration of the applied processing steps in this chapter, i.e., data integration procedure, retracking, and water storage estimations, is provided in Fig. 8.4.

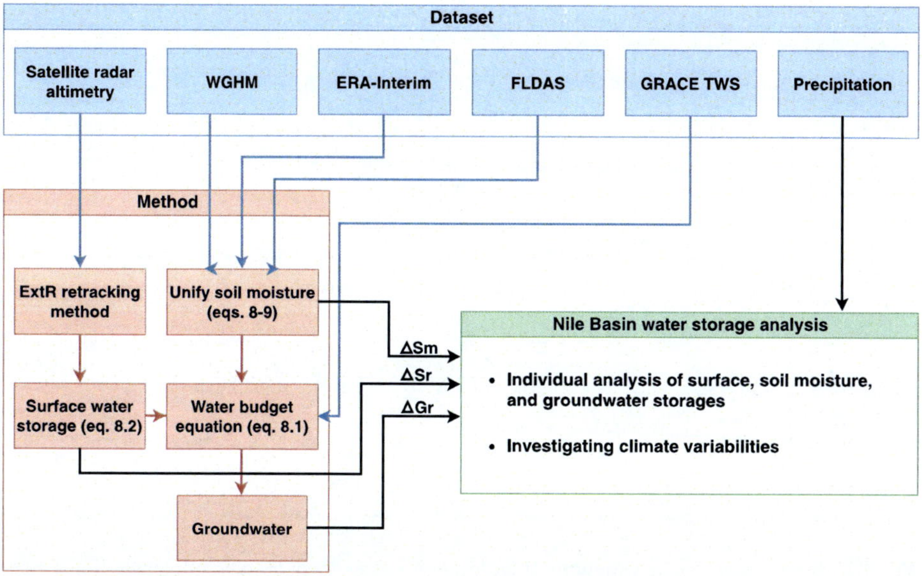

Fig. 8.4 A schematic illustration of the applied methodology. The figure shows how the various stored water compartments; surface water, soil moisture and groundwater of the NRB are generated from multi-satellite and multi-models. *Source* [65]

8.5 Compartmental Change Analysis

8.5.1 River Height Variations

First, a review of the altimetry-derived river level height time series (Fig. 8.5) is undertaken to understand the river's fluctuations with time. Thus, water lever time series is calculated for each virtual station within the three study regions of Fig. 8.2. The average of the stations' level height variations after removing the long-term mean in each region is presented in Fig. 8.5. Also, the trend for different parts of time series are plotted (cf. black solid lines in Fig. 8.5). As can be seen from Fig. 8.5, despite some similar behaviours such as positive trends after 2007 for all regions, the river height fluctuations differ from region to region. Between 1992–2002, water levels rose in the Upper and Central Nile regions, while in contrast, over the same period of time, water level fell at a rate of 0.01 m/year in the Lower Nile region. Moreover, unlike the Upper and Central Nile regions, the Lower Nile region experienced water level fall in 2002. The Upper Nile region shows remarkably larger variations possibly due to higher rainfall in the region.

A decrease in the Upper Nile's water levels between 2002 and 2006 is consistent with the findings of [9, 97], where excessive dam construction (e.g., expansion of the Nalubaale Dam to include Kiira Dam) contributed to the fall. A similar negative trend is observed in the Lower Nile (0.76 average correlation coefficient between

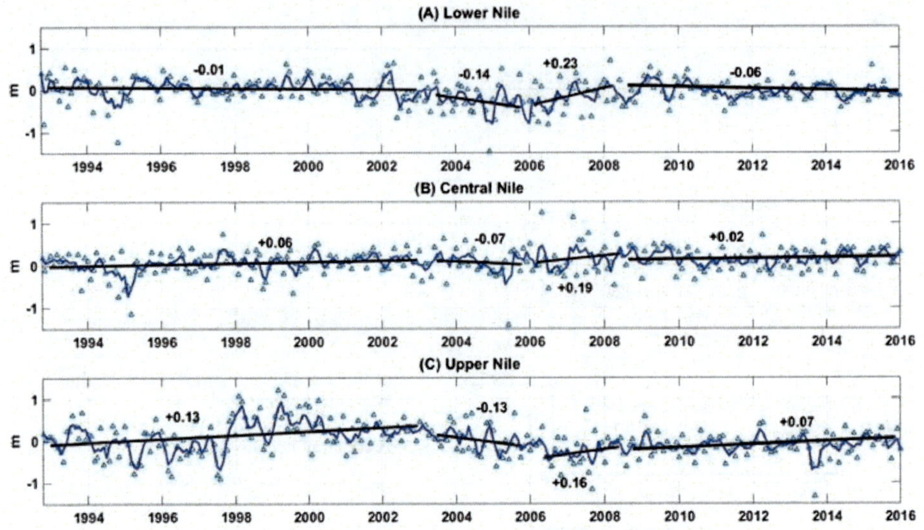

Fig. 8.5 River level height variations (blue triangles) and their 60-day smoothed (for a better presentation) time series (blue lines) for each region. Variation rate (m/year) are reported above the fitted lines (black lines). Note that long-term average height levels are removed from each time series. *Source* [65]

water levels of the Upper and Lower Nile) and Central Nile (0.68 on average). These correlation coefficients depicts the effects of Upper Nile's water management policies (e.g., dam constructions) on the other regions. The small change in the Central Nile during the period 2002–2006 compared to the Upper Nile could indicate other factors (e.g., climatical) since the Blue Nile comes from the Ethiopian Highlands and as such was not impacted by the expansion of the dam in Uganda. This can explain the lower correlation coefficient between water level variations in Upper Nile and the Central Nile in comparison to the Lower Nile. The rate of fall in 2002–2006 in the lower Nile is higher than the upper and Central Nile as a result of the possible combined effects of anthropogenic, see e.g., [13, 95] for example on irrigation (Upper Nile) and climatic (Central Nile) origins.

After 2006, water levels rose in all the three regions until 2008 at different rates (i.e., 0.16 m/year for Upper Nile region, 0.19 m/year for Central Nile and 0.23 m/year for the Lower Nile region). The main reasons for this are positive precipitation rates (cf. Figure. 8.6), and the fact that the excessive exploitation of the White Nile for hydroelectric purposes by Uganda reduced, see, e.g., [9, 97]. The differences between water level rise rates over the three regions could be attributed to climatic impacts. For example, the ENSO rainfall of 2007 [82] brought heavy rainfall to the East African region and caused water level increase in the Nile region. In addition, La Niña impacts, which caused drought in Ethiopia led to a smaller water level increase in the Central Nile compared to the Lower Nile, where the Blue Nile contribution resulted in a larger increase. Same positive trends can also be seen after 2008 over Upper and Central Niles. The positive trend is smaller over Central Nile possibly due to the fact that the White Nile's waters are lost in the Sudd-wetland [12], hence, the diminishing effect of the increase can be seen from the Upper Nile to the Central Nile. Lower Nile, on the other hand, depicts a negative water level variation rate for the same period. The Lower Nile reflects already reduced water level variation rate of the Central Nile plus withdrawal effects for irrigation purposes (cf. Fig. 8.2) for irrigated areas, see also [13, 95]. This could also explain larger negative trends between 2002 and 2006, where water use aggravates the excessive dam construction in Upper and Central Niles (see, e.g., Fig. 8.2). Moreover, as it will be shown in Fig. 8.6, negative precipitation rates for the periods of 1992–2002 and 2002–2006 result in surface water decline over the same periods in Lower Nile.

8.5.2 *Water Storages: Surface, Soil Moisture and Groundwater*

Following [46], surface water storage is derived from water level fluctuations over the NRB. The average surface storage time series for the Upper, Central, and Lower Nile regions are shown in Fig. 8.6. The average time series of precipitation and GRACE TWS are also shown in the figure. It can be seen from the figure that the time series generally follow water level variation patterns of Fig. 8.5. The Upper Nile's time series depicts larger variations with various trends (see also the Lower Nile with a negative trend) unlike the Central Nile. Similar to the water level variation time series in Fig. 8.5, various smaller (short-term) trends can be seen, particularly for the Upper and Lower Nile regions. The negative trend in surface water seen before 2002 in the Lower Nile does not exist in the Upper or Central Nile regions possibly due to extensive usage in activities such as irrigations see, e.g., [95].

Overall, the largest fluctuations are observed in the Upper Nile mainly due to high precipitation. TWS changes naturally follow the precipitation and hence a similar pattern can also be seen. This is followed by the Central Nile, which shows larger variations in both precipitation and TWS time series compared to the Lower Nile (a region with the least precipitation). As with water level variations (cf. Fig. 8.5), negative surface water storage trends exist in all the three regions (see Fig. 8.6) between 2002 and 2006 due to similar reasons discussed in Sect. 8.5.1. This, however, is followed by positive trends in all the regions thanks to the two strong rainfall anomalies in 2006–2007 and 2010–2011 [13]. These trends are also evident in TWS variations over the Central and Upper Nile. Nevertheless, it can be seen that the TWS variation over the Lower Nile is generally negative, which can be attributed to larger water usages in the region as discussed in Sect. 8.5.1. Low precipitation during this period can also be responsible for some parts of this TWS negative changes.

The impacts of precipitation on TWS variations and also surface storage changes can be seen in Fig. 8.6. This is evident from strong rise and fall in both surface water storage and TWS variations' time series followed by rainfall pattern. For example, 1997–1998 El Niño event [69] contributed to more seasonal rainfall over Upper Nile,

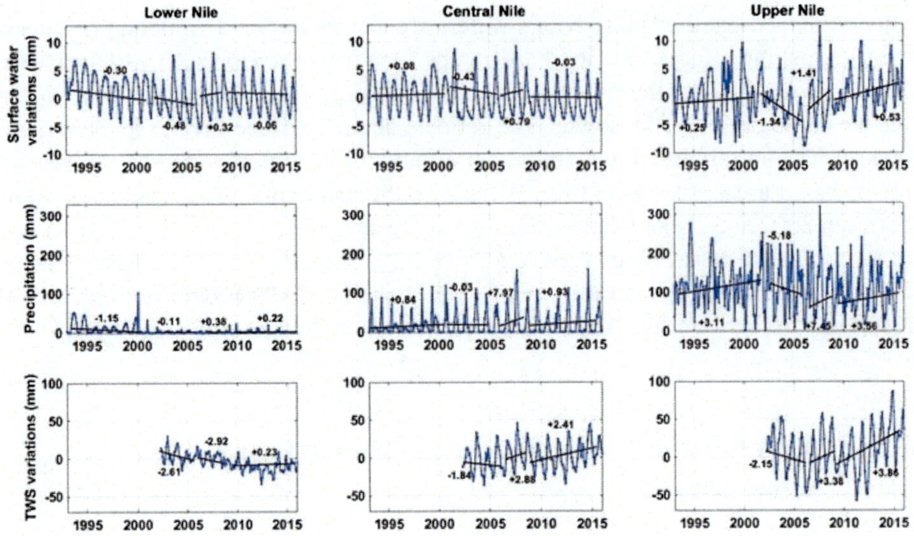

Fig. 8.6 Average surface water storage, precipitation, and TWS variations over the Lower, Central, and Upper Nile regions. The temporal coverage of TWS is limited to 2002 to 2016 subject to the availability of GRACE data. *Source* [65]

and correspondingly surface storage increase as indicated in Fig. 8.6. This, along with positive IOD event, led to highly unstable air and catastrophic floods over Equatorial East Africa [49], which also caused water level increase of Equatorial Lakes, see also [23, 75]. Strong precipitation in 2000 largely affected surface storage, especially over Upper and Central Niles. McSweeney et al., [74] revealed the annual precipitation increases during the 1990s and 2000s, as well as a drought period of 2002–2003. The impact of this drought can clearly be seen in TWS and surface water variations (cf. Fig. 8.6). The 2007 ENSO rainfall [82] has the same impact on both surface water and TWS time series. The 2011 drought over the basin [8, 13] is found to remarkably impacts on TWS and surface water storage. Drought conditions in Central and Lower Niles during 2015 can be attributed to an intense El Niño year suggested by [71].

The results of TCA are presented in Fig. 8.7, which shows the average estimated soil moisture variation from FLDAS-NOAH, WGHM, and ERA-Interim over the Lower, Upper, and Central Nile regions. Furthermore, using the surface water storage, soil moisture, and TWS changes, groundwater changes calculated based on Eq. 8.1 are also presented in Fig. 8.7. Following the patterns of precipitation and TWS time series in Fig. 8.6, smaller soil moisture and groundwater variations exist over the Lower Nile and to a lesser degree over the Central Nile compared to the Upper Nile while no considerable trend is observed in soil moisture variations. It is worth mentioning that seasonal variabilities for Lower Nile time series have considerably lower magnitudes in comparison to other regions. This small variability, which has also recently been shown by [6, 26] can largely be explained by lower precipitation rate (cf. Fig. 8.6) over Lower Nile. Groundwater changes exhibit short-term and long-

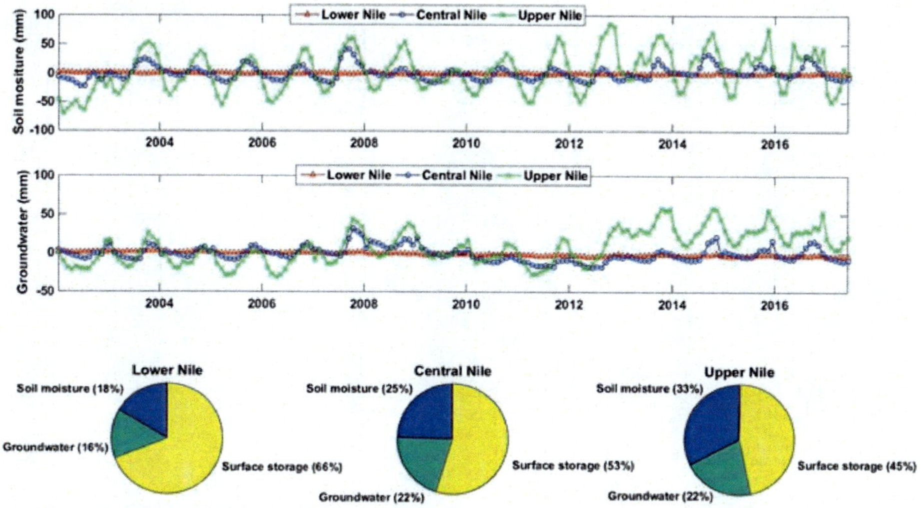

Fig. 8.7 Average soil moisture and groundwater variations over the Lower, Central, and Upper Nile. Larger variations are clearly seen over the Upper Nile and to a lesser degree over the Central Nile. Contrary to the soil moisture time series, trends can be observed for groundwater time series, especially for the Lower Nile. *Source* [65]

term trends over different regions. Negative trends can be seen between 2002 and 2006 generally over the Upper and Central Nile, followed by remarkable increases, likely due to the same reasons explained earlier, see also [9]. More importantly, a negative groundwater trend is dominant over the Lower Nile. This trend exists over the entire study period regardless of precipitation trends, which shows considerable groundwater depletion over the region. The rate of this decline, however, is found to be larger after 2008. As explained, reducing water controls in the Upper Nile after 2006 and the impact of the 2007 ENSO likely caused groundwater increase over the Central and Upper Nile and smaller negative trend over the Lower Nile. Nevertheless, this effect is found to be degraded by 2008 resulting in negative trends (with a higher rate for the Lower Nile) in all groundwater time series between 2008 and 2012. Increasing amount of precipitation after 2012 caused groundwater to rise in both Upper and Central Niles. This high rainfall has the same impact on soil moisture variation between 2012 and 2016.

For each water storage compartment (surface water storage, groundwater, and soil moisture variations), the time series are averaged over each grid point for the period of 2002 to 2016 to generate the spatial pattern maps displayed in Fig. 8.8. It can clearly be seen that the Lower Nile depicts negative variations in the surface storage and groundwater. The Central Nile on the other hand does not show considerable change in most of the cases. Larger variations in terms of amplitude are found in the Upper Nile, thus confirming the previous findings. Besides negative groundwater changes in Egypt indicating huge usage, cf. [13, 95], Sudan and South Sudan also show

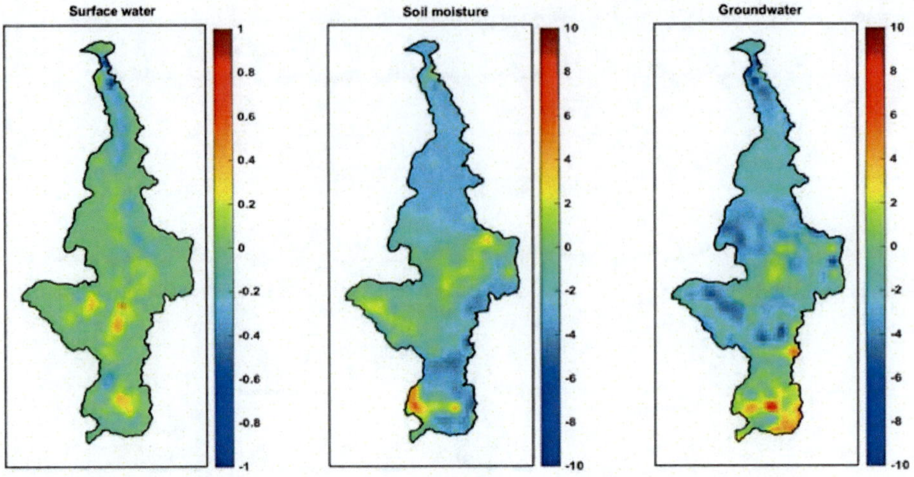

Fig. 8.8 Temporally averaged surface water storage, soil moisture, and groundwater over the Nile Basin for the period 2002–2016. *Source* [65]

considerable decline. In terms of surface water storage, the Upper Nile generally depicts positive variations.

To better compare the water storage changes within the various compartments, Fig. 8.9 show surface storage and groundwater trends neglecting those of soil moisture that indicated no considerable change in Fig. 8.7. Negative trends are observed for both surface water and groundwater storages over the Lower Nile, with the latter being more prominent thus corroborating the conclusions of [13, 95]. No considerable surface storage trend is found for the Central Nile while the Upper Nile depicts positive values. Negative trends can also be seen in some parts of the Central Nile (e.g., in some parts of Sudan and Ethiopia) and in the Lower Nile (mostly in Egypt). In contrast, most parts of the Upper Nile shows positive groundwater trends.

8.5.3 Compartmental Changes: Impacts of Climate Variability

In order to further understand the interactions of precipitation, river level heights, and TWS with climate variabilities, e.g., [21], their correlation coefficients with ENSO and IOD climate variability indices are calculated for each region and presented in Table 8.3. Those of IOD are found to be generally lower than ENSO's. Table 8.3 show the highest correlation coefficient to be between ENSO and precipitation especially for the Upper Nile. For the correlations between ENSO and TWS, the highest value is also achieved in the Upper Nile probably due to the strong connection between precipitation and TWS over the region, see e.g., [12]. For all variables (water level, precipitation, and TWS), the smallest correlation coefficients are achieved in the

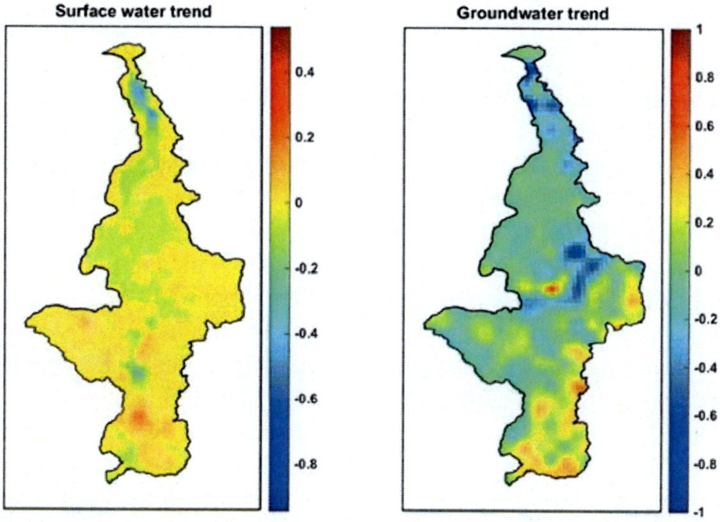

Fig. 8.9 Surface water storage and groundwater trend over the Nile Basin for the period of 2002 to 2016. Trends of soil moisture that indicated no considerable change in Fig. 8.7 are neglected here. *Source* [65]

Table 8.3 Correlation coefficients (at 0.05 significant level) between the river level heights, precipitation, and TWS time series for each region and climate variabilities of ENSO and IOD. Note that cross-correlation is applied to account for lag differences between the time series. *Source* [65]

Climate index	Variable	Lower Nile	Central Nile	Upper Nile
ENSO	Water level	0.47	0.52	0.58
	Precipitation	0.61	0.68	0.72
	TWS	0.58	0.64	0.69
IOD	Water Level	0.41	0.44	0.45
	Precipitation	0.52	0.54	0.59
	TWS	0.43	0.48	0.55

Lower Nile, a factor which can be attributed to very limited precipitation on the one hand, and high human impacts on water storages on the other hand e.g., [95]. Comparing ENSO's correlation coefficients between water levels and TWS, larger values are obtained with TWS. This can be explained by the fact that contrary to the water level fluctuations, non-climatic impacts on TWS are smaller. Considering the entire NRB, a higher correlation coefficient between the river level heights and climate indexes are found in the Central Nile (0.52 for ENSO). The addition of the Blue Nile in this area could possibly explain this higher correlation coefficient in comparison to the other regions.

8.6 Concluding Remarks

This chapter analyzed the Nile River water level fluctuations and total water storage (TWS) compartments (surface water, soil moisture, and groundwater storage) using multi-mission satellite products, as well as land surface models. The association between these variables and climate variabilities are also investigated using precipitation and ENSO/IOD time series (see also [82]). This is crucial for water management policies of the Nile Basin Authority that manages the water resources on behalf of the eleven countries whose livelihood is highly dependent on the Nile for various aspects, e.g., water supply, agriculture, and industry, e.g., [33]. Importantly, it is shown that water usages over the Upper Nile can largely impact the Nile's water variations in Central Nile and particularly Lower Nile, as seen e.g., between 2002–2006 [9]. Excessive dam and barrage constructions, e.g., by Uganda, Rwanda, Kenya, and Ethiopia should be controlled due to their immediate influence on water availability on other parts, e.g., Egypt, which relies on the Nile. It was also found that climatic impacts can also be very important over the entire basin, such as frequent droughts over Central and Lower Nile, and flood events over the Upper Nile. The following summarizes the major outcomes of the study.

- A considerable long-term (2002–2016) negative groundwater trend is found in the Lower Nile (Egypt) signifying a potential depletion. The rate of decline is seen to increase rapidly from 2008 despite increase in precipitation and TWS time series, thus signifying the possibility of human usage, e.g., for irrigation purposes. Smaller soil moisture and groundwater variations exist over the Lower Nile and to a lesser degree over the Central Nile compared to the Upper Nile. While no considerable trend is observed for soil moisture variations, groundwater changes exhibit short-term and long-term trends over different regions. Negative trends are found between 2002 and 2006 over the Upper and Central Nile.
- The Upper Nile, the headwaters of the White Nile, depicts large water level variations compared to the Central (region covering the Blue Nile) and the Lower Nile (Egypt and Sudan). In general, a negative trend is found for water level variation in the Lower Nile (with the highest for the period 2002 – 2006) in contrast to the Central and Upper Nile.
- Stronger correlation between the river level variations and precipitation exist in the Central Nile compared to the Upper and Lower Nile regions over the study period. The contribution of the Blue Nile (originating from the Ethiopian Highlands) appears to cushion the Central Nile.
- Larger precipitation and TWS variations exist in the Upper Nile and to a lesser degree over the Central Nile, which can explain larger water storage fluctuation in these regions compared to the Lower Nile. Contrary to the Upper and Central Niles, negative trends are found for TWS variations over the Lower Nile.
- In addition to the trends, several strong impacts of precipitation, e.g., the 2007 ENSO rains, are also observed leading to strong rise and fall in both surface water storage and TWS variations time series.

- Large correlation coefficient between the precipitation and ENSO is found with an average of 0.67 indicating that precipitation and correspondingly surface water in the Nile Basin follows the global climate variability. ENSO had the highest correlation coefficients with the three variables over the Central and Upper Nile in comparison to IOD.

References

1. Aboulela H (2012) Contribution of satellite altimetry data in the environmental geophysical investigation of the Northern Egyptian Continental Margin. Int J Geosci 3(3):431–442. https://doi.org/10.4236/ijg.2012.33047
2. Agrawal AG, Petersen LR (2003) Human immunoglobulin as a treatment for West Nile virus infection. J Infect Dis 188:1–4. https://doi.org/10.1086/376871
3. Ahmed AA, El Daw AK (2004) An overview on cooperation of trans-boundary water: case of the Nile River Basin, the 2nd Regional Arab water conference Egypt., Cairo
4. Ahmed AA, Ismail UAE (2008) Sediment in the Nile River System, Consultancy Study requested by UNESCO
5. Ahmed AE, Roghim S, Saleh A (2014) Poverty determinants in South Sudan: the case of renk county. Asian J Agric Ext Econ Sociol 4(2):125–136; Article no. AJAEES.2015.013, ISSN: 2320–7027, https://doi.org/10.9734/AJAEES/2015/12724
6. Ahmed M, Abdelmohsen K (2018) Quantifying modern recharge and depletion rates of the Nubian Aquifer in Egypt. Surv Geophys 39:729. https://doi.org/10.1007/s10712-018-9465-3
7. Alsdorf DE, Rodríguez E, Lettenmaier DP (2007) Measuring surface water from space. Rev Geophys 45:RG2002. https://doi.org/10.1029/2006RG000197
8. AMCEN (2011) Addressing climate change challenges in Africa: a practical guide towards sustainable development. African Ministerial Conference on Environment (AMCEN), Addis Ababa
9. Awange JL, Sharifi MA, Ogonda G et al (2008) The falling lake victoria water level: GRACE, TRIMM and CHAMP satellite analysis of the lake basin. Water Res Manag 22:775. https://doi.org/10.1007/s11269-007-9191-y
10. Awange JL, Forootan E, Kusche J, Kiema JBK, Omondi PA, Heck B, Fleming K, Ohanya SO, Gonçalves RM (2013) Understanding the decline of water storage across the Ramser-Lake Naivasha using satellite-based methods. Adv Water Res 60:7–23. ISSN 0309–1708, https://doi.org/10.1016/j.advwatres.2013.07.002
11. Awange JL, Anyah R, Agola N, Forootan E, Omondi P (2013) Potential impacts of climate and environmental change on the stored water of Lake Victoria Basin and economic implications. Water Res Res 49:8160–8173. https://doi.org/10.1002/2013WR014350
12. Awange JL, Gebremichael M, Forootan E, Wakbulcho G, Anyah R, Ferreira VG, Alemayehu T (2014) Characterization of Ethiopian mega hydrogeological regimes using GRACE, TRMM and GLDAS datasets. Adv Water Res 74:64–78. ISSN 0309–1708, https://doi.org/10.1016/j.advwatres.2014.07.012
13. Awange JL, Forootan E, Kuhn M, Kusche J, Heck B (2014) Water storage changes and climate variability within the Nile Basin between 2002 and 2011. Adv Water Res 73:1–15. https://doi.org/10.1016/j.advwatres.2014.06.010
14. Awange JL, Ferreira VG, Forootan E, Khandu Andam-Akorful SA, Agutu NO, He XF (2016) Uncertainties in remotely sensed precipitation data over Africa. Int J Climatol 36:303–323. https://doi.org/10.1002/joc.4346

15. Awulachew SB, Demissie SS, Hagos F, Erkossa T, Peden D (2012) Water management intervention analysis in the Nile Basin. In: Awulachew SB, Smakhtin V, Molden D, Peden D (eds) The Nile River Basin: water, agriculture, governance and livelihoods. Routledge, Oxon, UK, pp 292–311
16. Ayana EK, Awulachew Bekele S (2008) Comparison of irrigation performance based on management and cropping types. In Awulachew, Seleshi Bekele; Loulseged, Makonnen; Yilma, Aster Denekew (Comps.). Impact of irrigation on poverty and environment in Ethiopia: draft proceedings of the symposium and exhibition, Addis Ababa, Ethiopia, 27–29 (2007) Colombo. Sri Lanka, International Water Management Institute (IWMI), pp 14–26
17. Barnston AG, Livezey RE (1987) Classification, seasonality, and persistence of low-frequency atmospheric circulation patterns. Mon Weather Rev 115:1083–1126
18. Becker M, Lovel W, Cazenave A, Güntner A, Crétaux JF (2010) Recent hydrological behavior of the East African great lakes region inferred from GRACE. Comptes Rendus Geosci Satell Altimetry Rainfall Observ. https://doi.org/10.1016/j.crte.2009.12.010
19. Benada JR (1997) PODAAC Merged GDR (TOPEX/POSEIDON) Generation B User's Handbook, Version 2.0. JPL D-11007. Jet Propulsion Laboratory, California Institute of Technology, Pasadena
20. Berry PAM, Garlick JD, Freeman JA, Mathers EL (2005) Global inland water monitoring from multi-mission altimetery. Geophys Res Lett 32:L16401. https://doi.org/10.1029/2005GL022814
21. Beyene T, Lettenmaier DP, Kabat P (2010) Hydrologic impacts of climate change on the Nile River Basin: implications of the 2007 IPCC scenarios. Clim Change 100(3–4):433–461. https://doi.org/10.1007/s10584-009-9693-0
22. Birkett CM (1995) The contribution of TOPEX/POSEIDON to the global monitoring of climatically sensitive lakes. J Geophys Res 25:25–179
23. Birkett C, Murtugudde R, Allan T (1999) Indian Ocean climate event brings floods to East Africa's lakes and the Sudd Marsh. GRL 26(8):1031–1034
24. Birkett CM, Mertes LAK, Dunne T, Costa MH, Jasinski MJ (2002) Surface water dynamics in the amazon basin: application of satellite radar altimetry. J Geophys Res 107 https://doi.org/10.1029/2001JD000609
25. Boergens E, Dettmering D, Schwatke C, Seitz F (2016) Treating the hooking effect in satellite altimetry data: a case study along the Mekong River and its tributaries. Remote Sens 8:91. https://doi.org/10.3390/rs8020091
26. Bonsor HC, Shamsudduha M, Marchant BP, MacDonald AM, Taylor RG (2018) Seasonal and decadal groundwater changes in African sedimentary aquifers estimated using GRACE products and LSMs. Remote Sens 10:904
27. Brown GS (1977) The average impulse response of a rough surface and its applications. IEEE Trans Antennas Propag 25:67–74
28. Calmant S, Seyler F, Cretaux JF (2008) Monitoring continental surface waters by satellite altimetry. Surv Geophys 29. https://doi.org/10.1007/s10712-008-9051-1
29. Camberlin P (2009) Nile Basin Climates. Dumont, Henri J. The Nile: origin, environments, limnology and human use. Springer, pp 307–333, Monographiae Biologicae
30. Cretaux J-F, Jelinski W, Calmant S, Kouraev A, Vuglinski V, Bergé Nguyen M, Gennero M-C, Nino F, Abarca Del Rio R, Cazenave A, Maisongrande P (2011) SOLS: a Lake database to monitor in Near Real Time water level and storage variations from remote sensing data. J Space Res Adv. https://doi.org/10.1016/j.asr.2011.01.004
31. Cheng MK, Tapley BD (2004) Variations in the Earth's oblateness during the past 28 years. J Geophys Res Solid Earth 109:B09402. https://doi.org/10.1029/2004JB003028
32. Conway D (2002) Extreme rainfall events and lake level changes in East Africa: recent events and historical precedents. In: Odada EO, Olago DO (eds) The East African Great Lakes: limnology, palaeolimnology and biodiversity, vol 12. Advances in Global Change Research. Kluwer, Dordrecht, pp 63–92
33. Conway D (2017) Water resources: future Nile river flows. Nat Clim Change 7(5), 319–320. ISSN 1758–678X

34. Davis CH (1995) Growth of the greenland ice sheet: a performance assessment of altimeter retracking algorithms. IEEE Trans Geo-Sci Remote Sens 33(5):1108–1116. https://doi.org/10.1109/36.469474
35. Davis CH (1997) A robust threshold retracking algorithm for measuring ice-sheet surface elevation change from satellite radar altimeters. IEEE Trans Geosci Remote Sens 35(4). doi: https://doi.org/10.1109/36.602540
36. Dee DP, Uppala SM, Simmons AJ, Berrisford P, Poli P, Kobayashi S, Andrae U, Balmaseda MA, Balsamo G, Bauer P, Bechtold P, Beljaars ACM, van de Berg L, Bidlot J, Bormann N, Delsol C, Dragani R, Fuentes M, Geer AJ, Haimberger L, Healy SB, Hersbach H, Hólm EV, Isaksen L, Kållberg P, Köhler M, Matricardi M, McNally AP, Monge-Sanz BM, Morcrette J-J, Park B-K, Peubey C, de Rosnay P, Tavolato C, Thépaut J-N, Vitart F (2011) The ERA-Interim reanalysis: configuration and performance of the data assimilation system. QJR Meteorol Soc 137:553–597. https://doi.org/10.1002/qj.828
37. Döll P, Kaspar F, Lehner B (2003) A global hydrological model for deriving water availability indicators: model tuning and validation. J Hydrol 207:105–134
38. Döll P, Müller Schmied H, Schuh C, Portmann FT, Eicker A (2014) Global-scale assessment of groundwater depletion and related groundwater abstractions: combining hydrological modeling with information from well observations and GRACE satellites. Water Res Res 50. https://doi.org/10.1002/2014WR015595
39. Elshamy ME, Seierstad IA, Sorteberg A (2009) Impacts of climate change on Blue Nile flows using bias-corrected GCM scenarios. Hydrol Earth Syst Sci 13:551–565. https://doi.org/10.5194/hess-13-551-2009
40. FAO Land and Water Division (1997) Irrigation potential in Africa. A basin approach, Food and Agriculture Organization of the United Nations, Rome (Italy)
41. Forootan E, Kusche J (2012) Separation of global time-variable gravity signals into maximally independent components. J. Geod. 86:477. https://doi.org/10.1007/s00190-011-0532-5
42. Forootan E, Kusche J (2013) Separation of deterministic signals using independent component analysis (ICA). J. Stud. Geophys. Geod. 57:17. https://doi.org/10.1007/s11200-012-0718-1
43. Forootan E, Rietbroek R, Kusche J, Sharifi MA, Awange J, Schmidt M, Omondi P, Famiglietti J (2014) Separation of large scale water storage patterns over Iran using GRACE, altimetry and hydrological data. J Remote Sens Environ 140:580–595. https://doi.org/10.1016/j.rse.2013.09.025
44. Forootan E, Khandu Awange J, Schumacher M, Anyah R, van Dijk A, Kusche J (2016) Quantifying the impacts of ENSO and IOD on rain gauge and remotely sensed precipitation products over Australia. Remote Sens Environ 172:50–66. https://doi.org/10.1016/j.rse.2015.10.027
45. Frappart F, Calmant S, Cauhope M, Seyler F, Cazenave A (2006) Preliminary results of envisat ra-2-derived water levels validation over the amazon basin. Remote Sens Environ 100(2):252–264
46. Frappart F, Papa F, Famiglietti JS, Prigent C, Rossow WB, Seyler F (2008) Interannual variations of river water storage from a multiple satellite approach: A case study for the Rio Negro River basin. J Geophys Res 113:D21104. https://doi.org/10.1029/2007JD009438
47. Frappart F, Papa F, Güntner A, Werth S, da Silva JS, Tomasella J, Seyler F, Prigent C, Rossow WB, Calmant S, Bonnet MP (2011) Satellite-based estimates of groundwater storage variations in large drainage basins with extensive floodplains. Remote Sens Environ 115(6), 1588–1594. ISSN 0034–4257, https://doi.org/10.1016/j.rse.2011.02.003
48. Fu LL, Christensen E, Yamarone CA (1994) TOPEX/POSEIDON mission overview. J Geophys Res 99. https://doi.org/10.1029/94JC01761
49. Goddard L, Graham NE (1999) Importance of the Indian Ocean for simulating rainfall anomalies over eastern and Southern Africa. J Geophys Res 104:19099
50. Goita K, Diepkile AT (2012) Radar altimetry of water level variability in the Inner Delta of Niger River. In: IEEE, International Geoscience and Remote Sensing Symposium (IGARSS), Munich, Germany, pp 5262–5265, https://doi.org/10.1109/IGARSS.2012.6352422

51. Gomez-Enri J, Vignudelli S, Quartly G, Gommenginger C, Benveniste J (2009) Bringing satellite radar altimetry closer to shore. In: Remote sensing, SPIE newsroom. doi: https://doi.org/10.1117/2.1200908.1797
52. Gruber A, Dorigo WA, Crow W, Wagner W (2017) Triple collocation-based merging of satellite soil moisture retrievals. IEEE Trans Geosci Remote Sens 55(12):6780–6792. https://doi.org/10.1109/TGRS.2017.2734070
53. Hassan AA, Jin S (2014) Lake level change and total water discharge in East Africa Rift Valley from satellite-based observations. In: Global and planetary change, vol 117. pp 79–90. ISSN 0921–8181, https://doi.org/10.1016/j.gloplacha.2014.03.005
54. Hwang CW, Kao YH (2010) A preliminary analysis of lake Level and Water Storage Changes over Lakes Baikal and Balkhash from Satellite Altimetry and Gravimetry. Terrestrial, Atmospheric and Oceanic Sciences 22:97–108. https://doi.org/10.3319/TAO.2010.05.19.01(TibXS)
55. Ismail SS, Samuel MG (2011) Response of river Nile dredging on water levels. In: Fifteenth international water technology conference, IWTC-15 2011, Alexandria, Egypt
56. Khaki M, Forootan E, Sharifi MA (2014) Satellite radar altimetry waveform retracking over the Caspian Sea. Int J Remote Sens 35(17):6329–6356. https://doi.org/10.1080/01431161.2014.951741
57. Khaki M, Forootan E, Sharifi MA, Awange J, Kuhn M (2015) Improved gravity anomaly fields from retracked multimission satellite radar altimetry observations over the Persian Gulf and the Caspian Sea. Geophys J Int 202(3):1522–1534. https://doi.org/10.1093/gji/ggv240
58. Khaki M, Hoteit I, Kuhn M, Awange J, Forootan E, van Dijk AIJM, Schumacher M, Pattiaratchi C (2017) Assessing sequential data assimilation techniques for integrating GRACE data into a hydrological model. Adv Water Res. 107. ISSN 301–3160309–1708, https://doi.org/10.1016/j.advwatres.2017.07.001
59. Khaki M, Schumacher M, Forootan J, Kuhn M, Awange E, van Dijk AIJM (2017) Accounting for spatial correlation errors in the assimilation of GRACE into hydrological models through localization. Adv Water Res 1 August 2017. ISSN 0309–1708, doi: https://doi.org/10.1016/j.advwatres.2017.07.024
60. Khaki M, B, Ait-El-Fquih I, Hoteit E, Forootan J, Awange and M, Kuhn (2017) A Two-update ensemble Kalman filter for land hydrological data assimilation with an uncertain constraint. J Hydrol 555:447–462. https://doi.org/10.1016/j.jhydrol.2017.10.032
61. Khaki M, Forootan E, Kuhn M, Awange J, Papa F, Shum CK (2018) A study of Bangladesh's sub-surface water storages using satellite products and data assimilation scheme, accepted in Adv Water Res
62. Khaki M, Forootan E, Kuhn M, Awange J, van Dijk AIJM, Schumacher M, Sharifi MA (2018) Determining water storage depletion within Iran by assimilating GRACE data into the W3RA hydrological model. Adv Water Res 114:1–18
63. Khaki M, Forootan E, Kuhn M, Awange J, Longuevergne L, Wada W (2018) Efficient basin scale filtering of GRACE satellite products. Remote Sens Environ 564:175–190. ISSN 0022-1694, https://doi.org/10.1016/j.rse.2017.10.040
64. Khaki M, Ait-El-Fquih B, Hoteit I, Forootan E, Awange J, Kuhn M (2018) Unsupervised ensemble Kalman filtering with an uncertain constraint for land hydrological data assimilation. J Hydrol 564. ISSN 175–190:0022–1694. https://doi.org/10.1016/j.jhydrol.2018.06.080
65. Khaki M, Awange J, Forootan E, Kuhn M (2018) Understanding the association between climate variability and the Nile's water level fluctuations and water storage changes during 1992–2016. Sci Total Environ 645:1509–1521. https://doi.org/10.1016/j.scitotenv.2018.07.212
66. Khandu Forootan E, Schumacher M, Awange JL, Müller Schmied H (2016) Exploring the influence of precipitation extremes and human water use on total water storage (TWS) changes in the Ganges-Brahmaputra-Meghna River Basin. Water Res Res 52:2240–2258. https://doi.org/10.1002/2015WR018113
67. Kizza M, Rodhe A, Xu CY, Natle KH, Halldian S (2009) Temporal rainfall variability in the Lake Victoria Basin in east Africa during the twentieth century Theor. Appl Climatol 98:119–135

68. Lee H, Shum CK, Yi Y, Ibaraki M, Kim JW, Braun A, Kuo CY, Lu Z (2009) Louisiana wetland water level monitoring using retracked TOPEX/POSEIDON altimetry. Marine Geod 32:284–302. https://doi.org/10.1080/01490410903094767
69. Lyon B, Mason SJ (2007) The 1997–98 summer rainfall season in Southern Africa. Part I: observations. J Climate 20:5134–5148. https://doi.org/10.1175/JCLI4225.1
70. Martin TV, Zwally H, Brenner AC, Bindschadler RA (1983) Analysis and retracking of continental ice sheet radar altimeter waveforms. J Geophys Res 88(C3):1608. https://doi.org/10.1029/JC088iC03p01608
71. Massachusetts Institute of Technology (2017) Climate change predicted to increase Nile flow variability: Climate change could lead to overall increase in river flow, but more droughts and floods, study shows. www.sciencedaily.com Retrieved from 22 June 2018 from www.sciencedaily.com/releases/2017/04/170424141236.htm
72. Mayer-Gürr T, Zehentner N, Klinger B, Kvas A (2014) ITSG-Grace2014: a new GRACE gravity field release computed in Graz. - in: GRACE Science Team Meeting (GSTM), Potsdam am: 29.09.2014
73. McNally A, Arsenault K, Kumar S, Shukla S, Peterson P, Wang S, Funk C, Peters-Lidard CD, Verdin JP (2017) A land data assimilation system for sub-Saharan Africa food and water security applications. Sci Data 4, Article number: 170012. doi: https://doi.org/10.1038/sdata.2017.12
74. McSweeney C, Lizcano G, New M, Lu X (2010) The UNDP climate change country profiles: improving the accessibility of observed and projected climate information for studies of climate change in developing countries. Bull Am Meteorol Soc 91:157–166
75. Mistry VV, Conway D (2003) Remote forcing of East African rainfall and relationships with fluctuations in levels of Lake Victoria. Int J Climatol 23(1):67
76. Muala E, Mohamed YA, Duan Z, van der Zaag P (2014) Estimation of reservoir discharges from Lake Nasser and Roseires Reservoir in the Nile basin using satellite altimetry and imagery data. Remote Sens 6:7522–7545
77. Multsch S, Elshamy ME, Batarseh S, Seid AH, Frede H-G, Breuer L (2017) Improving irrigation efficiency will be insufficient to meet future water demand in the Nile Basin. J Hydrol: Reg Stud 12:315–330. ISSN 2214–5818, doi: https://doi.org/10.1016/j.ejrh.2017.04.007
78. Nazemosadat MJ, Cordery I (2000) On the relationships between ENSO and autumn rainfall in Iran. Int J Climatol 20:47–61
79. Nunzio J (2013) Conflict on the Nile: the future of transboundary water disputes over the world's longest river. Future Directions International, Dalkeith
80. Nurutami MN, Hidayat R (2016) Influences of IOD and ENSO to Indonesian rainfall variability: role of atmosphere-ocean interaction in the indo-pacific sector. Proc Environ Sci 33:196–203 (2016). ISSN 1878–0296, doi: https://doi.org/10.1016/j.proenv.2016.03.070
81. Omondi P, Awange JL, Ogallo LA, Okoola RA, Forootan E (2012) Decadal rainfall variability modes in observed rainfall records over East Africa and their relations to historical sea surface temperature changes. J Hydrol 464–465:140–156
82. Omondi P, Awange JL, Ogallo LA, Ininda J, Forootan E (2013) The influence of low frequency sea surface temperature modes on delineated decadal rainfall zones in Eastern Africa region. Adv Water Res 54:161–180
83. Papa F, Durand F, Rossow WB, Rahman A, Bala S (2010) Satellite altimeter-derived monthly discharge of the Ganga-Brahmaputra River and its seasonal to interannual variations from (1993) to 2008. J Geophys Res 115:C12013. https://doi.org/10.1029/2009JC006075
84. Papa F, Frappart F, Malbeteau Y, Shamsudduha M, Vuruputur V, Sekhar M, Ramillien G, Prigent C, Aires F, Pandey RK, Bala S, Calmant S (2015) Satellite-derived surface and sub-surface water storage in the Ganges-Brahmaputra River Basin. J Hydrol: Reg Stud 4:15–35. Part A. ISSN 2214–5818, doi: https://doi.org/10.1016/j.ejrh.2015.03.004
85. Prigent C, Matthews E, Aires F, Rossow WB (2001) Remote sensing of global wetland dynamics with multiple satellite data sets. Geophys Res Lett 28:4631–4634. https://doi.org/10.1029/2001GL013263

86. Prigent C, Papa F, Aires F, Rossow WB, Matthews E (2007) Global inundation dynamics inferred from multiple satellite observations, 1993–2000. J Geophys Res 112:D12107. https://doi.org/10.1029/2006JD007847
87. Rao SA, Behara SK, Masumoto Y, Yamagata T (2002) Interannual subsurface variability in the tropical Indian Ocean with a special emphasis on the Indian Ocean Dipole. Deep Sea Res Part II 49:1549–1572
88. Samuel MG (2014) Limitations of navigation through Nubaria canal. Egypt. J Adv Res 5:147–155. https://doi.org/10.1016/j.jare.2013.01.006
89. Sandwell DT (1990) Geophysical applications of satellite altimetry. Rev Geophys Suppl 132–137
90. Schneider U, Becker A, Finger P, Meyer-Christoffer A, Rudolf B, Ziese M (2015) GPCC Full Data Reanalysis Version 7.0 at 1.0: Monthly Land-Surface Precipitation from Rain-Gauges built on GTS-based and Historic Data. https://doi.org/10.5676/DWD_GPCC/FD_M_V7_100
91. Schwatke C, Dettmering D, Bosch W, Seitz F (2015) DAHITI - an innovative approach for estimating water level time series over inland waters using multi-mission satellite altimetry. Hydrol Earth Syst Sci 19:4345–4364. https://doi.org/10.5194/hess-19-4345-2015
92. Seyler F, Calmant S, Santos da Silva J, Filizola N, Roux E, Cochonneau G, Vauchel P, Bonnet M-P (2008) Monitoring water level in large trans-boundary ungauged basins with altimetry: the example of ENVISAT over the Amazon basin. J Appl Remote Sens 7150:715017. https://doi.org/10.1117/12.813258
93. Siam MS, Eltahir EAB (2015) Explaining and forecasting interannual variability in the flow of the Nile River. Hydrol Earth Syst Sci 19:1181–1192. https://doi.org/10.5194/hess-19-1181-2015
94. Stoffelen A (1998) Toward the true near-surface wind speed: Error modeling and calibration using triple collocation. J Geophys Res 103(C4):7755–7766
95. Sultan M, Ahmed M, Sturchio N, Eugene Y, Milewski A, Becker R, Wahr J, Becker D, Chouinard K (2013) In: Pielke RA (ed) Assessment of the vulnerabilities of the Nubian sandstone fossil aquifer. North Africa. Understanding and addressing threats to essential resources. Elsevier, Climate Vulnerability, pp 311–333
96. Swenson S, Chambers D, Wahr J (2008) Estimating geocentervariations from a combination of GRACE and ocean model output. J Geophys Res 113:B08410. https://doi.org/10.1029/2007JB005338
97. Swenson S, Wahr J (2009) Monitoring the water balance of Lake Victoria, East Africa, from space. J Hydrol 370(1–4):163–176. https://doi.org/10.1016/j.jhydrol.2009.03.008
98. Taye MT, Ntegeka V, Ogiramoi NP, Willems P (2011) Assessment of climate change impact on hydrological extremes in two source regions of the Nile River Basin. Hydrol Earth Syst Sci 15:209–222. https://doi.org/10.5194/hess-15-209-2011
99. Trenberth KE (1990) Recent observed interdecadal climate changes in the Northern Hemisphere. Bull Am Meteor Soc 71:988–993
100. Tropical Rainfall Measuring Mission (TRMM) (2011) TRMM (TMPA/3B43) Rainfall Estimate L3 1 month 0.25 degree x 0.25 degree V7, Greenbelt, MD, Goddard Earth Sciences Data and Information Services Center (GES DISC), Accessed [Data Access Date]. https://disc.gsfc.nasa.gov/datacollection/TRMM_3B43_7.html
101. Tseng KH, Shum CK, Yi Y, Fok HS, Kuo CY, Lee H, Cheng X, Wang X (2013) Envisat altimetry radar waveform retracking of quasi-specular echoes over the ice-covered Qinghai lake. Terr Atmos Ocean Sci 24:615–627. https://doi.org/10.3319/TAO.2012.12.03.01(TibXS)
102. Uebbing B, Kusche J, Forootan E (2015) Waveform retracking for improving level estimations from Topex/Poseidon, Jason-1 and -2 altimetry observations over African lakes. IEEE Trans Geosci Remote Sens 53(4):2211–2224
103. Wingham DJ, Rapley CG, Griffiths H (1986) New techniques in satellite altimeter tracking systems. In: ESA proceedings of the 1986 international geoscience and remote sensing symposium (IGARSS 86) on remote sensing. Todays Solutions for Tomorrows Information Needs 3: 1339–1344

104. Woodward JC, Macklin MG, Krom MD, Williams MAJ (2007) The Nile: evolution. Quaternary River Environments and material fluxes. Geomorphology and Management. Wiley, Chichester, In Large Rivers, pp 261–292
105. Yang L, Lin M, Liu Q, Pan D (2012) A coastal altimetry retracking strategy based on waveform classification and sub-waveform extraction. Int J Remote Sens 33(24):7806–7819. https://doi.org/10.1080/01431161.2012.701350
106. Yilmaz MT, Crow WT, Anderson MC, Hain C (2012) An objective methodology for merging satellite- and model-based soil moisture products. Water Res Res 48:W11502. https://doi.org/10.1029/2011WR011682
107. Yin X, Nicholson SE (1998) The water balance of Lake Victoria. Hydrol Sci - J Sci Hydrol 43(5):789–811
108. Zakharova E, Kouraev A, Cazenave A (2006) Amazon river discharge estimated from the Topex/Poseidon altimetry. C R Geosci 338:188–196
109. Nile Basin Water Resources Atlas (2017). http://atlas.nilebasin.org/

Chapter 9
Modelling the Nile's Waters: Assimilation

The main objective of assimilating surface water storage derived from satellite radar altimetry into a land surface model is to improve its simulations over the Nile River Basin, a data deficient region. This will be done to explore the impact of altimetry-derived surface water storage for improving various water compartments over the basin.

—Khaki et al. [48]

9.1 Summary

Global hydrological models facilitate studying of water resources and their variations over time. The accuracies of these models are enhanced when combined with ever-increasing satellite remotely sensed data. Traditionally, these combinations are done via data assimilation approach, which permits the use of improved hydrological outputs to study regions with limited in-situ measurements such as the Nile Basin. This chapter,[1] based on the work of [48] aims at using the state-of-art satellite radar altimetry data to enhance a land-based hydrological model for studying water storage changes over the Nile Basin (see Fig. 9.1). Altimetry-derived surface water storage, for the first time, is assimilated into the model using the ensemble Kalman filter (EnKF) for the period of 2003–2016. Multiple datasets from ground measurements, as well as space observations, are used to evaluate the performance of the assimilated satellite altimetry data. Results indicate that the assimilation successfully improves model outputs, especially the surface water component. The process increases the

[1]This is an invited chapter from Dr. Mehdi Khaki, School of Engineering, University of Newcastle, Callaghan, New South Wales, Australia.

J. Awange, *The Nile Waters*,
https://doi.org/10.1007/978-3-030-64756-8_9

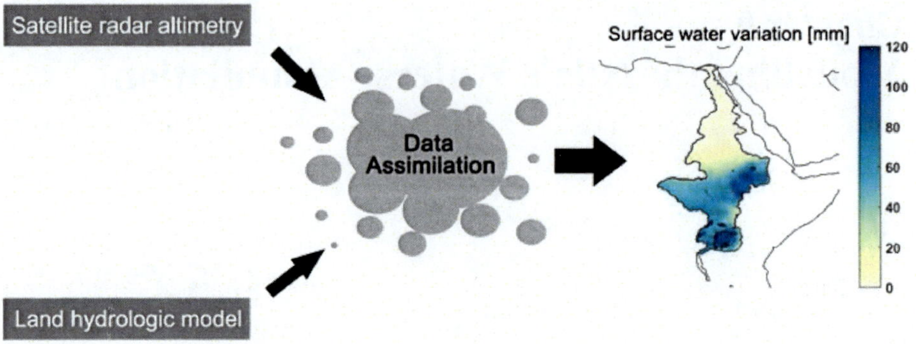

Fig. 9.1 Graphic abstract summarizing the chapters work. *Source* [48]

correlation between surface water storage changes and water level variations from satellite radar altimetry by 0.44 and reduces the surface water discharge root-mean-square errors (RMSE) by approximately 33%.

9.2 Land Surface Model: Assimilating Altimetry Products

The Nile River, one of the world's largest river systems, is a vital freshwater source for 11 Eastern Africa countries (see e.g., Fig. 2.5), supporting the region's various socio-economic activities and correspondingly the livelihood of over 300 million people [6, 8, 11, 27, 31, 57, 70]. It is, therefore, essential to accurately monitor surface and sub-surface water over the entire Nile Basin. This is very important for water management policies of the countries within the basin in order to analyse related climate impacts e.g., [9, 10, 17, 20, 58, 68, 75] and anthropogenic influences e.g., [2, 3, 19] on the basin's water resources. Such monitoring, however, is challenging due to the extent of the Nile Basin and its limited in-situ measurements both spatially and temporally. Land surface models are valuable tools to address these limitations by providing high-resolution information on various water storage compartments such as groundwater, soil moisture, and surface water [36, 73].

There are, however, a variety of factors that could negatively affect the models' performances over the Nile Basin. These include imperfect modelling of physical phenomena, insufficient and erroneous inputs, uncertain parameters, and simplifications applied during modelling, see e.g., [24, 39, 41, 59]. Attempting to address these shortcomings, assimilating satellite remotely sensed products into land surface models has been put forward. This approach has gained a lot of attention in recent years due to the fast development of several satellite missions providing unprecedented global information on various water components e.g., [5, 44, 47, 50, 63]. One important satellite mission over both ocean and land is the satellite radar altimetry, which measures sea surface height (SSH) variations [1, 38]. Altimetry products over land have successfully been used to measure surface variations of inland water bodies e.g., [13, 37] and rivers e.g., [12, 14, 49, 69], where they have been proven to enhance different hydrological models. For example, [29] integrated altimetry data from Environmental Satellite (ENVISAT) into MGB-IPH (Modelo de Grandes Bacias by Instituto de Pesquisas Hidraulicas) model using an optimization technique to improve its parameters over the Branco River Basin (in northern parts of the Amazon Basin). ENVISAT water level data were also used by [54] to enhance the performance of a rainfall-runoff model. The altimetry data over six virtual stations were assimilated into the model over the Brahmaputra River and improved daily discharge outputs. On a different effort, water level data from satellite Synthetic Aperture Radar (SAR) and digital terrain model (DTM) were assimilated into a coupled hydrological and hydrodynamic model to enhance its discharge estimates [55]. Biancamaria et al. [15] assimilated virtual Surface Water and Ocean Topography (SWOT) data to enhance Arctic river modelling and concluded that the approach reduced channel water depth both spatially and temporally, see also [25].

The main objective of the present chapter is to assimilate surface water storage derived from satellite radar altimetry into a land surface model to improve its simulations over the Nile Basin, a data deficient region. This will be done to explore the impact of altimetry-derived surface water storage for improving various water compartments over the basin. In so doing, [48] becomes one of the pioneering works to assimilate altimetry products into a land surface model. Furthermore, for the first time, multi-mission satellite altimetry SSH from TOPEX/Poseidon (T/P), Jason-1, and Jason-2 are used during the assimilation process. The SSH data, converted to surface water storage by [42] is assimilated into the model using a sequential filtering method, namely ensemble Kalman filter (EnKF). EnKF is a relatively cost-effective filter that does not require to record historical observations or minimizing a chosen complex cost function contrary to other techniques such as variational methods [21, 33, 67]. The filter has been widely used in hydrological studies to integrate satellite data products such as those from satellite soil moisture e.g., [22, 60, 77], evapotranspiration e.g., [30, 32, 72], and the Gravity Recovery And Climate Experiment (GRACE) mission, see also [45, 64, 76].

In what follows, brief information about the data and method are described in Sects. 9.3 and 9.4, respectively. Section 9.5 presents and discusses the results and a summary of outcomes is presented in Sect. 9.6.

9.3 Surface Water Storage

9.3.1 Surface Water Storage and Hydrological Model Products

Surface water storage products for data assimilation are derived from [42]. The data include monthly surface water storage changes from 2003 to 2016 on $1^\circ \times 1^\circ$ spatial scale. Multiple satellite radar altimetry are used to generate these products. For this purpose, SSH time series are produced over 62 virtual stations across the Nile from TOPEX/Poseidon (T/P), Jason-1 and -2 satellite altimetry missions. The time series are then improved using a retracking method called Extremum Retracking (ExtR) [37] by addressing the usual altimetry errors over the land, see e.g., [16, 38]. These SSH time series are then converted into surface water storage following [28]. The approach first employs a bilinear interpolation method to estimate grid point water levels and then estimates surface water storage changes between two consecutive months based on the water level changes, grids' geographical locations, and the percentage of inundation, see more details e.g., in [28, 42].

9.3.2 *Hydrological Model*

A global water balance system of the World-Wide Water Resources Assessment (W3RA) model [71] developed by the Commonwealth Scientific and Industrial Research Organisation (CSIRO) is used for the data assimilation experiment. The model simulates various water storage compartments such as surface water, soil moisture, and groundwater on independent grid points. W3RA further represents the interactions between water and energy fluxes, e.g., partitioning precipitation between net precipitation and evaporation, and correspondingly net precipitation between infiltration, excess and saturation excess surface runoffs. Soil moisture is represented at different depths based on the incoming and exiting soil drainage and root water uptake. The model relies on a variety of parameters including water holding capacity that relates greenness and groundwater recession, effective soil parameters, as well as the saturated area to catchment characteristics [71]. The forcing datasets here are acquired from Princeton University meteorological forcing data [62], which includes daily meteorological fields of minimum and maximum temperature, downwelling short-wave radiation, and precipitation.

9.3.3 *Validation Data*

Various datasets are used to assess data assimilation results. This includes both satellite and in-situ measurements. Estimated total water storage (TWS) with and without data assimilation are compared with those from the Gravity Recovery And Climate Experiment (GRACE) satellite mission. GRACE-derived TWS changes are derived from the monthly ITSG-Grace2014 gravity field model [52] up to degree and order 90 level 2 (L2). L2 products are converted to $1^\circ \times 1^\circ$ TWS changes following [74] after replacing low degrees (i.e. C10, C11, S11, and C20) with accurate coefficients [18, 66]. The products are also filtered to reduce colored/correlated noise and leakage errors using the Kernel Fourier Integration (KeFIn) filter [43].

In addition, satellite soil moisture data from the Advanced Microwave Scanning Radiometer for EOS (AMSR-E) and ESA's Soil Moisture Ocean Salinity (SMOS)

Earth Explorer mission are used to validate soil moisture estimates from the model. AMSR-E data with a daily $0.25^\circ \times 0.25^\circ$ spatial resolution is obtained from Level-3 land surface products [56] for the period of 2003 and 2011. SMOS data with temporal and spatial resolutions of three days and about 50 km, respectively, is acquired from Level 3 Centre Aval de Traitement des Donnees SMOS (CATDS) [23] between 2011 and 2016. Both soil moisture data products are rescaled to monthly $1^\circ \times 1^\circ$ for the validation purpose. Since satellite soil moisture data only measures top layer's few centimetres, it is compared with the model's top layer simulations.

In-situ water discharge measurements from multiple stations across the Nile Basin are used between 2003 and 2016 to evaluate the model's discharge outputs. The in-situ data are derived from the Jinja and Entebbe stations in Uganda (Ministry of Energy & Mineral Development), from literature [35, 61], as well as from different sources including Power Planning Associates (PPA, 2007) report, River Watch project (http://floodobservatory.colorado.edu/), as well as the Global Runoff Data Centre (GRDC; http://www.bafg.de/). Discharge time series is produced at each station and compared to that of the model simulation at the closest model grid points.

9.4 Assimilation: Approach and Validation

9.4.1 Approach

Sequential data assimilation scheme using ensemble Kalman filter (EnKF) [26] is applied to constraint the model surface storage (or system state) using the incoming observations from satellite radar altimetry. EnKF is based on the Bayesian filtering procedures to numerically conditions a prior probability density function (PDF) of the state ($\mathbf{x}$) based on observations ($\mathbf{y}$). This is done following two steps; (i) *forecast*, and (ii), *analysis* steps. The forecast step is applied to estimate the state PDF using a dynamical model and the analysis step is undertaken to update the forecast PDF according to the observations using Bayes' rule. These steps are repeated at each assimilation step when new observations are available.

In what follows, a general process for every assimilation step is described.

- *Forecast step.* Starting at time t, the state analysis at $t-1$, i.e. $\{\mathbf{x}_{t-1}^{a,(i)}\}_{i=1}^{n}$ (with n being ensemble number) is integrated with the dynamical model ($\mathcal{M}$) to achieve the current (the time with the next available observation) forecast state $\{\mathbf{x}_t^{f,(i)}\}_{i=1}^{n}$ by,
$$\mathbf{x}_t^{f,(i)} = \mathcal{M}(\mathbf{x}_{t-1}^{a,(i)}) + \nu^{(i)}, \quad i = 1, \ldots, n. \tag{9.1}$$
In Eq. 9.1, $\nu^{(i)}$ is a random sample from the Gaussian $\mathcal{N}(\mathbf{0}, \mathbf{Q}_t)$ with covariance matrix $\mathbf{Q}_t$.
- *Analysis step.* The estimated state forecast is updated using the observation $\mathbf{y}_t$. To this end, one first needs to estimate observation forecast $\{\mathbf{y}_t^{f,(i)}\}_{i=1}^{n}$ via,

$$\mathbf{y}_t^{f,(i)} = \mathbf{H}\mathbf{x}_t^{f,(i)} + \mathbf{w}^{(i)}; \quad i = 1, \cdots, n, \tag{9.2}$$

where $\mathbf{H}$ is the observation transition matrix that translates state vector into observation space. The observation process noise $\mathbf{w}^{(i)}$ is assumed to be Gaussian $\mathcal{N}(\mathbf{0}, \mathbf{R}_t)$ with observation error covariance $\mathbf{R}_t$. Here, the observation error covariance is assumed to be uncorrelated (diagonal) and linearly depend on observations at each grid point, i.e., $\mathbf{R}_t = (\alpha.\mathbf{y}_t)^2$ with $\alpha = 0.1$, e.g., [25, 53].
Next, the analysis step is applied using an EnKF update to obtain $\{\mathbf{x}_t^{a,(i)}\}_{i=1}^n$ following,

$$\mathbf{x}_t^{a,(i)} = \mathbf{x}_t^{f,(i)} + \underbrace{\mathbf{P}_{\mathbf{x}_t^f}\mathbf{H}^T[\mathbf{H}\mathbf{P}_{\mathbf{x}_t^f}\mathbf{H}^T + \mathbf{R}_t]^{-1}}_{\boldsymbol{K}_t^{(i)}}[\mathbf{y}_t - \mathbf{y}_t^{f,(i)}], \quad i = 1, \cdots, n. \tag{9.3}$$

$\boldsymbol{K}_t^{(i)}$ in Eq. 9.3 is called Kalman gain. $\mathbf{P}_{\mathbf{x}_t^f}$ is covariance matrix, which can be calculated from the previous analysis and forecast ensembles as,

$$\mathbf{P}_{\mathbf{x}_t^f} = (n-1)^{-1}\mathbf{S}_{\mathbf{x}_t^f}\mathbf{S}_{\mathbf{x}_t^f}^T, \tag{9.4}$$

with $\mathbf{S}_{\mathbf{x}_t^f}$ representing the perturbation matrix, which can be computed from the corresponding ensemble members minus the ensemble mean. Once the state is updated, it will be used as forecast state in the next assimilation cycle.

9.4.2 Validation

An experiment is designed to test the performance of data assimilation for incorporating altimetry-derived surface storage into the model. The model is run for the period of 2000–2003 to generate $n = 30$ initial ensemble by perturbing forcing products. To this end, 30% Gaussian multiplicative error is added to precipitation, 50 Wm^{-2} additive Gaussian error is applied to the shortwave radiation, and 2 °C Gaussian additive error is added to the temperature term. During the data assimilation process, two standard tuning techniques are implemented to increase the efficiency of EnKF. These include ensemble inflation $[1.1 - 1.3]$ and localization $[2° - 5°]$ radii following [40] to address ensemble inbreeding issue as well as applying the limited ensemble size, see, e.g., [4, 34]. The applied process of data assimilation to improve the model surface water storage using the altimetry-derived data products is presented in Fig. 9.2.

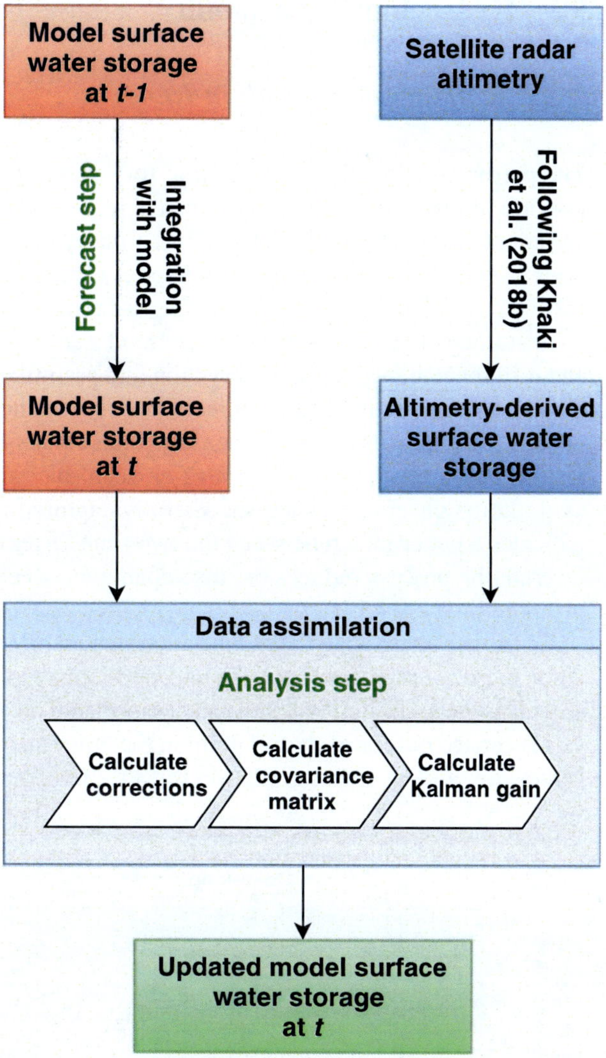

Fig. 9.2 Data assimilation steps for improving the model surface water storage using the altimetry-derived data products. *Source* [48]

9.4.3 Evaluation Metrics

To evaluate the assimilation results, three metrics including the Root-Mean-Squared Errors (RMSE), standard deviation (STD), and correlation are used. The time series from the open-loop run and data assimilation at the closest grid point to the discharge in-situ stations are used to compute the RMSE, STD, as well as correlation values with respect to the in-situ measurements. A significance test for the correlation coefficients using t-distribution is also undertaken. The estimated t-value and the distribution at 95% confidence level confidence level are then used to calculate the p-values for the correlations that are found to be under 5%, indicating that the coefficients are significant.

9.5 Hydroclimate Variability

9.5.1 Surface Assimilation Impacts

The performance of data assimilation for reflecting altimetry-derived surface storage into model simulations is analysed here. For this purpose, surface water storage variations with and without data assimilation are compared with assimilated observations. Figure 9.3a shows average time series of surface water variations over the basin for each case. The error time series, i.e. difference between the observation (derived from altimetry) and the open-loop run (model simulations without data assimilation) and difference between the observation and assimilation results are also depicted in the Fig. 9.3b, showing a better agreement between the assimilation results and observations. This results in smaller errors (blue error bar) compared to the open-loop results (black error bar). It can also be seen that surface water variabilities in the assimilation observations are successfully captured in the assimilation results. This indicates a better match between the two time series as a result of data assimilation. Overall, the process reduces the discrepancy between model and observational data by approximately 45%.

To further investigate the assimilation impact on surface storage estimates, correlation maps are produced between the open-loop results and assimilated observations as well as the assimilation results and assimilated observations. The correlation analysis is carried out for each grid point in the Nile Basins (Fig. 9.4). Higher correlation values are evident for assimilation results. It is found that the proposed approach

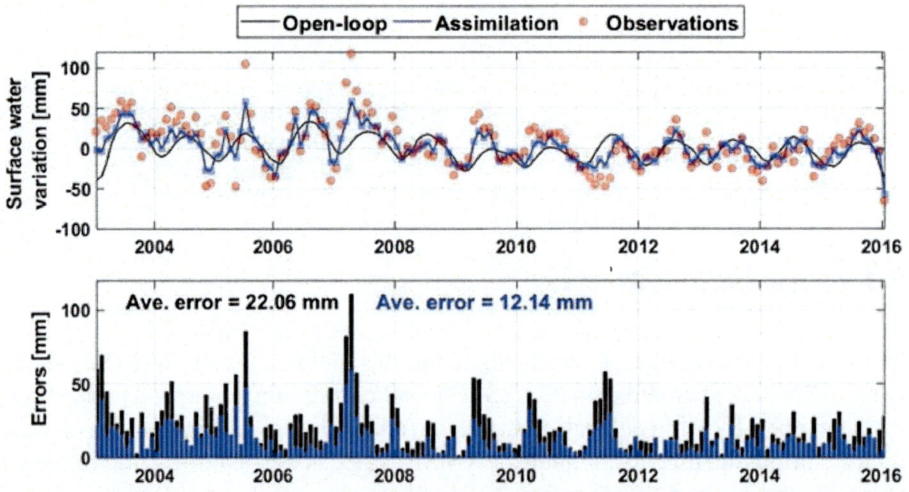

Fig. 9.3 Average surface water variations from the open-loop run (i.e., model simulation without data assimilation), data assimilation, and assimilated observations (**a**). Corresponding error values as different between observations and both the open-loop and assimilation results are presented in **b**. *Source* [48]

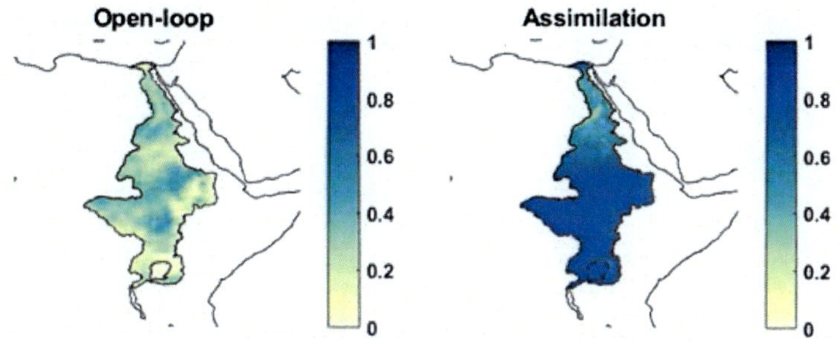

Fig. 9.4 Spatial correlation map between the model outputs (without (**a**) and with assimilation (**b**)) and altimetry-derived surface water observations. *Source* [48]

enhanced correlation from 0.34 (on average) to 0.78 (on average). Located in the southern parts of the basin, a considerable correlation increase can be seen over Lake Victoria. While better agreement can be seen between the observations and the assimilation results over the central to southern parts (Fig. 9.4a), no clear pattern can be observed for the open-loop results. The lower correlation values in the northern parts in the assimilation results (Fig. 9.4b) can be related to larger uncertainties over the areas with the less surface water availability (see Sect. 9.5.3 for more details). Nevertheless, the assimilation process effectively incorporates surface water storage information from observations into the model over the majority of the basin's area.

9.5.2 Validation of the Results

Remotely sensed data from GRACE and satellite soil moisture platforms are used to examine the model's output with and without data assimilation. Figure 9.5 shows the comparison between (i) the estimated surface storage from the model and GRACE observed satellite (Fig. 9.5a), (ii) estimated total water storage (TWS) from the model and those from GRACE (Fig. 9.5b), and (iii), estimated soil moisture from the model and those observed from the satellites (Fig. 9.5c). More details on the statistics are presented in Table 9.1. The comparison analyses are done using correlation analysis between the open-loop and assimilation results on the one hand and those of observations across the Nile Basin on the other hand. Correlation values are calculated for each point between simulations and observations and the results for each case, i.e., open-loop and data assimilation are displayed in Fig. 9.5. From Fig. 9.5, it can be observed that the assimilation results obtain the highest correlation values to satellite observations in all the three tests. This can also be seen from Table 9.1, where the assimilation results outperformed the open-loop results with smaller RMSE and STD values with respect to the satellite data. Importantly, a better agreement can be seen between the surface estimates (Fig. 9.5a) as well as TWS from data assimilation (Fig. 9.5b) and GRACE observations than the open-loop. This indicates the

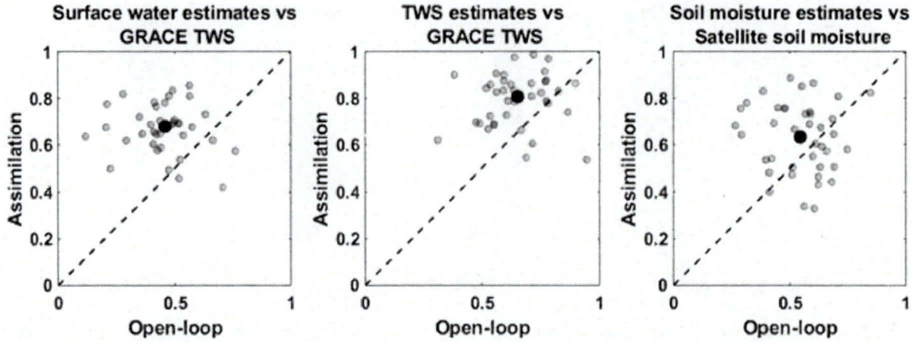

Fig. 9.5 Correlation values between GRACE observed TWS and estimated surface storage (**a**) and estimated TWS (**b**) from the open-loop run and data assimilation. Correlation values for the estimated soil moisture (**c**) are calculated with respect to satellite soil moisture products. *Source* [48]

positive impact of the applied data assimilation approach on the model. The higher correlation values for the TWS comparison analysis in Fig. 9.5b (see also Table 9.1) is largely related to the improved surface water component (in the model's TWS), which is updated during assimilation. It can also be seen that soil moisture estimates from data assimilation are more correlated to satellite soil moisture measurements. Smaller errors can also be found for the soil moisture estimates after data assimilation (cf. Table 9.1). This indicates the positive impact of assimilating surface storage into a non-assimilated variable (i.e., model top layer soil moisture).

Next, model discharge estimates are compared with in-situ measurements. This is particularly important due to the impact of surface water component on water discharge. The time series from the open-loop run and data assimilation at the closest grid point to the in-situ stations are used to compute the RMSE, STD, as well as correlation values with respect to the in-situ measurements. Table 9.2 summarises the statistics from the average results. The results indicate the superiority of data assimilation, i.e., the method reduces average RMSE and STD by 33% and 26%, respectively. Higher correlation to the in-situ measurements is also achieved from the application of data assimilation, i.e., with the average of 0.83 against 0.64 for the open-loop run. These results stress the capability of surface water data assimilation for improving the model's water storage estimates within the Nile Basin with considerable amount of water discharge, especially in the southern parts of the basin.

To further investigate the assimilation results, altimetry data at three virtual stations are used. Surface water storage from the open-loop run and data assimilation are compared with water level variation derived from satellite radar altimetry (Fig. 9.6). Note that, due to the different nature of these two products, standardize values (to have a mean of zero and a standard deviation of 1) are indicated in this figure. From Fig. 9.6, surface storage variations better match to SSH from altimetry. This is evident from the reported correlation values in the figure. Moreover, the assimilation results are found to better capture strong anomalies such as in 2009 in the basin's middle

Table 9.1 Statistics from evaluating estimated water storage compartments with respect to the satellite observations. *Source* [48]

	Evaluation case	RMSE (mm)	STD (mm)	Correlation
Open-loop	Surface water estimates versus GRACE TWS	–	–	0.47
	TWS estimates versus GRACE TWS	16.78	10.35	0.62
	Soil moisture estimates versus Satellite soil moisture	6.80	4.81	0.56
Assimilation	Surface water estimates versus GRACE TWS	–	–	0.67
	TWS estimates versus GRACE TWS	14.01	9.39	0.82
	Soil moisture estimates versus Satellite soil moisture	5.49	3.24	0.66

Table 9.2 Statistics from evaluating estimated water discharge with respect to the in-situ measurements. *Source* [48]

Method	RMSE (m^3/s)	STD (m^3/s)	Correlation
Open-loop	573.48	439.79	0.64
Assimilation	380.07	322.10	0.83
Improvement (%)	33.73	26.76	22.89

part, and also missing trends such as that of after 2014 in the basin's southern part. The better agreement between SSH time series and the assimilation results is clearer over the southern part of the basin and to a lesser degree in the middle part. This can be due to larger surface water uncertainties (as discussed in Sect. 9.5.3), lower surface water variations that can introduce errors in both altimetry and model products, as well as incorporation for non-climatic induced water changes imposed by human activities across the basin [42, 46], which make the modelling more difficult.

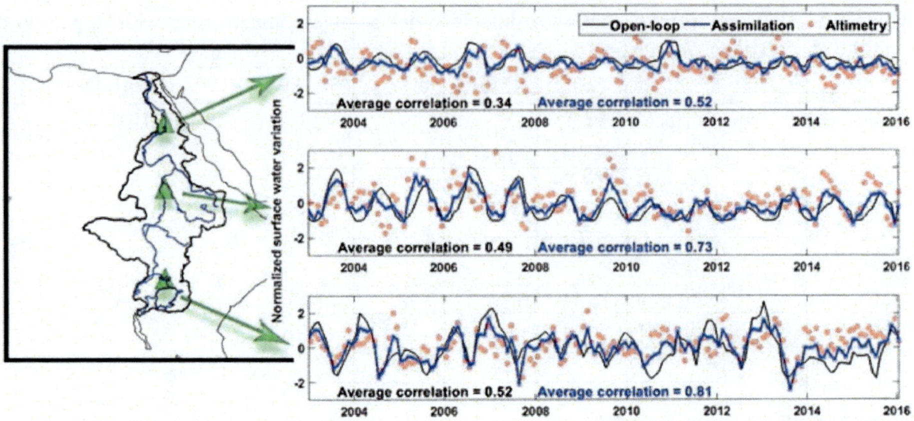

Fig. 9.6 Comparison between the surface water storage (from the open-loop run and data assimilation) and water level height variations produced by satellite radar altimetry over three virtual stations across the basin. *Source* [48]

9.5.3 Water Storage Changes

In this section, we explore the variations of various water storage compartments from the assimilation process over the Nile Basin. Figure 9.7 shows the average surface water variation, and also surface water storage trend and STD values. These are calculated for each grid point over the basin. STD values are calculated based on ensemble members from data assimilation that shows uncertainties in the results. These uncertainties are more dominant over the northern parts of the basin. As mentioned earlier, less amount of surface water availability and human-induced interferences pose challenges in both modelling and observing (from the space) water changes. These appear as a source of uncertainty in data assimilation. Figure 9.7 depicts larger variations in the southern parts of the basin, especially over Lake Victoria. Large surface water variabilities can also be seen over the eastern parts, where the Blue Nile exist. Over the central toward the northern parts, however, surface water changes are very small, which can also indicate less surface water availability. No significant trend values can also be observed in these areas. Over the central parts, on the other hand, negative surface water trend exists, which were partly shown by [7, 65]. Positive trends can be observed in some parts, such as Lake Victoria and southeast of the basin.

The changes of TWS, groundwater, soil moisture, and leaf area index (LAI) from the assimilation results are shown in Fig. 9.8. Similar to Fig. 9.7, the average water storage variations at each grid point is calculated and plotted in Fig. 9.8. The overall pattern of water changes in the various components of Fig. 9.8 follows the surface water pattern (cf. Fig. 9.7). Larger variabilities can be observed in the Nile's southern parts and to a lesser degree in the basin's central parts. This can be attributed to the availability of the Nile's main sources, i.e., Blue and White Niles, as well as Lake Victoria. The minimum amount of water storage changes belong to Nile Delta

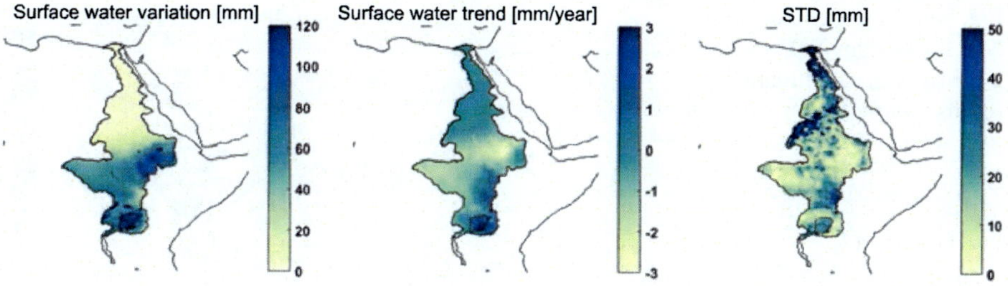

Fig. 9.7 Surface water spatial variation (**a**), trend (**b**), and STD (**c**). Surface variations are computed as temporal average over each grid point. Trend values represent average annual trend over each grid point for the period of 2003–2016. Both variations and trends are calculated from data assimilation results. STD values are computed based on the estimated surface water storage ensemble from data assimilation. *Source* [48]

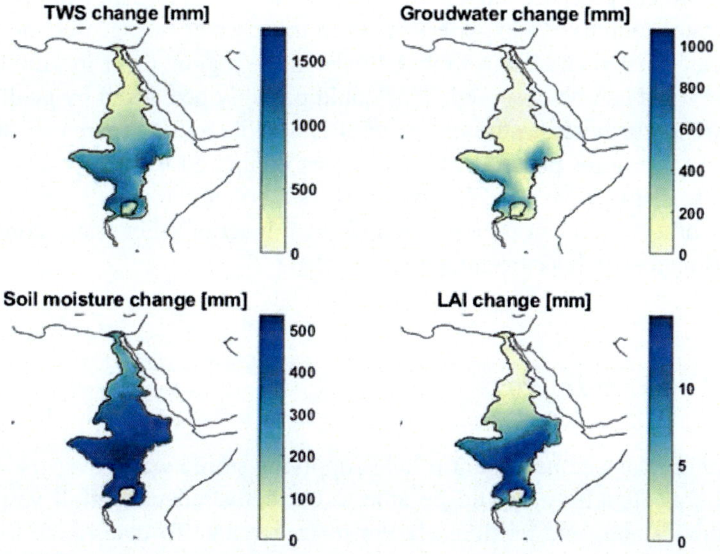

Fig. 9.8 Temporally averaged water storage components from data assimilation results. *Source* [48]

and many parts of Sudan. It can also be seen from Fig. 9.8b that a majority of the Nile Basin have limited groundwater storage resources, which corresponds to small groundwater variabilities. Soil moisture changes (Fig. 9.8c), however, demonstrate a much smoother pattern with more variabilities over the northern parts compared to groundwater and TWS changes (Fig. 9.8b, a, respectively). This can be explained by the fact that soil moisture mostly reflects rainfall patterns and correspondingly climatic changes. LAI changes (Fig. 9.8d), which largely shows vegetation changes display a considerable amount of changes in the southern parts of the basin with more vegetation cover.

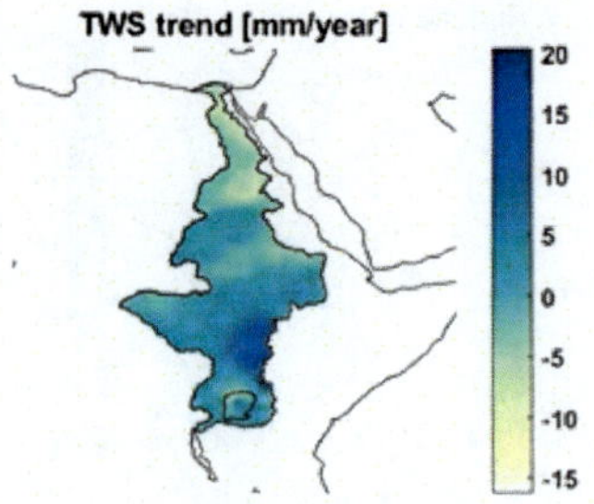

Fig. 9.9 TWS trend over the Nile Basin from 2003 to 2016 from the assimilation results. *Source* [48]

TWS trend map from the assimilation results is presented in Fig. 9.9. Trend values are estimated for each grid point for the period of 2003–2016. While no considerable trend can be noticed over the central Nile Basin, negative trends can be observed over the northern parts. Located in these areas, Egypt and Sudan, which already have limited water sources, are experiencing even more water storage decline. This TWS decline can cause many environmental risks in the region. Over the southeast parts, a positive trend can be observed. This could possibly be caused by positive surface storage trend attributed to the presence of Blue Nile and its increased storage (see also Fig. 9.7). Larger groundwater variations compared to other parts of the Nile Basin were displayed in Fig. 9.8. Nevertheless, no similar trends can be seen over Lake Victoria, which can be attributed to the impact of other water components in TWS estimation such as groundwater, see also [46].

9.5.4 Climatic Impacts

Precipitation plays an important role in supplying stored water over the Nile Basin. A successful modelling should be able to correctly reflect rainfall variations into the system's states, and particularly the surface water. To investigate this, average precipitation over a randomly selected point in the southern part of the basin (with larger water variability) is compared with the surface water changes (with and without data assimilation) at the same location. Figure 9.10 presents the results, as well as the correlation time series between the surface water estimates and precipitation. The calculated correlation value shows better agreement between the assimilation results and precipitation time series than that of the open-loop (without assimilation). This better match between the two time series is more evident, for example in 2009–2010, during 2012, and after 2014. Again, these time periods mainly correspond to extreme anomalies and missing trends. Data assimilation successfully incorporates these changes into the estimated surface water storage.

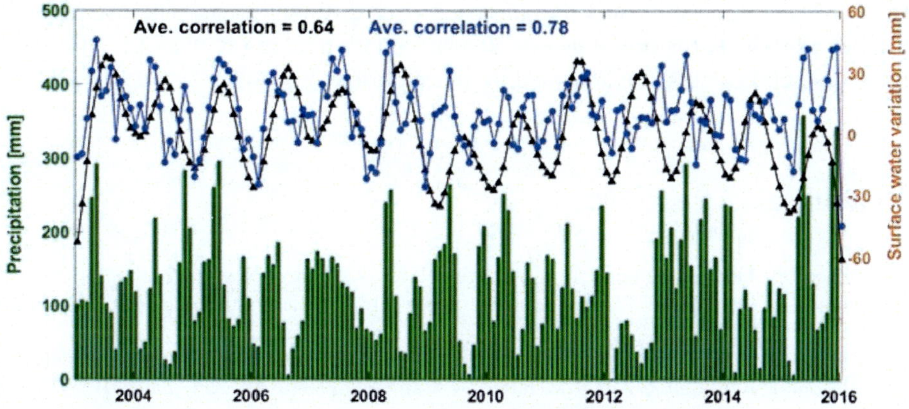

Fig. 9.10 Time evolution of surface water variation with (blue) and without (black) data assimilation over a randomly selected grid point in the southern part of the Nile Basin. Precipitation time series is also plotted (with green). Average correlation values between the precipitation and estimated surface water storage for the two cases are also presented. *Source* [48]

9.6 Concluding Remarks

The Nile River is a crucial source of fresh water for 11 riparian countries in Eastern Africa. This chapter offered an advanced water monitoring technique that combined a land hydrological model and satellite remote sensing data. The approach utilises data assimilation to integrate surface water storage data from satellite radar altimetry into the World-Wide Water Resources Assessment (W3RA) over the Nile River Basin. Evaluation of the assimilation results against validation datasets showed the capability of the proposed data assimilation scheme to improve the model outputs. It was found that

- The estimated surface water storage variations from data assimilation are more correlated to surface water variations from satellite altimetry than those of the open-loop run, e.g., 0.83 against 0.64.
- The assimilation process improved total water storage (TWS) when compared with the Gravity Recovery And Climate Experiment (GRACE) TWS products.
- Importantly, data assimilation indicated positive impacts on non-assimilated variables by showing a better match between the estimated soil moisture changes and those from satellite soil moisture products.
- Evaluation against in-situ measurements also demonstrated the improvements in model estimates after data assimilation. The method decreased the root-mean-squared errors (RMSE) of water discharge estimates by (approximately) 33%.

Based on the results, this chapter suggests that altimetry data can be effectively used over the land for improving hydrological models. Nevertheless, more investigations are required to test the approach over different areas. The usefulness of altimetry products for simultaneous data assimilation (along with assimilating other data products) can also be analysed in future efforts.

References

1. Acker J et al (2003) Remote sensing from satellites. Academic, New York
2. Ahmed AA, Ismail UAE (2008) Sediment in the Nile river system. Consultancy Study requested by UNESCO
3. Aman A (2014) Egypt seeks to halt Ethiopian dam, Al-Monitor: the pulse of the middle East. AlMonitor. Accessed 23 Apr 2014. http://www.almonitor.com/pulse/originals/2014/02/egypt-lobby-renaissance-dam-ethiopia.html
4. Anderson MC, Norman JM, Mecikalski JR, Otkin JA, Kustas WP (2007) A climatological study of evapotranspiration and moisture stress across the continental United States based on thermal remote sensing: 1. Model formulation. J Geophys Res 112 (D10117). https://doi.org/10.1029/2006JD007506
5. Awange J, Sharifi M, Baur O, Keller W, Featherstone W, Kuhn M (2009) GRACE hydrological monitoring of Australia: current limitations and future prospects. J Spat Sci 54(1):23–35
6. Awange JL, Forootan E, Kuhn M, Kusche J, Heck B (2014) Water storage changes and climate variability within the Nile basin between 2002 and 2011. Adv Water Resour 73:64–78. https://doi.org/10.1016/j.advwatres.2014.06.010
7. Awange JL, Gebremichael M, Forootan E, Wakbulcho G, Anyah R, Ferreira VG, Alemayehu T (2014) Characterization of Ethiopian mega hydrogeological regimes using GRACE, TRMM and GLDAS datasets. Adv Water Resour 74:64–78. ISSN 0309-1708. https://doi.org/10.1016/j.advwatres.2014.07.012
8. Awange J, Forootan E, Fleming K, Odhiambo GO (2015) Dominant patterns of water storage changes in the Nile basin during 2003–2013. In: Venkataraman Lakshmi (ed) Remote sensing of the terrestrial water cycle. John Wiley, New Jersey, pp 367–382. https://doi.org/10.1002/9781118872086.ch22.
9. Awange JL, Ferreira VG, Forootan E, Khandu, Andam-Akorful SA, Agutu NO, He XF (2016) Uncertainties in remotely sensed precipitation data over Africa. Int J Climatol 36:303–323. https://doi.org/10.1002/joc.4346
10. Awange JL, Hu KX, Khaki M (2019) The newly merged satellite remotely sensed, gauge and reanalysis-based multi-source weighted-ensemble precipitation: evaluation over Australia and Africa (1981–2016). Sci Total Environ 670:448–465. https://doi.org/10.1016/j.scitotenv.2019.03.148
11. Barnes J (2014) Cultivating the Nile: the everyday politics of water in Egypt. Duke University Press, Durham
12. Berry PAM, Garlick JD, Freeman JA, Mathers EL (2005) Global inland water monitoring from multi-mission altimetery. Geophys Res Lett 32:L16401. https://doi.org/10.1029/2005GL022814
13. Birkett CM (1995) The contribution of TOPEX/POSEIDON to the global monitoring of climatically sensitive lakes. J Geophys Res 25179–25204
14. Birkett CM, Mertes LAK, Dunne T, Costa MH, Jasinski MJ (2002) Surface water dynamics in the Amazon basin: application of satellite radar altimetry. J Geophys Res 107. https://doi.org/10.1029/2001JD000609
15. Biancamaria S, Durand M, Andreadis KM, Bates PD, Boone A, Mognard NM, Rodríguez E, Alsdorf DE, Lettenmaier DP, Clark EA (2011) Assimilation of virtual wide swath altimetry to improve Arctic river modeling. Remote Sens Environ 115(2):373–381. https://doi.org/10.1016/j.rse.2010.09.008

16. Brown GS (1977) The average impulse response of a rough surface and its applications. IEEE Trans Antennas Propag 25:67–74
17. Camberlin P (2009) Nile basin climates. In: Dumont, Henri J (ed) The Nile: origin, environments, limnology and human use. Monographiae Biologicae. Springer, Berlin, pp 307–333
18. Cheng MK, Tapley BD (2004) Variations in the Earth's oblateness during the past 28 years. J Geophys Res Solid Earth 109:B09402. https://doi.org/10.1029/2004JB003028
19. Consulate General of Ethiopia (2012) Grand Ethiopian Renaissance Dam, Los Angeles, CA
20. Conway D (2002) Extreme rainfall events and lake level changes in East Africa: recent events and historical precedents. In: Odada EO, Olago DO (eds) The East African great lakes: limnology, palaeolimnology and biodiversity. Advances in global change research, vol 12. Kluwer, Dordrecht, pp 63–92
21. Courtier P, Thépaut JN, Hollingsworth A (1994) A strategy for operational implementation of 4DVAR, using an incremental approach. Q J R Meteorol Soc 120:1367–1387
22. Crow WT, Wood EF (2003) The assimilation of remotely sensed soil brightness temperature imagery into a land surface model using ensemble Kalman filtering: a case study based on ESTAR measurements during SGP97. Adv Water Resour 26(2):137–149
23. Draper CS, Mahfouf J-F, Walker JP (2009) An EKF assimilation of AMSR-E soil moisture into the ISBA land surface scheme. J Geophys Res 114:D20104. https://doi.org/10.1029/2008JD011650
24. Dumedah G, Walker JP (2014) Assessment of land surface model uncertainty: a crucial step towards the identification of model weaknesses. J Hydrol Part B 519:1474–1484. https://doi.org/10.1016/j.jhydrol.2014.09.015
25. Emery CM, Biancamaria S, Boone A, Ricci S, Rochoux MC, Pedinotti V, David CH (2019) Assimilation of wide-swath altimetry observations to correct large-scale river routing model parameters. Earth Syst Sci Discuss Hydrol. https://doi.org/10.5194/hess-2019-242 (In review)
26. Evensen G (1994) Sequential data assimilation with a nonlinear quasi-geostrophic model using Monte Carlo methods to forecast error statistics. J Geophys Res 99:10143–10162
27. Fielding L, Najman Y, Millar I, Butterworth P, Garzanti E, Vezzoli G, Barfod D, Kneller B (2018) The initiation and evolution of the River Nile. Earth Planet Sci Lett 489:166–178. https://doi.org/10.1016/j.epsl.2018.02.031
28. Frappart F, Papa F, Famiglietti JS, Prigent C, Rossow WB, Seyler F (2008) Interannual variations of river water storage from a multiple satellite approach: a case study for the Rio Negro river basin. J Geophys Res 113:D21104. https://doi.org/10.1029/2007JD009438
29. Getirana A (2010) Integrating spatial altimetry data into the automatic calibration of hydrological models. J Hydrol 387(3–4):244–255. https://doi.org/10.1016/j.jhydrol.2010.04.013
30. Guerschman JP, Van Dijk A, Mattersdorf G, Beringer J, Hutley LB, Leuning R, Pipunic RC, Sherman BS (2009) Scaling of potential evapotranspiration with MODIS data reproduces flux observations and catchment water balance observations across Australia. J Hydrol 369:107–119. https://doi.org/10.1016/j.jhydrol.2009.02.013
31. Hamza W, Mason S (2005) Water availability and food security challenges in Egypt. In: Hamdy A, Monti R (eds) Food security under water scarcity in the Middle East: problems and solutions. CIHEAM, Bari, pp 249–259 (Options Méditerranéennes : Série A. Séminaires Méditerranéens; n. 65)
32. Hartanto IM, van der Kwast J, Alexandridis TK, Almeida W, Song Y, van Andel SJ, Solomatine DP (2017) Data assimilation of satellite-based actual evapotranspiration in a distributed hydrological model of a controlled water system. Int J Appl Earth Obs Geoinformation 57:123–135. https://doi.org/10.1016/j.jag.2016.12.015
33. Hoteit I, Triantafyllou G, Petihakis G (2005) Efficient data assimilation into a complex, 3-D physical-biogeochemical model using partially-local Kalman filters. Ann Geophys (European Geosciences Union) 23(10):3171–3185
34. Houtekamer PL, Mitchell HL (2001) A sequential ensemble Kalman filter for atmospheric data assimilation. Mon Weather Rev 129(1):123–137
35. Ismail SS, Samuel MG (2011) Response of river Nile dredging on water levels. In: 15th international water technology conference, IWTC-15 2011, Alexandria, Egypt

36. Ji L, Gong P, Wang J, Shi J, Zhu Z (2018) Construction of the 500-m resolution daily global surface water change database (2001–2016). Water Resour Res 54:10270–10292. https://doi.org/10.1029/2018WR023060
37. Khaki M, Forootan E, Sharifi MA (2014) Satellite radar altimetry waveform retracking over the Caspian Sea. Int J Remote Sens 35(17):6329–6356. https://doi.org/10.1080/01431161.2014.951741
38. Khaki M, Forootan E, Sharifi MA, Awange J, Kuhn M (2015) Improved gravity anomaly fields from retracked multimission satellite radar altimetry observations over the Persian Gulf and the Caspian Sea. Geophys J Int 202(3):1522–1534. https://doi.org/10.1093/gji/ggv240
39. Khaki M, Hoteit I, Kuhn M, Awange J, Forootan E, van Dijk AIJM, Schumacher M, Pattiaratchi C (2017) Assessing sequential data assimilation techniques for integrating GRACE data into a hydrological model. Adv Water Resour 107:301–316. https://doi.org/10.1016/j.advwatres.2017.07.001
40. Khaki M, Schumacher M, Forootan E, Kuhn M, Awange J, van Dijk AIJM (2017) Accounting for spatial correlation errors in the assimilation of GRACE into hydrological models through localization. Adv Water Resour 108:99–112. https://doi.org/10.1016/j.advwatres.2017.07.024
41. Khaki M, Forootan E, Kuhn M, Awange J, van Dijk AIJM, Schumacher M, Sharifi MA (2018) Determining water storage depletion within Iran by assimilating GRACE data into the W3RA hydrological model. Adv Water Resour 114:1–18. https://doi.org/10.1016/j.advwatres.2018.02.008
42. Khaki M, Awange J, Forootan E, Kuhn M (2018) Understanding the association between climate variability and the Nile's water level fluctuations and water storage changes during 1992–2016. Sci Total Environ 645:1509–1521. https://doi.org/10.1016/j.scitotenv.2018.07.212
43. Khaki M, Forootan E, Kuhn M, Awange J, Longuevergne L, Wada W (2018) Efficient basin scale filtering of GRACE satellite products. Remote Sens Environ 204:76–93. https://doi.org/10.1016/j.rse.2017.10.040,.
44. Khaki M, Hoteit I, Kuhn M, Forootan E, Awange J (2019) Assessing data assimilation frameworks for using multi-mission satellite products in a hydrological context. Sci Total Environ 647:1031–1043. https://doi.org/10.1016/j.scitotenv.2018.08.032
45. Khaki M, Awange J (2019) The application of remotely sensed products to enhance model-derived water storage changes over South America. Sci Total Environ 647:1557–1572. https://doi.org/10.1016/j.scitotenv.2018.08.079
46. Khaki M, Awange J (2019) Improved remotely sensed satellite products for studying Lake Victoria's water storage changes. Sci Total Environ 652:915–926. https://doi.org/10.1016/j.scitotenv.2018.10.279
47. Khaki M, Ait-El-Fquih B, Hoteit I (2020) Calibrating land hydrological models and enhancing their forecasting skills using an ensemble Kalman filter with one-step-ahead smoothing. J Hydrol 584:124708. https://doi.org/10.1016/j.jhydrol.2020.124708
48. Khaki M, Awange J (2020) Altimetry-derived surface water data assimilation over the Nile basin. Sci Total Environ. https://doi.org/10.1016/j.scitotenv.2020.139008
49. Langhorst T, Pavelsky TM, Frasson RPDM, Wei R, Domeneghetti A, Altenau E et al (2019) Anticipated improvements to water surface DEMs from the surface water and ocean topography mission. Front Earth Sci 7:102. https://doi.org/10.3389/feart.2019.00102
50. Lettenmaier DP, Alsdorf D, Dozier J, Huffman GJ, Pan M, Wood EF (2015) Inroads of remote sensing into hydrologic science during the WRR era. Water Resour Res 51:7309–7342. https://doi.org/10.1002/2015WR017616
51. Massari C, Brocca L, Tarpanelli A, Moramarco T (2015) Data assimilation of satellite soil moisture into rainfall-runoff modelling: a complex recipe. Remote Sens 7:11403–11433
52. Mayer-Gürr T, Zehentner N, Klinger B, Kvas A (2014) ITSG-Grace2014: a new GRACE gravity field release computed in Graz. In: GRACE science team meeting (GSTM). Potsdam am: 29.09.2014
53. Mazzoleni M, Noh SJ, Lee H, Liu Y, Seo DJ (2018) Real-time assimilation of streamflow observations into a hydrological routing model: effects of model structures and updating methods. Hydrol Sci J 63(3):386–407. https://doi.org/10.1080/02626667.2018.1430898

54. Michailovsky CI, Milzow C, Bauer-Gottwein P (2013) Assimilation of radar altimetry to a routing model of the Brahmaputra river. Water Resour Res 49:4807–4816. https://doi.org/10.1002/wrcr.20345
55. Neal J, Schumann G, Bates P, Buytaert W, Matgen P, Pappenberger F (2009) A data assimilation approach to discharge estimation from space. Hydrol Process 23:3641–3649. https://doi.org/10.1002/hyp.7518
56. Njoku EG (2004) AMSR-E/Aqua Daily L3 surface soil moisture, interpretive parameters, QC EASE-Grids. Version 2. NASA National Snow and Ice Data Center Distributed Active Archive Center, Boulder, Colorado, USA. https://doi.org/10.5067/AMSR-E/AE_LAND3.002
57. Nile Basin Water Resources Atlas (2017) http://atlas.nilebasin.org/
58. Omondi PA, Awange JL, Forootan E, Ogallo LA, Barakiza R, Girmaw GB, Fesseha I, Kululetera V, Kilembe C, Mbati MM, Kilavi M, King'uyu SM, Omeny PA, Njogu A, Badr EM, Musa TA, Muchiri P, Bamanya D, Komutunga E (2014) Changes in temperature and precipitation extremes over the Greater Horn of Africa region from 1961 to 2010. Int J Climatol 34:1262–1277. https://doi.org/10.1002/joc.3763
59. Orth R, Dutra E, Pappenberger F (2016) Improving weather predictability by including land surface model parameter uncertainty. Mon Weather Rev 144:1551–1569. https://doi.org/10.1175/MWR-D-15-0283.1
60. Reichle RH, McLaughlin D, Entekhabi D (2002) Hydrologic data assimilation with the ensemble Kalman filter. Mon Weather Rev 130(1):103–114
61. Samuel MG (2014) Limitations of navigation through Nubaria canal, Egypt. J Adv Res 5:147–155. https://doi.org/10.1016/j.jare.2013.01.006
62. Sheffield J, Goteti G, Wood EF (2006) Development of a 50-year high-resolution global dataset of meteorological forcings for land surface modeling. J Clim 19(13):3088–3111
63. Sheffield J, Wood EF, Pan M, Beck H, Coccia G, Serrat-Capdevila A, Verbist K (2018) Satellite remote sensing for water resources management: potential for supporting sustainable development in data-poor regions. Water Resour Res 54:9724–9758. https://doi.org/10.1029/2017WR022437
64. Shokri A, Walker JP, van Dijk AIJM, Pauwels VRN (2019) On the use of adaptive ensemble Kalman filtering to mitigate error misspecifications in GRACE data assimilation. Water Resour Res 55:7622–7637. https://doi.org/10.1029/2018WR024670
65. Sultan M, Ahmed M, Sturchio N, Eugene Y, Milewski A, Becker R, Wahr J, Becker D, Chouinard K (2013) Assessment of the vulnerabilities of the Nubian sandstone fossil aquifer, North Africa. In: Pielke RA (ed) Climate vulnerability: understanding and addressing threats to essential resources. Elsevier, Amsterdam, pp 311–333
66. Swenson S, Chambers D, Wahr J (2008) Estimating geocentervariations from a combination of GRACE and ocean model output. J Geophys Res 113:B08410. https://doi.org/10.1029/2007JB005338
67. Talagrand O, Courtier P (1987) Variational assimilation of meteorological observations with the adjoint vorticity equation-Part 1. Theory. Q J R Meteorol Soc 113:1311–1328
68. Taye MT, Ntegeka V, Ogiramoi NP, Willems P (2011) Assessment of climate change impact on hydrological extremes in two source regions of the Nile river basin, Hydrol. Earth Syst Sci 15:209-222. https://doi.org/10.5194/hess-15-209-2011
69. Tseng KH, Shum CK, Yi Y, Fok HS, Kuo CY, Lee H, Cheng X, Wang X (2013) Envisat altimetry radar waveform retracking of quasi-specular echoes over the ice-covered Qinghai lake. Terr Atmos Ocean Sci 24:615–627. https://doi.org/10.3319/TAO.2012.12.03.01(TibXS)
70. Tutwiler R (2010) Personal communication. MEST 430 The Nile: environment, irrigation and civilization. American University in Cairo
71. van Dijk AIJM, PeñArancibia JL, Wood EF, Sheffield J, Beck HE (2013) Global analysis of seasonal streamflow predictability using an ensemble prediction system and observations from 6192 small catchments worldwide. Water Resour Res 49:2729–2746. https://doi.org/10.1002/wrcr.20251
72. van Dijk AIJM, Schellekens J, Yebra M, Beck HE, Renzullo LJ, Weerts A, Donchyts G (2018) Global 5km resolution estimates of secondary evaporation including irrigation through satellite

data assimilation. Hydrol Earth Syst Sci 22:4959–4980. https://doi.org/10.5194/hess-22-4959-2018
73. Wada Y, Wisser D, Bierkens M (2014) Global modeling of withdrawal, allocation and consumptive use of surface water and groundwater resources. Earth Syst Dyn 5(1):15–40. https://doi.org/10.5194/esd-5-15-2014
74. Wahr JM, Molenaar M, Bryan F (1998) Time variability of the Earth's gravity field: hydrological and oceanic effects and their possible detection using GRACE. J Geophys Res 108(B12):30205–30229. https://doi.org/10.1029/98JB02844
75. Yin X, Nicholson SE (1998) The water balance of Lake Victoria. Hydrol Sci J 43(5):789–811
76. Zaitchik BF, Rodell M, Reichle RH (2008) Assimilation of GRACE terrestrial water storage data into a land surface model: results for the Mississippi river basin. J Hydrometeorol 9(3):535–548. https://doi.org/10.1175/2007JHM951.1
77. Zheng W, Zhan X, Liu J, Ek M (2018) A preliminary assessment of the impact of assimilating satellite soil moisture data products on NCEP global forecast system. Adv Meteorol 12 p. https://doi.org/10.1155/2018/7363194. Article ID 7363194

Chapter 10
Hydrometeorological Drought: Impact on Nile

The Greater Horn of Africa (GHA) shares many common experiences with the rest of the world when it comes to impacts of drought. However, the impacts of drought in some Member States of the GHA region often appear to have more adverse effects on sustainable development than elsewhere in similar developing countries. For example, for the southern Africa region, drought in Botswana is regarded as a learning opportunity to improve/achieve water security.
—Mpelasoka et al. [9]

10.1 Summary

Spatio-temporal patterns of hydrological droughts over the Greater Horn of Africa (GHA) are explored based on total water storage (TWS) changes derived from time-variable gravity field solutions of Gravity Recovery And Climate Experiment (GRACE, 2002–2014), together with those simulated by Modern Retrospective Analysis for Research Application (MERRA, 1980–2014). These hydrological extremes are then related to meteorological drought events estimated from observed monthly precipitation products of Global Precipitation Climatology Center (GPCC, 1979–2010) and Tropical Rainfall Measuring Mission (TRMM, 1998–2014). The major focus of this chapter based on the work of [4] lies on the application of spatial Independent Component Analysis (sICA) to extract distinguished regions with similar rainfall and total water storage (TWS; surface, groundwater, soil moisture and vegetation water) with similar overall trend and seasonality. Rainfall and TWS are used to estimate Standard Precipitation Indices (SPIs) and Total Storage Deficit Indices (TSDIs), respectively that are employed to characterize frequency and intensity of hydro-meteorological droughts over GHA. Significant positive (negative) changes in monthly rainfall over Ethiopia (Sudan) between 2002 and 2010 leading to a significant increase in TWS over the central GHA region were noted in both MERRA and GRACE TWS (2002–2014). However, these trends were completely reversed in the long-term (1980–2010) records of rainfall (GPCC) and TWS (MERRA). The

J. Awange, *The Nile Waters*,
https://doi.org/10.1007/978-3-030-64756-8_10

four independent hydrological sub-regions extracted based on the sICA (i.e., Lake Victoria Basin, Ethiopia–Sudanese border, South Sudan, and Tanzania) indicated fairly distinct temporal patterns that matched reasonably well between precipitation and TWS changes. While meteorological droughts were found to be consistent with most previous studies in all sub-regions, their impacts are clearly observed in the TWS changes resulting in multiple years of extreme hydrological droughts. Correlations between SPI and TSDI were found to be significant over Lake Victoria Basin, South Sudan, and Tanzania. The low correlations between SPI and TSDI over Ethiopia are likely related to inconsistency between TWS and precipitation signals. Further, hydrological droughts in these regions were found to be significantly associated with Indian Ocean Dipole (IOD) events while El Niño Southern Oscillation (ENSO) played a secondary role.

10.2 Greater Horn of Africa's Drought

The people of the semi-arid region of Greater Horn of Africa (GHA) largely depend on rain-fed agriculture and livestock that is increasingly coming under threat from the intensified frequency and severity of drought events over the past decades e.g., [29]. Since most crops in GHA are planted during rainy seasons, i.e., March–May (MAM) and October–December (OND), food security is increasingly coming under jeopardy given the fact that these seasons too are now reported to experience drought episodes [36]. This is further exacerbated by the continued warming of the Indian-Pacific Oceans possibly linked to anthropogenic influences or multi-decadal climate variability, which has been shown to contribute to more frequent droughts in GHA over the past 30 years during the spring and summer seasons, see e.g., [26, 35, 53].

Even with the reality of the threat posed by droughts in the GHA, see e.g., [35], effective drought monitoring in this region is, however, challenged by multiple factors including: inadequate spatial coverage and hydro-meteorological observations (e.g., precipitation, temperature, soil moisture, and groundwater storage) that are not readily available beyond the local meteorological services, poor communication frameworks, and national red tapes in accessing data.

Furthermore, the large spatial extent of GHA makes the collection of hydrological data quite challenging and hence the reason why most studies e.g., [51] have focused largely on meteorological droughts. This view is supported e.g., by [40] who pointed out that the lack of reliable hydro-meteorological data hold up development of effective real-time drought monitoring and early warning systems in the region. To account for various drought conditions over GHA, therefore, an integrated drought index that involves various meteorological and hydrological variables is desirable e.g., the African Drought Monitor(ADM[1]) project [46]. This, therefore, highlights the significant role played by satellite remote sensing in data deficient regions.

[1] http://stream.princeton.edu/AWCM/WEBPAGE/interface.php?locale=en.

To this extent, the Gravity Recovery and Climate Experiment (GRACE) [48] satellite remote sensing mission (see Chap. 3), launched in 2002, offers the possibility to characterize droughts at a wider spatial coverage albeit for shorter time span (from 2002 onwards). The value of GRACE-derived total water storage (TWS) changes for drought studies estimated through Total Storage Deficient Index (TSDI), has been demonstrated e.g., in [10, 11, 17, 18, 27, 32–34, 49, 56].

Although GRACE has the capability to offer a wider spatial coverage, spatio-temporal analysis of drought in GHA using a single areal-average as a representation of the entire GHA region is challenging. This is mainly due to the meteorological observations that are influenced by spatial inhomogeneity of climate over the GHA caused by uneven topography, north-south migration of Inter-tropical Convergence Zone (ITCZ), and spatially varying influences of large-scale climate events among others, see e.g., [51].

In this chapter, as opposed to [12, 14–16] that focused mainly on TWS changes in Lake Victoria Basin, Nile River Basin (NRB) and Ethiopia, GRACE satellites and sICA method [23–25] are employed for the first time to characterize meteorological and hydrological droughts in the entire GHA. The novelty is that for the first time, the long-term (1980–2014) meteorological and hydrological drought in GHA is extracted using the Standardized Precipitation Index (SPI) and TSDI, respectively. The spatio-temporal variability of SPI and TSDI are also compared and a comprehensive interpretation of various drought events across the GHA is provided.

Specifically, the chapter (i) applies sICA to partition the GHA into various sub-regions based on precipitation and TWS changes, (ii) employs GRACE (2002–2014) and Modern Retrospective Analysis for Research Application (MERRA, 1980–2014) TWS changes to derive hydrological drought indices for the four independent sub-regions (as computed in i above), (iii) uses observed precipitation products, Global Precipitation Climatology Center (GPCC, 1979–2010) and Tropical Rainfall Measuring Mission (TRMM, 1998–2014) to compute remotely sensed SPI, and (iv), examines the relative influence of ENSO and IOD events on hydro-meteorological droughts (reflected in TSDI and SPI) within the GHA sub-regions.

The remainder of the chapter is organized as follows. In Sect. 10.3, a general climatological background of GHA is presented while Sect. 10.4 presents an overview of the data as well as the analysis methods used. In Sect. 10.5, the drought indicators used are described, while in Sect. 10.6, the results are analysed and interpreted. The chapter is concluded in Sect. 10.9.

10.3 GHA: Climatological Background

The GHA, a semi-arid region, is made up of Burundi, Djibouti, Ethiopia, Eritrea, Kenya, Rwanda, Somalia, South Sudan, Sudan, Tanzania and Uganda. Long rains over equatorial GHA region mostly occur during March–May (MAM, Fig. 10.1b) while short rains occur in October–December (OND, Fig. 10.1d) corresponding to the migration of the intertropical convergence zone (ITCZ) from south to north and vice versa [36]. Furthermore, Ethiopia, South Sudan, Sudan, and parts of Uganda experience a single rainy season from June–September (JJAS, Fig. 10.1c).

Apart from its seasonal differences, rainfall variability over the GHA is closely associated with the large-scale regional and global circulations such as ENSO, fluctuation of the Indian Ocean and Atlantic Ocean SST, and moisture fluxes over the Congo region [35, 41, 50, 54]. Mean temperature patterns over GHA follow the annual rainfall pattern.

10.4 Hydrometeorological Products

The hydrometeorological products used in this chapter include monthly precipitation products obtained from global gridded rain gauge and near-global satellite-based estimates, as well as total water storage (TWS) changes from a reanalysis model and Gravity Recovery and Climate Experiment (GRACE) time-variable gravity field solutions.

10.4.1 Precipitation Products

1. *GPCC v6*: The Global Precipitation Climatology Center (GPCC) provides gridded precipitation products at various temporal and spatial resolutions derived from up to 67,000 quality controlled station data around the world between 1901 and 2010 [45]. This chpater used monthly precipitation data at $0.50° \times 0.50°$ (latitude × longitude) spatial resolution covering the period 1979–2010. GPCC products have been found to be consistent with other rainfall products in Africa, see e.g., [3, 13]. The distribution of rain gauges over the GHA region, however, is rather sparse over certain areas such as Somalia, Sudan, and Tanzania, and as such, may not provide accurate representation of the precipitation variability over these regions.
2. *TRMM 3B43*: To complement the GPCC v6 data, monthly precipitation estimates from the Tropical Rainfall Measuring Mission (TRMM) Multi-satellite Precipitation Analysis (TMPA) [28] version 7 for the period 1998 to 2014 were used. These products at $0.25° \times 0.25°$ spatial resolution were interpolated to $0.50° \times 0.50°$ to make them consistent with those of GPCC. Awange et al. [3, 13] assessed various precipitation products over Africa covering 2003–2010, from which TRMM 3B43 (hence forth referred to as TRMM) indicated similar skills compared to GPCC v6.

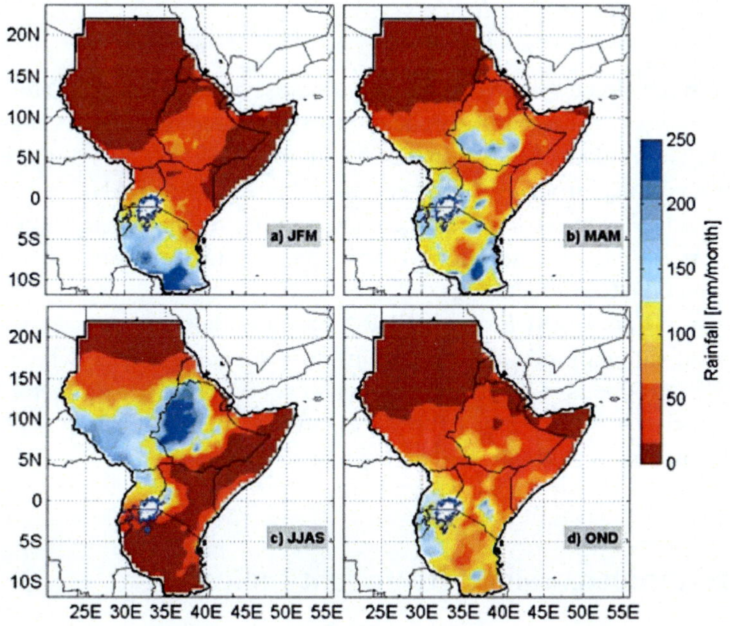

Fig. 10.1 Spatial variability of seasonal rainfall over GHA from 1979–2010 based on GPCC v6 data: **a** January–March (JFM), **b** March–May (MAM), **c** June–September (JJAS), and **d** October–December (OND). Note that the signals over Lake Victoria has been masked out in the analysis to avoid spurious trends due to varying lake levels. *Source* [4]

10.4.2 Water Storage Change Products

The reprocessed GRACE Level-2 RL05a time-series for the period 2002 to 2014 from the Gravity Recovery and Climate Experiment GRACE, e.g., [48] was used. The GRACE data is provided by GeoForschungsZentrum (GFZ, Potsdam) [21] as sets of (approximately) monthly fully normalized geopotential spherical harmonic coefficients up to degree and order 90 (see, ftp://podaac.jpl.nasa.gov/allData/grace/L2/).

GRACE degree one (C_{10}, C_{11}, S_{10}) coefficients were replaced by those from [19], and the more accurate degree two coefficients derived from satellite laser ranging (SLR) [20] replaced C_{20}. A commonly used non-isotropic de-correlation filter DDK3 [22, 30] was applied to reduce the high degree/order correlated errors, which are manifested as stripes in the spatial domain. This filter removes most of north-south stripes and seems to be suitable to process GFZ RL05a products over the GHA. To be consistent with GPCC, the time-variable spherical harmonic coefficients were converted to $0.50° \times 0.50°$ (latitude $\times$ longitude) TWS grids using the approach of [52]. The selected spatial grid is optimistic for GRACE products whose spatial resolution is $\sim 1°$.

10.4.3 Reanalysis Products

GRACE TWS estimates were compared with those simulated by the Modern Retrospective Analysis for Research Application(MERRA) [44]. MERRA reanalysis are run globally at a relatively high spatial resolution ($0.67° \times 0.50°$) and are available as monthly products from 1979-present. Monthly MERRA-LAND TWS estimates for the period 1980 to 2014 were processed by filtering in the spectral domain and then converted to a grid resolution of $0.50° \times 0.50°$ to be consistent with those of GRACE and GPCC products.

Filtering of both GRACE and MERRA products, however, causes some damping of the signal amplitude and might introduce spatial leakages, which should be restored by introducing a multiplicative scaler (or a gridded) gain factor e.g., [31]. A multiplicative scale factor of 1.05 was obtained from the MERRA TWS and was uniformly applied to both GRACE and reanalysis data used.

10.5 Hydrometeorological Drought Indices

To assess the hydro-meteorological drought events, two drought indicators were analysed: (i) the Standard Precipitation Index (SPI) estimated from monthly precipitation products of GPCC and TRMM, and (ii), the Total Storage Deficit Index (TSDI) deduced from MERRA and GRACE. Drought conditions were assessed over various sub-regions of the GHA extracted using the spatial independent component analysis approach discussed in Sect. 10.5.3.

10.5.1 Standardized Precipitation Index (SPI)

The Standardized Precipitation Index (SPI) is a widely-used meteorological drought indicator based on probability distribution of long-term rainfall time-series developed by [37, 38] to provide a spatially and temporally invariant measure of rainfall deficit (or excess) over a variety of accumulation timescales (e.g., 3, 6, 12, 24 months). SPI is derived by fitting a parametric cumulative probability distribution (CDF) of e.g., γ-distribution to the precipitation time-series, which are then transformed to obtain the SPI using the inverse normal (Gaussian) distribution function, e.g., [51] (Table 10.1). SPI provides the value of standard anomalies from the median indicating negative for drought and positive for wet conditions (see details in Table 10.1). SPI is computed for the 12-month accumulation time periods to capture the long-term trend of meteorological droughts over the GHA.

Table 10.1 Various categories of droughts and wet conditions based on the SPI Index [37, 51]. *Source* [4]

SPI	Category
+2.0 and above	Extreme wet
+1.5 to +1.99	Very wet
+1.0 to +1.49	Moderately wet
+0.99 to −0.99	Normal
−1.0 to −1.49	Moderate drought
−1.5 to −1.99	Severe drought
−2 and below	Extreme drought

10.5.2 Total Storage Deficit Index (TSDI)

The Total Storage Deficit Index (TSDI) is a renamed version of Soil Moisture Deficient Index (SMDI) [39] by [56]. TWS changes, which comprise information of the changes in all water components (surface, groundwater, soil moisture, biomass, ice and snow) are used to compute TSDI. The procedure to estimate TSDI (based on GRACE data) involves an estimation of the Total Storage Deficit as (TSD%) [56]:

$$TSD(k, j-2001) = \frac{TWS(k,j) - Mean(TWS(k,:))}{Max(TWS(k,:)) - Min(TWS(k,:))} \times 100, \qquad (10.1)$$
$$k = 1, 2, \ldots, 12, j = 2002, 2003, \ldots, 2014$$

where $TWS(k, j)$ is the k'th month of total water storage (anomaly) in each year ($j = 2002, 2003, \ldots, 2014$) obtained from GRACE. For MERRA, the left hand side of Eq. 10.1 becomes $TSD(k, j-1979)$ and $j = 1980, 1981, \ldots, 2014$. $Mean(TWS(k,:))$, $Max(TWS(k,:))$, and $Min(TWS(k,:))$ are respectively the mean, maximum, and minimum of the k'th month of TWS over the study period (for GRACE covering 2002–2014 and for MERRA 1980–2014). The TSD values in Eq. 10.1 can be stored in a vector TSD_i ($i = 1, 2, \ldots,$ length of TWS time series). Consequently, TSDI can be computed as

$$TSDI_i = p \times TSDI_{i-1} + q \times TSD_i, \qquad (10.2)$$

where the drought severity and duration factors p and q respectively are defined from the cumulative TSD plot (e.g., Fig. 10.2) based on the following relation [56]:

$$p = 1 - \frac{m}{m+b}, \qquad q = -\frac{C}{m+b}. \qquad (10.3)$$

In Eq. 10.3, C represents drought intensity, which is obtained from the best fit line of the cumulative TSD (from the drought monograph) during the dryness period

Table 10.2 Various categories of droughts and wet conditions based on the TSDI Index [43, 56]. *Source* [4]

TSDI	Category
+4.0 and above	Extreme wet
+3.0 to +3.99	Very wet
+2.0 to +2.99	Moderate wet
+1.99 to −1.99	Normal
−2.0 to −2.99	Moderate drought
−3.0 to −3.99	Severe drought
−4 and below	Extreme drought

(see e.g., line 3 in Fig. 10.2). In general, C can take on any of the four drought classifications values (−4.0 for extreme drought, −3.0 for severe drought, −2.0 for moderate drought, and −1.0 for mild drought) defined according to [43]. The other elements of the regression line are the slope m and the intercept b of the best fit of the drought monograph for the dryness period of the cumulative TSD curve.

Using an example, we illustrate how the TSDI values of this chapter are estimated. Figure 10.2a shows a TSD (%) plot over Ethiopia derived from GRACE TWS changes based on Eq. 10.1 between 2004 and 2013. A prolonged drought pattern from March 2004 to November 2006 (33 months) has been detected with the TSD values reaching as low as 66%. Assuming this scenario, the cumulative TSDs in Fig. 10.2b are plotted to derive the parameters in Eqs. 10.2–10.3. Using the linear regression parameters during the declining TWS period (i.e., 2004–2007), TSDI was computed for the period of decline. The slope ($m = -31.54$) and y-intercept ($b = -20.51$) parameters derived from the cumulative TSD based on 33 months from March 2004 to November 2006 were subsequently used to calculate the critical parameters p and q. Therefore, $C = -3$ is the category of drought event as can be seen in Fig. 10.2. Correspondingly p and q were derived as 0.394 and 0.058, respectively. $TSDI_{1,1}$ was assumed to be 2% of TSD_1 in line with [56]. The TWS droughts, categorized based on the "C" values of Fig. 10.2, are reported in Table 10.2. This procedure was used to compute TSDI from GRACE and MERRA TWS changes in order to explore hydrological droughts over the GHA for the entire study period (1980–2014).

10.5.3 Spatial Independent Component Analysis

The spatial independent component analysis (sICA) [24] approach was applied to extract statistically independent modes of precipitation and TWS changes. Note that sICA patterns are essentially regions with similar seasonality (and not typical spatial patterns of variability) and as such, there could be significant bias in the analysis from the wettest stations. As the estimated modes are statistically independent, they can

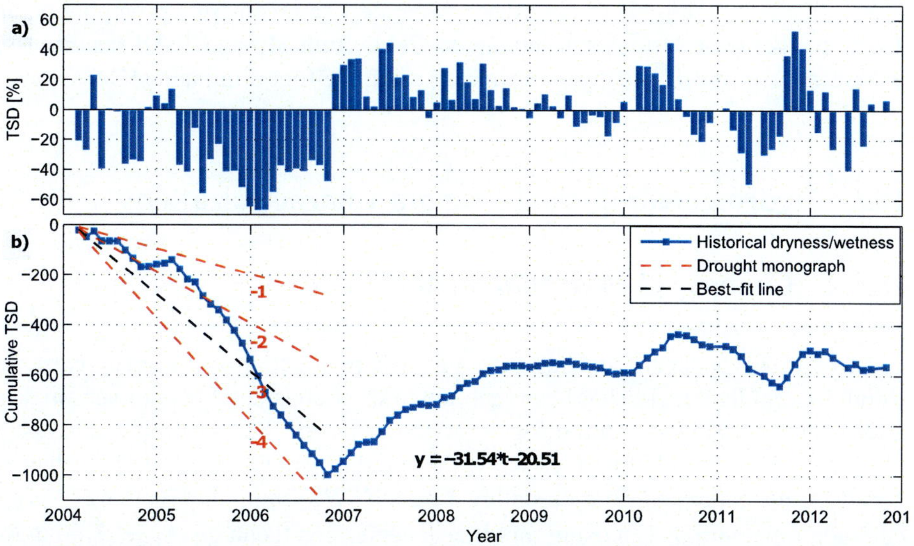

Fig. 10.2 Total storage deficit over eastern GHA region from 2004 to 2013 (**a**), and cumulative total storage deficit (%) over the same period (**b**). Considering the horizontal line of zero in the Figure (**b**) to represent near normal condition, following [43] classification, the interval from near normal to severe dryness was divided into four equal intervals, with the part above the best fit line having two lines (1 and 2) and the best fit is shown by line 3. To complete the [43] classification, a fourth line 4 was plotted. All the four lines were drawn at equal intervals to define the extreme (Line 4), severe (Line 3), moderate (Line 2), and mild (Line 1) drought conditions. *Source* [4]

be separately analysed without considering other modes, see an application of sICA over the Nile River Basin (NRB) in [14]. Suppose $\mathbf{X}(t, s)$ is a gridded time series of precipitation or TWS changes after removing their dominant annual and semi-annual cycles. sICA decomposes the time-series of $\mathbf{X}$ into j spatial $\mathbf{S}_j$ and temporal $\mathbf{A}_j$ modes as:

$$\mathbf{X}(t, s) = \mathbf{A}_j(t)\ \mathbf{S}_j(s), \tag{10.4}$$

where t is the time and s represents the grid points. By applying sICA the rows of $\mathbf{S}$ are statistically as independent as possible, and the columns of $\mathbf{A}$ are their corresponding temporal evolutions. The use of sICA as opposed to principal component analysis (PCA) or its rotated version (RPCA) makes sense since it helps to identify (or localize) specific hydrological areas impacted by droughts or extreme wet conditions over GHA, which is not possible when using PCA [57] or RPCA. The details of sICA estimation are reported in [22]. Here, only the first four dominant modes of both precipitation and TWS changes were found to be statistically significant and consequently were considered separately to reconstruct precipitation and TWS changes over various sub-regions of GHA. Both SPI and TSDI (as discussed in Sects. 10.5.1 and 10.5.2) were estimated using the four dominant sICA modes.

10.6 Hydrometeorological Drought Characterization

10.6.1 Changes in Precipitation and TWS

Changes in TWS depend on the variability of precipitation and climate extremes. Rainfall over GHA region has been reported to show considerable inter-seasonal and inter-annual variability over the past few decades beside the long-term changes e.g., [3, 9, 13, 35, 42, 54]. Figure 10.3 shows the variability of rainfall and TWS changes. Trends were estimated from monthly rainfall/TWS anomalies (y) over various common time periods (T) using multilinear regression technique ($y = \hat{\beta}_0 + \hat{\beta}_1 T + \hat{\beta}_2.\sin(2\pi T) + \hat{\beta}_3.\cos(2\pi T) + \hat{\beta}_4.\sin(4\pi T) + \hat{\beta}_5.\cos(4\pi T) + \epsilon(T)$), where $\hat{\beta}_0$ is a constant term, $\hat{\beta}_1$ represents the linear trend, and the remaining terms represent the annual and semi-annual signals, ϵ represents the error term. Only changes that are significant at 95% confidence level based on student's t-test are shown.

Figure 10.3a shows significant decline (up to 20 mm/decade) in rainfall over southern and northern Sudan as well as increase in rainfall over central Ethiopia between 2002 and 2014 based on monthly TRMM precipitation estimates. Equivalent increase (decrease) in TWS over central Ethiopia (eastern Kenya, Somalia, and northern Sudan) are noted in MERRA (Fig. 10.3b). GRACE also indicated decreasing trends over southern Sudan and eastern Kenya (similar to rainfall) as well as anomalously large positive trends (up to 20 mm/year) over the Lake Victoria region (Fig. 10.3c). However, a completely opposite sign is observed in the long-term records of GPCC precipitation products between 1980 and 2010 (Fig. 10.3d; i.e., multidecadal oscillations may be influencing the sign of the trends), which indicate significant negative (positive) changes over Ethiopia and Tanzania (Sudan and Somalia). This might have lead to an overall negative trend in TWS over the northern parts of GHA during the same period as shown by MERRA in Fig. 10.3e.

Figure 10.4 shows the area-averaged seasonal anomalies of TWS based on MERRA (1980–2014) and GRACE (2002–2014) data. Substantial inter-seasonal variations in TWS over the past 30–40 years are observed. Besides decadal trends, prolonged periods of negative TWS anomalies in the late1980 s to early 1990s and late 1990s and early 2000s are observed. Although, GRACE data covers a very short time period, the inter-seasonal variations are visible in all the seasons. While the TWS changes in 2009 may have occurred as a result of the historical 2009–2011 drought, TWS changes between 2002 and 2006 at a rate of 6.20 mm/month seen in GRACE data has been linked to other processes such as anthropogenic, evaporations, and rainfall deficiency [14, 16].

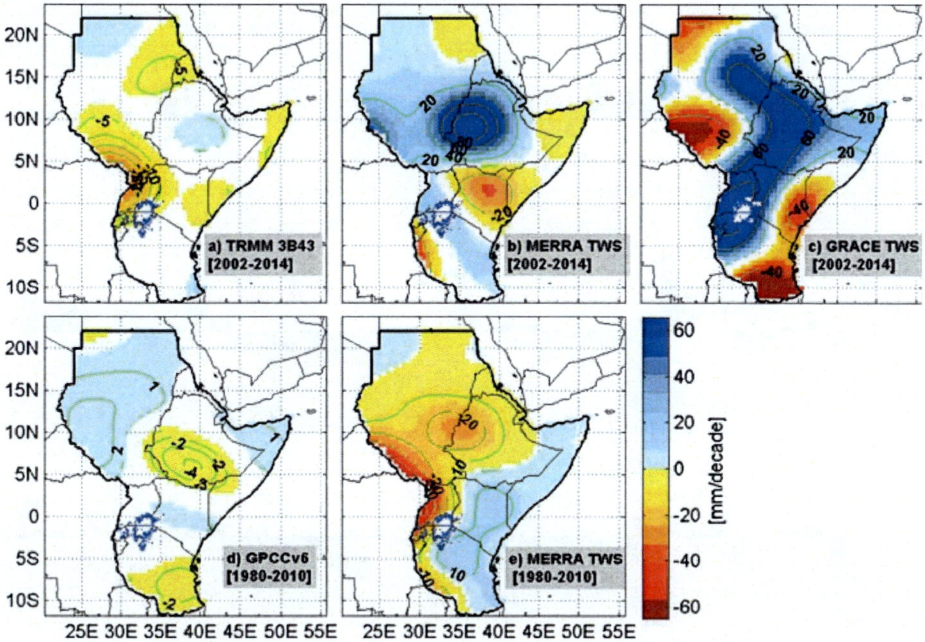

Fig. 10.3 Spatial variability of linear trend over GHA based on a full complement of monthly data at 95% confidence level from 2002–2014 for (**a**) TRMM rainfall, (**b**) MERRA TWS, and (**c**), GRACE TWS. Long term variability are also assessed from 1980–2010 for (**d**) GPCC rainfall, and (**e**), MERRA TWS. Note that signals over Lake Victoria were not included (i.e., they are masked). *Source* [4]

To assess the overall relationship between rainfall and GRACE TWS changes over the region, the temporal correlations are computed and presented in Fig. 10.5, together with their lags. High correlations between rainfall and TWS (with a 2-month time lag) observed in Fig. 10.5a indicate that TWS changes are primarily driven by precipitation. Areas of low correlation (over central GHA region) show a time-lag of about one month (see, Fig. 10.5d). This low correlations could be related to seasonality, where the rainfall is usually bimodal over these regions.

MERRA TWS changes on its part show considerably better correlations with rainfall indicating that TWS changes simulated by MERRA are mainly driven by precipitation (Fig. 10.5b). It should be noted that MERRA does not simulate changes in surface water and groundwater, and therefore, does not represent the integrated changes in TWS as observed by GRACE satellites [44]. The large-scale high correlation patterns observed in Fig. 10.5b are further illustrated by the time lags, where the majority of the region indicated a lag of only one month. Further, the correlations between GRACE and MERRA (Fig. 10.5c) closely resemble those of rainfall and GRACE, although over a different time lag (see, Fig. 10.5f).

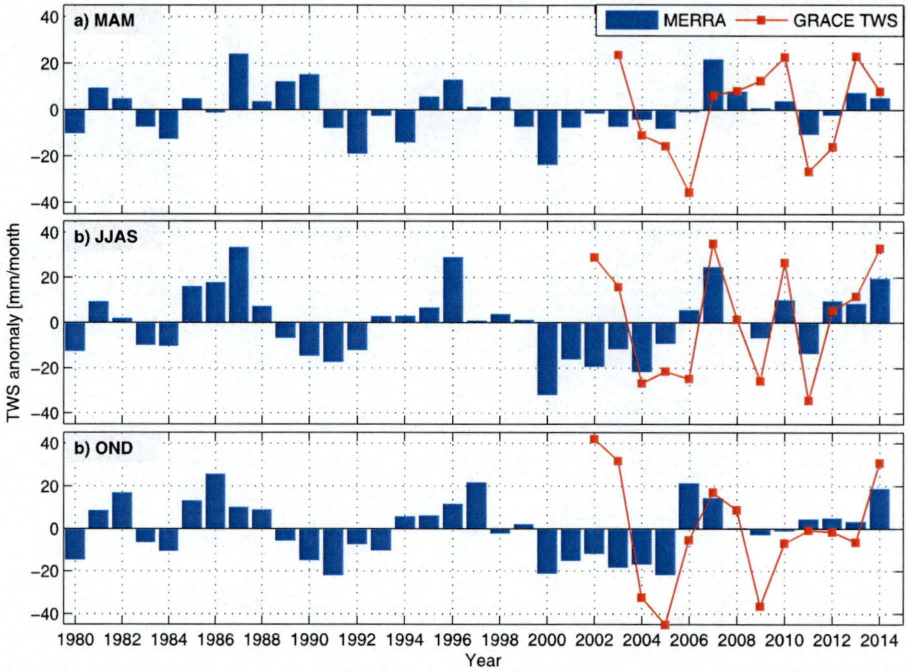

Fig. 10.4 Regional-averaged seasonal anomalies of TWS over the GHA region based on MERRA (1980–2014) and GRACE (2002–2014). The seasonal averages are shown for **a** MAM, **b** JJAS, and **c** OND corresponding to Fig. 10.1b–d. *Source* [4]

10.6.2 Spatio-Temporal Drought Patterns over GHA

To study the temporal drought patterns over the GHA, sICA was applied on the non-seasonal rainfall/TWS anomalies to derive region-specific patterns of temporal variability over the period 1979 to 2014. Figure 10.6 shows the standard deviations derived from the first four leading modes of rainfall variability, together with their corresponding temporal patterns (ICs). The first independent mode of rainfall variability based on GPCC indicates the largest anomalies over central GHA (Kenya, parts of Ethiopia and Somalia), accounting for 23.8% of the variability (Fig. 10.6a). The second mode (20.2%) is localized over central Ethiopia indicating a maximum standard deviation of about 30 mm (Fig. 10.6b) while the third mode (20.0%) is concentrated over Tanzania (Fig. 10.6c). The fourth mode (13.3%) is localized over northern Ethiopia and Sudan (Fig. 10.6d). TRMM showed similar spatial patterns to GPCC (Fig. 10.6e–h) except that the second and third modes were inter-changed (cf. that of GPCC in Figs. 10.6b and 10.6c). The maximum variability in the third mode was found to be localized completely over Ethiopia (Figs. 10.6g–h).

The corresponding temporal patterns in Fig. 10.6i–l, indicate considerable inter-annual variability and their temporal variations appeared to be considerably different from each other signifying the complexity of the region. Note that the ICs are only

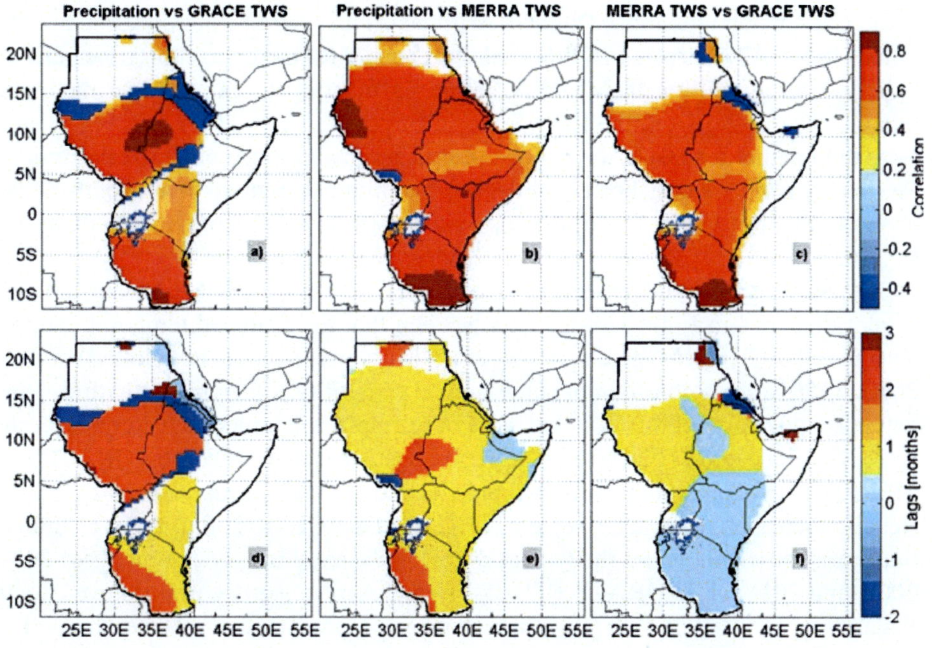

Fig. 10.5 Temporal correlations (with lags in months) between monthly rainfall and TWS changes for the common period 2002–2014. Correlations and lags were computed between a&d; monthly precipitation estimates of TRMM and TWS changes derived from GRACE Level 2 data, b&e; monthly precipitation estimates of TRMM and TWS changes simulated MERRA, and c&f; GRACE and MERRA TWS changes over the entire GHA region. Note that the values which are not significant at 95% were not shown. *Source* [4]

shown for the period 1998 to 2014 in order to highlight the differences between the two precipitation products.

The four dominant modes of TWS variability are shown in Fig. 10.7 with both MERRA (1980–2014) and GRACE (2002–2014) indicating distinct patterns of maximum variability over various parts of the GHA region. In order to highlight the differences in temporal evolutions between MERRA and GRACE, the ICs are plotted from 2002 to 2014 only. MERRA TWS showed the largest anomalies over western Ethiopia (explaining about 30% of the variability; Fig. 10.7a) while the second mode (18.2%) is localized over western Tanzania (Fig. 10.7b). The third independent mode showed maximum variations over the northern Lake Victoria region accounting for 17.3% of the variability (Fig. 10.7c). The fourth mode (14.3%) is mainly concentrated over South Sudan (Fig. 10.7d).

GRACE showed maximum variability (∼27%) over the Lake Victoria region (Fig. 10.7e). Its second mode (16%) indicates maximum variations between Sudan and Ethiopia (Fig. 10.7f) while its third mode (∼14%) is localized over South Sudan (Fig. 10.7g). The fourth mode (13.8%), although not very distinctive in its spatial pattern, shows a similar temporal pattern to that of MERRA over Tanzania. It should be noted here that the spatial patterns of maximum variability shown by MERRA do not exactly match with those from GRACE data. This could be possibly due to limitations in MERRA (errors in TWS modelling) and GRACE data (short time period). One obvious limitation in MERRA lies in its inability to capture the maximum variations resulting from the Lake Victoria Basin (see, e.g., sICA 1 in Fig. 10.7e). Note though that for GHA, the regions of maximum variations often overlap among the four leading modes. For example, the maximum variations over South Sudan shown by MERRA (Fig. 10.7d) matche quite well with sICA 3 of GRACE data (see, Fig. 10.7g).

In this way, the spatial patterns of sICA from GRACE and MERRA were matched to plot their ICs in Fig. 10.7i–l. It can be seen that the matched temporal patterns between GRACE and MERRA are very close. The corresponding ICs plotted between 2002 and 2014 show considerably inter-annual variability, indicating the hydrological extremes over parts of the GHA region. For example, the temporal patterns over the Lake Victoria region (Fig. 10.7e) are reproduced reasonably well by GRACE and MERRA (Fig. 10.7i), indicating a rapid decline in TWS from 2002 to 2006 and steadily increasing thereafter, see e.g., [16, 47]. GRACE (IC2, Fig. 10.7j, Ethiopia–Sudanese border) indicates the typical hydrological extremes, i.e., droughts of 2004, 2009, and 2011 while GRACE (IC3) fairly representes the dry periods over South Sudan (Fig. 10.7k). GRACE (IC4), which is mainly localized over the eastern GHA region (Fig. 10.7l) indicates a rapid decline of TWS from 2002 to 2006 and 2007 to 2010.

Based on the sICA results in Figs. 10.6 and 10.7, in Table 10.3, GHA is classified into four sub-regions in the order of sICA-derived patterns of GRACE TWS in Fig. 10.7e–h. Precipitation and TWS anomalies reconstructed from the individual spatial and temporal modes provide rainfall and TWS changes with anomalies mainly localized over these four regions. Four time-series were calculated for each dataset based on the four leading sICA modes to generate SPI and TSDI indices used to study the hydrological extremes over the period 1980 to 2014. Since the seasonal component from both precipitation and TWS changes have already been removed before applying the sICA, it was not necessary to use Eq. 10.1 to compute the TSD anomalies. Instead, the TWS anomalies were converted to TSD (in %) by dividing them with the range ($TWS_{max} - TWS_{min}$). While SPI is straight forward (Fig. 10.8), the calculation of TSDI index is rather complicated depending on the period of drought events, which is important for deriving the slope and y-intercept parameters from the cumulative TSDs. By analyzing the individual time-series, the TSDIs are estimated for all the four regions and plotted in Fig. 10.9. For example, TSDI (of GRACE TWS) over Lake Victoria region (Region 1) was calculated from cumulative TSDs between June 2006 and November 2006 while the TSDI over South Sudan (Region 3) was calculated for August 2009 to May 2010 (see Table 10.3).

Figure 10.8 shows the 12-month SPI indices calculated from GPCC (1979–2010) and TRMM (1998–2014) over the four regions as indicated in Table 10.3 and Fig. 10.7e–h. These are (a) Region 1 covering the Lake Victoria Basin, (b) Region 2, including the Ethiopia-Sudan border, (c) Region 3 covering South Sudan, and (d) Region 4, including Tanzania and parts of Ethiopia. The first 12 months of GPCC (i.e., 1979) and TRMM (i.e., 1998) were removed during the SPI computation. Table 10.4 summarizes the duration and intensity of severe to extreme meteorological drought

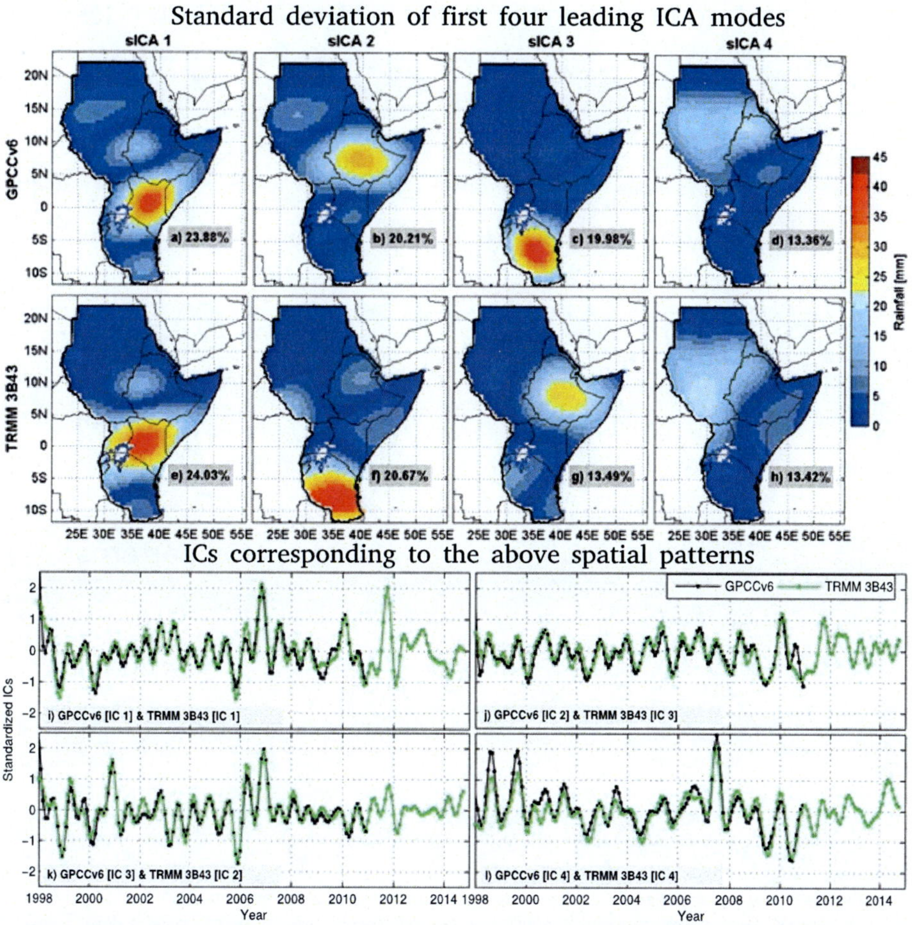

Fig. 10.6 The first four dominant modes of rainfall variability over the GHA region based on (a–d) GPCC (1979–2010), and (e–h) TRMM (1998–2014). The corresponding temporal patterns (or ICs) are shown in the bottom panels (i–l), where a low-pass filter is applied to enhance their interpretation. (i) Temporal patterns corresponding to a & e, (j) b& g, (k) c & f, and (l) d & h. *Source* [4]

episodes for the four sub-regions between 1980 and 2014. The SPI indices over Region 1 in Fig. 10.8a (corresponding to Fig. 10.6a–e) shows extreme drought events in 1984, 1999–2000, 2009, and 2011, consistent with the findings in [35]. The most notable droughts over Ethiopia and eastern Sudan (Region 2, Table 10.3) include the extreme drought events of 2009 and some moderate droughts in 1984, 1986, 1992, 1994, and 2002 (Fig. 10.8b). SPI patterns over South Sudan (Fig. 10.8c, see also, Table 10.3) indicate prolonged droughts from 1982 to early 1985 and between 2009 and 2010. The SPI indices over Tanzania (Region 4, Fig. 10.8d) indicate considerable variability with severe/extreme droughts lasting from one to two years (e.g., 1988–1989, 1997–1998, 2003–2004, and 2005–2006).

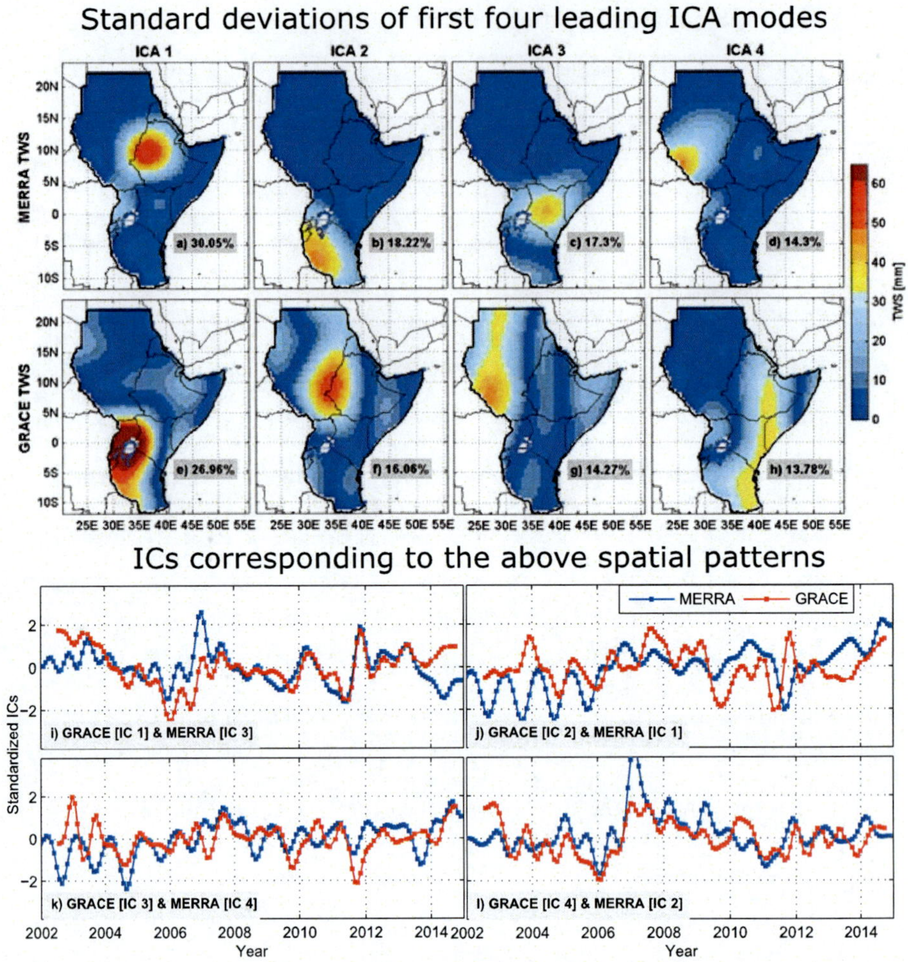

Fig. 10.7 The first four dominant modes of TWS variability over the GHA region based on (a–d) MERRA (1980–2014) and (e–h) GRACE data (2002–2014). The corresponding temporal patterns (or ICs) are shown in the bottom panels (e–h), where a 6-month low-pass filter is applied to enhance their interpretation. (i) Temporal patterns corresponding to c & e, (j) a& f, (k) d & g, and (l) b & h. *Source* [4]

Although there were indications of decreasing rainfall from 1980–1982, it did not lead to extreme droughts over Tanzania. The region exhibited only a moderate drought from 2010–2011, which caused devastating affects in other parts of the GHA [50, 55]. The two precipitation products, GPCC and TRMM were found to be fairly consistent across all the four regions between 1998 and 2010 differed slightly over Region 3 (over South Sudan, Fig. 10.8c), where rainfall variability of GPCC was highest over central Sudan and northern Ethiopia (see, Fig. 10.6d) in contrast to South Sudan as indicated by TRMM (Fig. 10.6g).

Figure 10.9 shows the TSDIs of the four regions (see Table 10.3). Major hydrological drought events between 1980 and 2014 based on TWS changes derived from

Table 10.3 Four sub-regions classified according to the first four leading modes of precipitation and TWS. *Source* [4]

Region	Area	GPCC v6	TRMM 3B43	MERRA	GRACE
Region 1	Kenya Uganda Rwanda Burundi Northern Tanzania	ICA1	ICA1	ICA3	ICA1
Region 2	Ethiopia Sudan	ICA2	ICA3	ICA1	ICA2
Region 3	South Sudan	ICA4	ICA4	ICA4	ICA3
Region 4	Tanzania Eastern Ethiopia	ICA3	ICA2	ICA2	ICA4

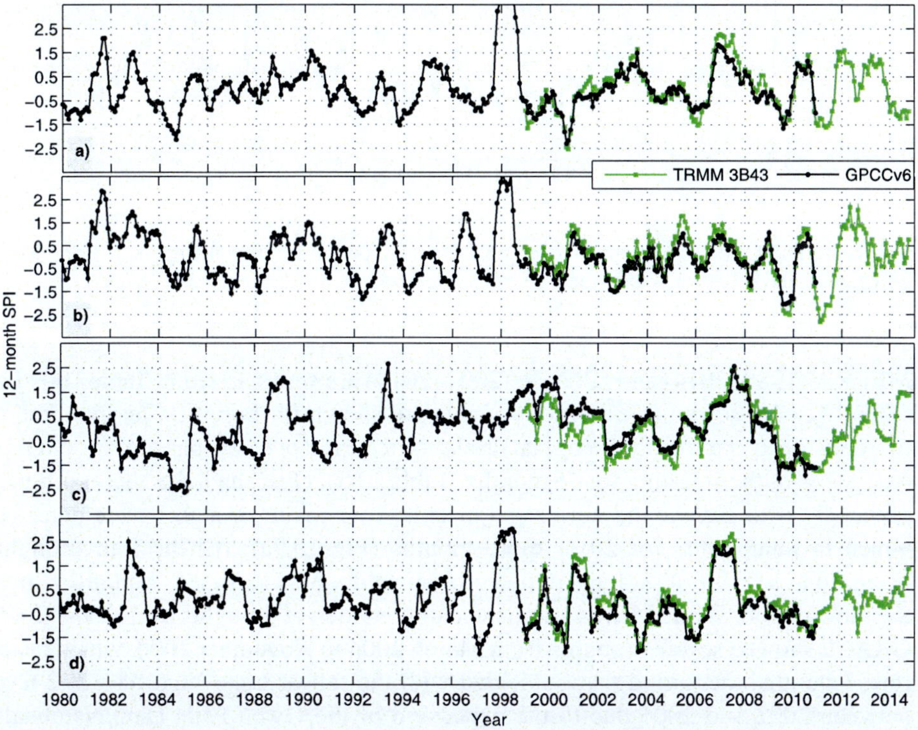

Fig. 10.8 12-month SPI indices estimated from monthly precipitation products of GPCC (1979–2010) and TRMM (1998–2014) over the four homogenous regions classified based on sICA (see also, Table 10.3). **a** Region 1, **b** Region 2, **c** Region 3, and **d** Region 4. Note that SPI was not calculated for the first 12 months in both datasets. *Source* [4]

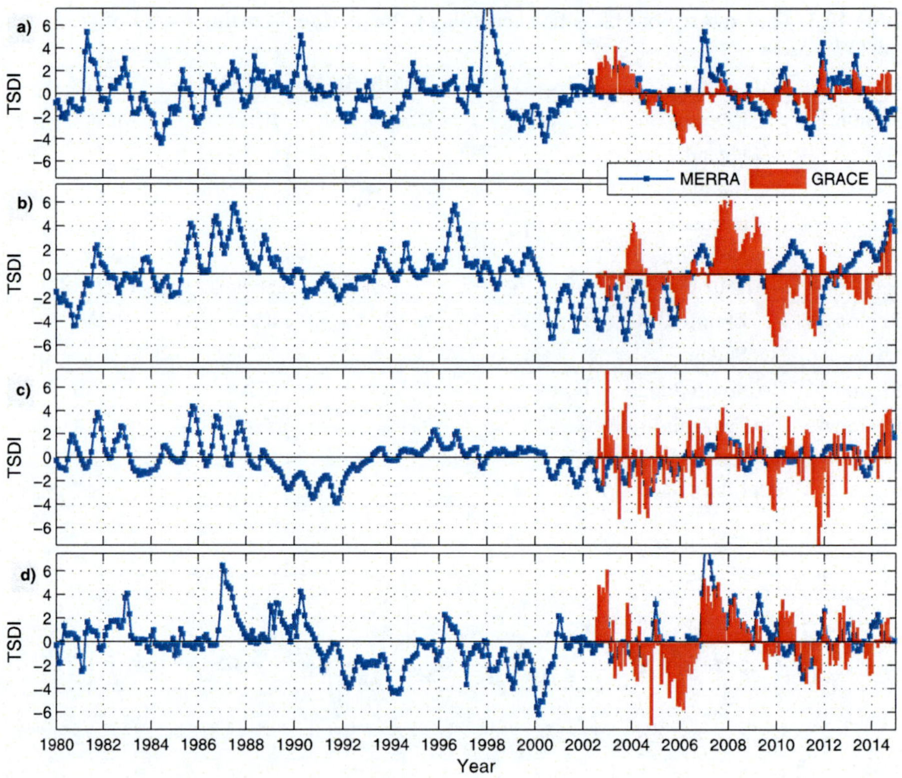

Fig. 10.9 TSDI indices computed based on sICA-derived modes over **a** Region 1, **b** Region 2, **c** Region 3, and **d** Region 4 (see Table 10.3). *Source* [4]

MERRA (1980–2014) and GRACE (2002–2014) are summarized in Table 10.5. For MERRA, only severe to extreme droughts are shown for brevity. Clearly, three or more severe to extreme hydrological droughts occurred over various regions of GHA with some of them lasting for 75 months (Table 10.5). Over the Lake Victoria Basin, especially over Kenya and northern Tanzania, four major droughts were detected, which in total lasted for 20 or more months (Fig. 10.9a). All the four droughts occurred as a result of prolonged meteorological dry episodes (see, Table 10.4) over the same region. While MERRA indicated only moderate droughts from 2004–2006, GRACE showed severe droughts from March 2004 to November 2006, which lasted for 33 months, the period that coincided with the fall of Lake Victoria water level between 2002 and 2006 due to the expansion of the Owen Falls Dam/Nalubaale, see e.g., [2, 16]. From Table 10.4 (see also, Fig. 10.8d), it can be seen that a severe meteorological drought also occurred from October further exacerbating the TWS decline. However, the extreme meteorological droughts of 2009–2010 and 2010–2011 (see, Table 10.4) only caused moderate hydrological droughts over the region (Table 10.5).

Over Ethiopia and Sudan (Region 2), very different scenarios of TWS variability between GRACE and MERRA were observed especially over 2002 to 2014, where GRACE data showed large positive TSDIs between 2007 and 2009 (2007 and 2011), which were not represented well by MERRA (negative; Fig. 10.9b). MERRA also indicates prolonged droughts from April 2000 to June 2006 lasting for about 75 months (Fig. 10.9b and Table 10.5), which could be spurious. However, it is also observed that two meteorological droughts occurred (1999–2000, and 2002–2004), which might have affected the TWS variability in MERRA. Extreme hydrological droughts have been found from July 2011-July 2012 (MERRA) and September 2009-October 2011 (GRACE) in response to the prolonged meteorological drought conditions over the region (see, Fig. 10.9b). MERRA also indicated extreme droughts between 1990 and 1993, which was again, in response to the meteorological droughts that occurred from 1991–1992. The region also experienced the largest TWS decline (extreme hydrological drought) in 2009–2010 and 2010–2011, corresponding to the historical drought events from 2009–2011. These events were well-captured by the two precipitation products (Fig. 10.8b) and GRACE-derived TWS changes (Fig. 10.9b). During this two periods, SPI dropped below −2.5 while the TSDI was round −6.0 during the peak period (i.e., December to February 2009–2010). It is also interesting to observe that both the meteorological droughts eventuated from rainfall deficits during the OND season.

TSDI patterns over Region 3 representing the South Sudan region are quite complicated as both GRACE and MERRA indicate high intra-seasonal variability (especially observed in GRACE) but the extreme drought of 2011–2012 was well captured by GRACE (Fig. 10.9c). This could have occurred as a result of prolonged meteorological droughts that last for 39 months from May 2009 to July 2012. Note that MERRA also indicated prolonged hydrological droughts between July 1988 and December 1992. Region 4 covering Tanzania and parts of Ethiopia experienced several instances of droughts with varying intensities between 1980 and 2014 (Fig. 10.9d). The most notable drought events shown by MERRA are the prolonged droughts that occurred in the early 1990, the late 1990s, and in 2011. Similar to the Lake Victoria Basin (Region 1), this region also exhibited significant decline in TWS from 2003 to mid-2006 leading to a prolonged (and extreme) hydrological drought over the region. It should be noted here that the region also suffered two meteorological droughts (2003–2004 and 2005–2006) during the period (see, Fig. 10.8d). The 2010–2011 extreme drought was also captured very well by GRACE data.

Figure 10.10 illustrates the spatial evolution of the 2009–2010 drought computed as a sum of rainfall/TWS over three important seasons. It is observed that the entire GHA experienced anomalously low rainfall during OND 2010 (Fig. 10.10a), which remained relatively dry during MAM 2011 over Kenya, Ethiopia, and South Sudan (Fig. 10.10b) before recovering to positive changes. TWS anomalies were mostly negative in the southern and western parts of GHA during OND 2010 (Fig. 10.10d) but their magnitudes further increased during MAM 2011 (Fig. 10.10e). High negative anomalies shifted to Region 3 during JJAS 2011 but the majority of the regions still indicated negative anomalies (Fig. 10.10f) as a result of deficit rainfall in the two preceding months.

Table 10.4 Severe to extreme meteorological droughts based on the SPI in the four regions (see, Table 10.3) of GHA from 1979 to 2014. *Source* [4]

Region	Periods	Duration	SPI [Max]	Class
Region 1	Jul 1983–Feb 1985	20 months	−2.13	Extreme
	Apr 1993–Sep 1994	18 months	−1.54	Severe
	Nov 1998–Mar 2002	43 months	−2.31	Extreme
	Mar 2004–Jul 2006	29 months	−1.53	Severe
	Jun 2008–Nov 2009	19 months	−1.66	Severe
	Nov 2010–Sep 2011	12 months	−1.62	Severe
Region 2	Apr 1984–Apr 1985	13 months	−2.13	Extreme
	Jan 1986–Jul 1988	28 months	−1.60	Severe
	Oct 1991–Dec 1992	15 months	−1.84	Severe
	Nov 1993–Nov 1994	13 months	−1.59	Severe
	Apr 2002–Jan 2004	22 months	−1.52	Severe
	Feb 2009–Feb 2010	13 months	−2.50	Extreme
	Oct 2010–Oct 2011	13 months	−2.80	Extreme
Region 3	Jul 1982–Sept 1985	39 months	−2.57	Extreme
	Apr 1982–May 1985	38 months	−2.61	Extreme
	Mar 2009–July 2012	41 months	−2.03	Extreme
Region 4	Dec 1987–Mar 1989	16 months	−1.54	Severe
	Nov 1996–Sep 1997	9 months	−2.18	Extreme
	Nov 1998–Oct 2000	24 months	−2.14	Extreme
	Mar 2003–Jul 2006	41 months	−2.10	Extreme
	Dec 2008–Jul 2011	32 months	−1.52	Severe

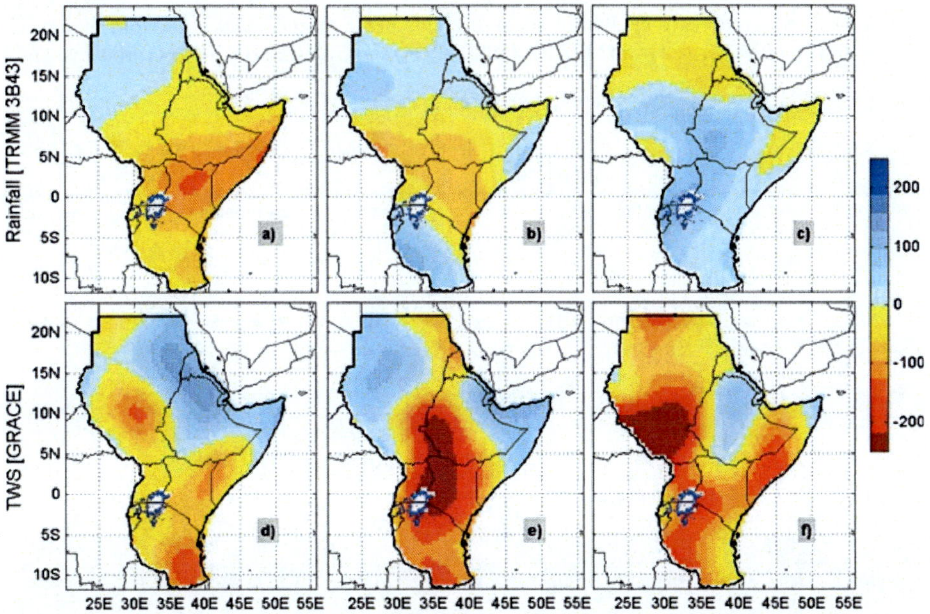

Fig. 10.10 Seasonal rainfall anomaly and TWS changes (mm/season) over the GHA region during **a–d** October–December 2010, **b–e** March–May 2011, and **c–f** June–September 2011. *Source* [4]

It is observed that almost all the hydrological droughts over the GHA region resulted from extreme (and/or prolonged) meteorological droughts with both SPI and TSDI indicating consistent temporal variability during the period 1980–2014. To illustrate the relationship between SPI and TSDI, the SPI and TSDI values were plotted based on TRMM precipitation estimates (1998–2014) and GRACE-derived TWS changes (2002–2014) for the period 1999 to 2014 in Fig. 10.11, with TSDI values scaled to the SPI values. The temporal patterns of SPI and TSDI are closely matched for Regions 1, 3, and 4 while the major drought events of 2009–2010 and 2010–2011 over Region 2 were also represented well by SPI and TSDI, where hydrological droughts have occurred almost 5–6 months after the meteorological drought.

The correlations between SPI and TSDI were computed for the period 2002–2014 (Table 10.6). Consistent with Fig. 10.11, significant correlations (at 95% confidence interval) were found between SPI and TSDI during the period with a correlation of 0.44 (at one month lag), 0.52 (at 3 month lag), and 0.59 (at one month lag) for Region 1, Region 3, and Region 4, respectively. The correlation between SPI and TSDI for Region 2 was not significant. There exists a high seasonal correlation over the region between rainfall and TWS changes (see, Fig. 10.5a). The correlations between MERRA-derived TSDI and GRACE-derived TSDI were found to be significant over all the four regions with a lag of up to 1 month (see, Table 10.6). In quantifying changes in TWS and the resulting hydrological drought events (as indicated by TSDI)

Table 10.5 Severe to extreme hydrological droughts based on the TSDI in the four regions (see Table 10.3) of GHA derived from MERRA (1980–2014) and GRACE (2002–2014) products. *Source* [4]

	MERRA (1980–2014)				GRACE (2002–2014)			
Region	Periods	Duration	TSDI [Max]	Category	Periods	Duration	TSDI [Max]	Category
Region 1	Nov 1983–May 1987	43 months	−4.33	Severe	Sep 2004–Nov 2006	27 months	−4.31	Severe
	Mar 1992–Mar 1996	49 months	−4.05	Extreme	Jun 2008–Feb 2010	21 months	−2.08	Moderate
	Feb 2000–Nov 2002	34 months	−3.34	Severe	Sept 2010–Sep 2011	13 months	−2.58	Moderate
	Oct 2010–May 2012	20 months	−3.68	Severe				
Region 2	Jan 1980–Jul 1981	19 months	−5.02	Extreme	Mar 2004–Nov 2006	33 months	−3.36	Severe
	Apr 2000–Jun 2006	75 months	−6.25	Extreme	Sep 2009–Oct 2011	26 months	4.60	Extreme
	Jul 2011–Jul 2012	13 months	−4.72	Extreme				
Region 3	Dec 1998–Mar 1993	52 months	−5.93	Extreme	May 2004–Nov 2006	31 months	−3.26	Extreme
	Jul 2000–Mar 2002	21 months	−3.9	Severe	Jun 2009–Sep 2011	28 months	−4.40	Extreme
	Jun 2000–Feb 2006	45 months	−4.81	Extreme				
Region 4	Feb 1991–Mar 1996	62 months	−5.84	Extreme	Mar 2004–Oct 2006	32 months	−5.70	Extreme
	Jan 1997–Jan 2001	49 months	−6.93	Extreme	Nov 2011–Jan 2014	27 months	−3.98	Severe
	Nov 2005–Nov 2006	13 months	−3.47	Severe				
	May 2010–Nov 2011	19 months	−3.17	Severe				

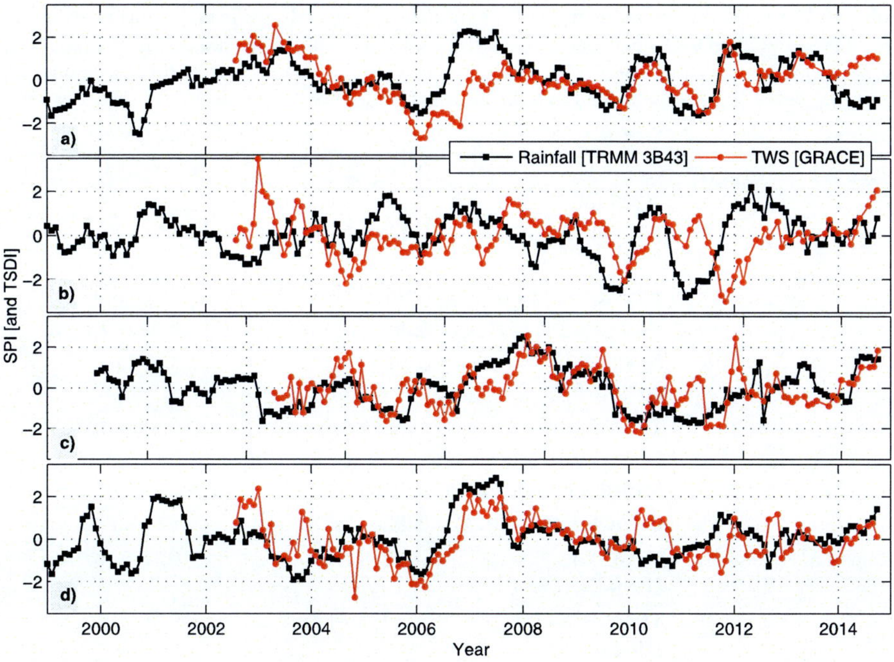

Fig. 10.11 SPI and TSDI indices of TRMM precipitation product and GRACE TWS changes for the period 1999 to 2014, i.e., **a** region 1, **b** region 2, **c** region 3, and **d** region 4. The TSDI indices were normalized to the SPI to highlight their temporal pattern over the four regions. *Source* [4]

Table 10.6 Correlations between TSDI derived from GRACE TWS changes and SPI derived from TRMM precipitation estimates between 2002 and 2014. Correlations between GRACE and MERRA TSDIs are also shown to indicate the skills of MERRA with respect to TWS changes derived from GRACE observations. Correlation coefficients that are significant at 95% confidence level are indicated in bold. *Source* [4]

Region	Region 1	Region 2	Region 3	Region 4
SPI and TSDI (GRACE)	**0.44** (1 month)	0.24 (2 month)	**0.52** (3 month)	**0.59** (1 month)
MERRA versus GRACE (TSDI)	**0.47** (1 month)	**0.51** (0 month)	**0.46** (1 month)	**0.57** (0 month)

in GHA, contributions due to evapotranspiration and increased use of freshwater (groundwater and surface water) in the region should be taken into account.

The study further explored global teleconnection relationships to SPI and TSDI indices. Table 10.7 shows the correlation coefficients and time lags (in months) between SPI (and TSDI) indices and Niño3.4 and the Dipole Mode Index (DMI) over various time periods between 1979 to 2014. Both SPI and TSDI (GRACE) of Region 1 are significantly correlated with ENSO and IOD and are more related to IOD than ENSO (with a maximum correlation of 0.53 between TRMM and DMI).

Table 10.7 Correlations (significant at 95% confidence level) and lags (in brackets) between rainfall (and TWS) and ENSO/IOD between 1979–2014. Negative months' lags indicate ENSO/IOD occurs before rainfall or variations of TWS and vice versa. Correlation coefficients that are significant at 95% confidence level are indicated in bold. *Source* [4]

Region	Products	ENSO (Niño3.4)	IOD (DMI)
Region 1	GPCC v6 (1979–2010)	**0.33** (−1)	**0.48** (−1)
	TRMM 3B43 (1998–2014)	**0.44** (−1)	**0.53** (−1)
	MERRA (1980–2014)	−0.23 (5)	−0.18 (6)
	GRACE (2002–2014)	**0.38** (−6)	**−0.42** (−6)
Region 2	GPCC v6 (1979–2010)	**0.44** (−1)	**0.53** (−1)
	TRMM 3B43 (1998–2014)	−0.26 (−6)	−0.07 (−2)
	MERRA (1980–2014)	−0.26 (1)	−0.24 (1)
	GRACE (2002–2014)	**0.48** (−1)	**0.57** (−5)
Region 3	GPCC v6 (1979–2010)	−0.23 (1)	−0.18 (−6)
	TRMM 3B43 (1998–2014)	−0.26 (−1)	−0.20 (−1)
	MERRA (1980–2014)	**−0.37** (3)	**−0.44** (3)
	GRACE (2002–2014)	−0.21 (2)	**−0.34** (−6)
Region 4	GPCC v6 (1979–2010)	−0.23 (−1)	**0.39** (−2)
	TRMM 3B43 (1998–2014)	−0.20 (−6)	**0.34** (−1)
	MERRA (1980–2014)	−0.22 (7)	**−0.34** (6)
	GRACE (2002–2014)	**0.34** (−6)	**0.41** (−6)

SPI (GPCC) and TSDI (GRACE) also show significant correlations with ENSO (IOD) while both SPI and TSDI over Region 4 show significant correlations with IOD with the highest correlation of 0.41 for the GRACE-derived TSDI. SPI and TSDI over Region 3 (representing South Sudan) show the least correlation against ENSO (IOD) indicating that rainfall and TWS variability over South Sudan was not associated with large-scale climate events as the region is located further inland compared to the other three regions.

It is observed that MERRA product shows low but opposite correlations with ENSO (IOD) compared to GRACE, which could be another limitation of the reanalysis product. Furthermore, IOD is seen to be the leading driver of rainfall and TWS variability especially over the eastern GHA region with the highest impact over Tanzania and Lake Victoria (see, Table 10.7). This is consistent with previous studies, which reported that the East African climate is highly governed by the Indian Ocean SST variability e.g., [9, 35, 50, 53].

10.7 Characterizing Agricultural Drought

Heavy reliance of East Africa (EA) on rain-fed agriculture makes it vulnerable to drought-induced famine. Yet, most research on EA drought focuses on meteorological aspects with little attention paid on agricultural drought impacts. Agricultural droughts are defined by Awange [5–8] as soil water deficiency, plant water stress, and reduced biomass and yield. The inadequacy of in-situ rainfall data across EA has also hampered detailed agricultural drought impact analysis. Recently, however, there has been increased data availability from remote sensing (rainfall, vegetation condition index—VCI, terrestrial water storage—TWS), reanalysis (soil moisture and TWS), and land surface models (soil moisture).

To support agricultural drought monitoring from diverse indicators, it is imperative to identify and provide information on the most effective agricultural drought indicator or a combination of indicators for the EA region [1]. Therefore, Agutu et al. [1] employed these products to characterise EA droughts between 1983 and 2013 in terms of severity, duration, and spatial extent. Furthermore, the capability of these products to capture agricultural drought impacts was assessed using maize and wheat production data. Their results showed that while all products were similar in drought characterisation in dry areas, the similarity of Climate Hazards Group InfraRed Precipitation with Station data (CHIRPS) and Global Precipitation Climatology Centre (GPCC) extended over the whole EA. CHIRPS and GPCC also identified the highest proportion of areas under drought followed closely by soil moisture products whereas VCI had the least coverage. Drought onset was marked first by a decline/lack of rainfall, followed by VCI/soil moisture, and then TWS. VCI indicated drought lag at 0–4 months following rainfall while soil moisture and TWS products had variable lags vis-à-vis rainfall. Global Land Data Assimilation System (GLDAS) mischaracterized the 2005–2006 drought vis-à-vis other soil moisture products. Based on the annual crop production variabilities explained, Agutu [1] identified CHIRPS, GPCC, Famine Early Warning Systems Network (FEWS NET) Land Data Assimilation System (FLDAS), and VCI as suitable for agricultural drought monitoring/characterization in the East African region for the study period (1983–2013). Finally, GLDAS explained the lowest percentages of the Kenyan and Ugandan annual crop production variances. In general, Agutu et al. [1]:

- characterized drought and assessed utility of drought indicators in East Africa (EA), which included the Lake Victoria Basin, the mother of the Nile.
- ordered drought detection as first rainfall, VCI/moisture then TWS,
- VCI was found to have 0–4 months of lag in detecting drought cycle with respect to rainfall,
- ranked the suitability of agricultural drought indicators for EA as: CHIRP, GPCC, FLDAS, and VCI
- found that the land surface models' products generally performed inconsistently over EA.

The assessment of the products are presented in Figs. 10.12, 10.13, 10.14, and their correlation with maize and wheat presented in Fig. 10.15. These findings are impor-

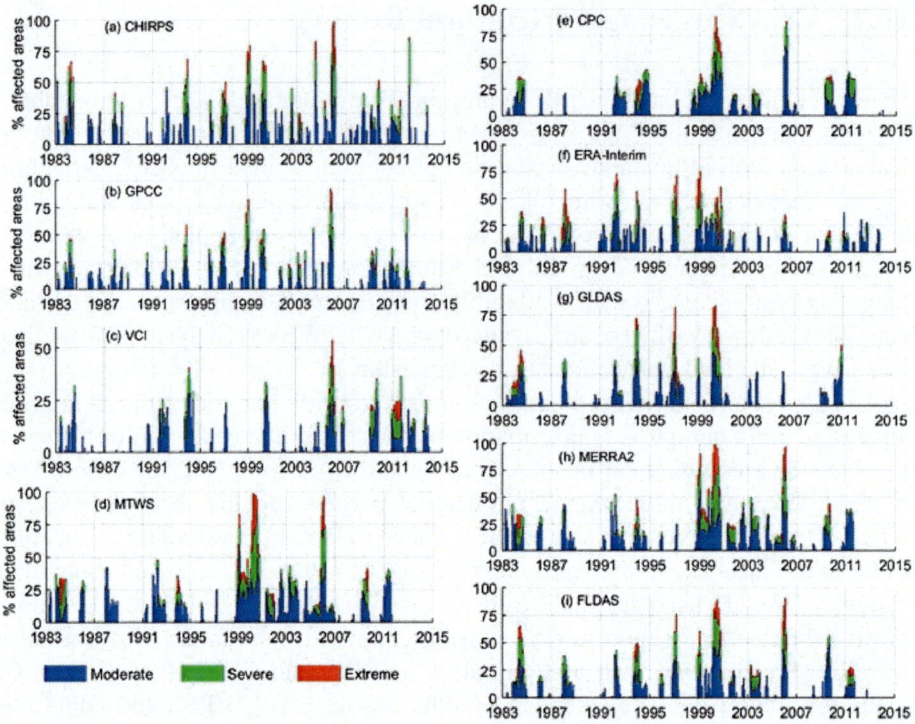

Fig. 10.12 Percentage of area affected by various drought intensities during the period 1983–2013. Percentage areas are computed by integrating the regional spatial and temporal patterns then determining percentage of pixels under each drought category (see Agutu et al. [1]). The rainfall products have the highest percentage areas under drought followed closely by soil moisture products and finally the lowest percentage areas are by VCI. In addition to the soil moisture products having different percentage areas under drought, Climate Prediction Center (CPC) soil moisture is consistent with MERRA-2, GLDAS is consistent with ERA-interim while FLDAS is in between. *Source* Agutu et al. [1]

tant for the gauge data deficient EA region as they provide alternatives for monitoring agricultural drought.

10.8 Influence of Coupled Ocean-Atmosphere Phenomena

Drought-like humanitarian crises in the Greater Horn of Africa (GHA) are increasing despite recent progress in drought monitoring and prediction efforts. Notwithstanding these efforts, there remain challenges stemming from uncertainty in drought prediction, and the inflexibility and limited buffering capacity of the recurrent impacted systems. The complexity of the interactions of El Niño/Southern Oscillation (ENSO), Indian Ocean Dipole (IOD), Interdecadal Pacific oscillation (IPO) and North Atlantic Oscillation (NAO), arguably remains the main source of uncertainty in drought prediction. To develop practical drought risk parameters that potentially can guide invest-

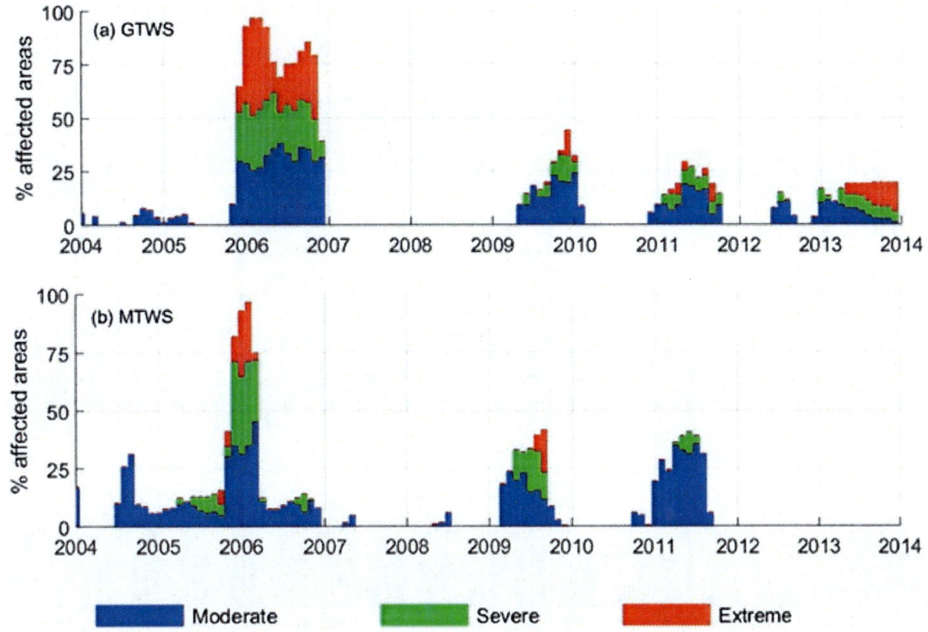

Fig. 10.13 Comparison of performance between GRACE-TWS (GTWS) and Merra-TWS (MTWS) in terms of percentage of areas affected by various drought categories. Percentage areas are computed as in Fig. 10.12. They have consistent performance, with GTWS having a lag in drought detection probably due to groundwater that is lacking in MTWS. *Source* Agutu et al. [1]

ment strategies and risk-informed planning, Mpelesoka et al. [9] quantified drought characteristics that underpin drought impacts management and highlighted the following:

- Zonality in drivers of drought occurrence over the GHA.
- Probabilities of drought-year occurrences range from 10 to 40%.
- Most droughts last 14 and 24 months for the 12- and 24-month timescales.
- Drought areal-extent decadal trends range from 0.6%, Tanzania to 3.7%, Ethiopia.

Drought characteristics that include probability of drought-year occurrences, durations, areal-extent and their trends over 11 decades (1903–2012) were derived from the Standardized Precipitation Index (SPI). Transient probability of drought-year occurrences, modelled on Beta distribution, across the region ranged from 10 to 40%, although most fall within 20–30%. For more than half of the drought events, durations of up to 4, 7, 14 and 24 months for the 3-, 6-, 12- and 24-month timescales were evident, while 1 out of 10 events persisted for up to 18 months for the short timescales, and up to 36 months or more for the long timescales. Apparently, only drought areal-extent showed statistically significant trends of up to 3%, 1%, 3.7%, 2.4%, 0.7%, 0.3% and 0.6% per decade over Sudan, Eritrea, Ethiopia, Somalia, Kenya, Uganda and Tanzania, respectively. Since there was no evidence of significant changes in drought characteristics, the peculiarity of drought-like crises in the

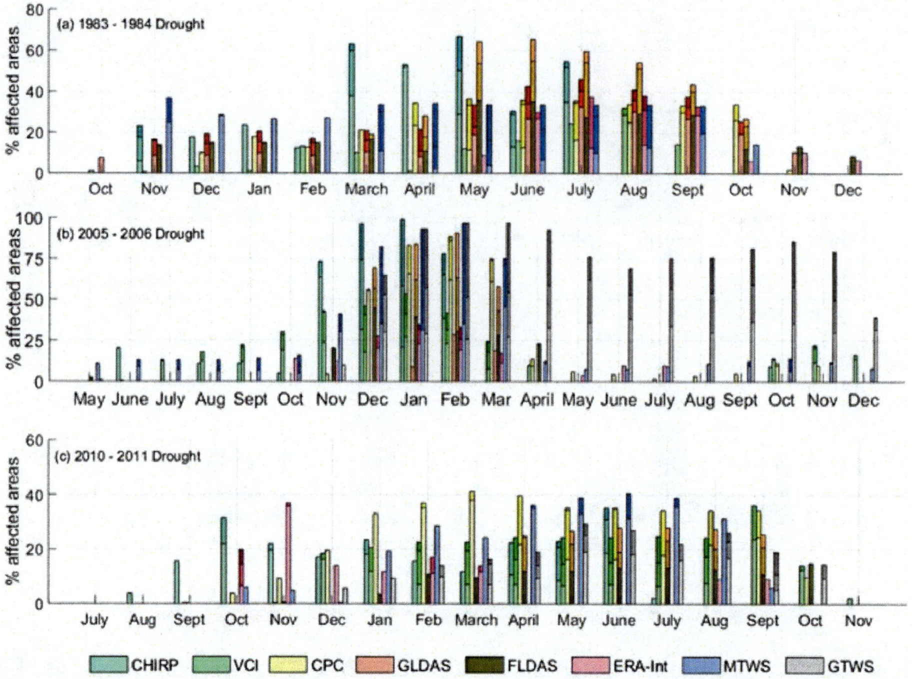

Fig. 10.14 Percentage of areas affected by various drought intensities during the 1983–1984, 2005–2006, and 2010–2011 drought episodes. Each bar has up to 3 colour grades (gradients) representing from bottom moderate, severe, and extreme droughts at the top. Percentage areas are computed as in Fig. 10.12 but only for the duration of drought. VCI has a lag of about 2–3 months in identifying the drought cycle in relation to CHIRPS. The rest of the products have inconsistent lags in relation to rainfall across the three drought episodes. *Source* Agutu et al. [1]

GHA was attributed by [9] (at least in part) to unaccounted for systematic rainfall reduction. This highlighted the importance of distinguishing drought impacts from those associated with new levels of aridity. In principle drought is a temporary phenomenon while aridity is permanent, a difference that managers and decision-makers should be more aware [9]. For details on the study, refer to Mpelasoka et al. [9], while a summarized abstract of the work is presented in Fig. 10.16.

10.9 Concluding Remarks

The results presented in this chapter found significant changes in long-term and decadal rainfall based on monthly estimates from GPCC and TRMM over 1979 to 2014. Long-term rainfall changes between 1980 and 2010 indicated significant positive (negative) changes over Sudan (Ethiopia and Tanzania) while changes from 2002 to 2014 showed completely opposite trends over these two regions.

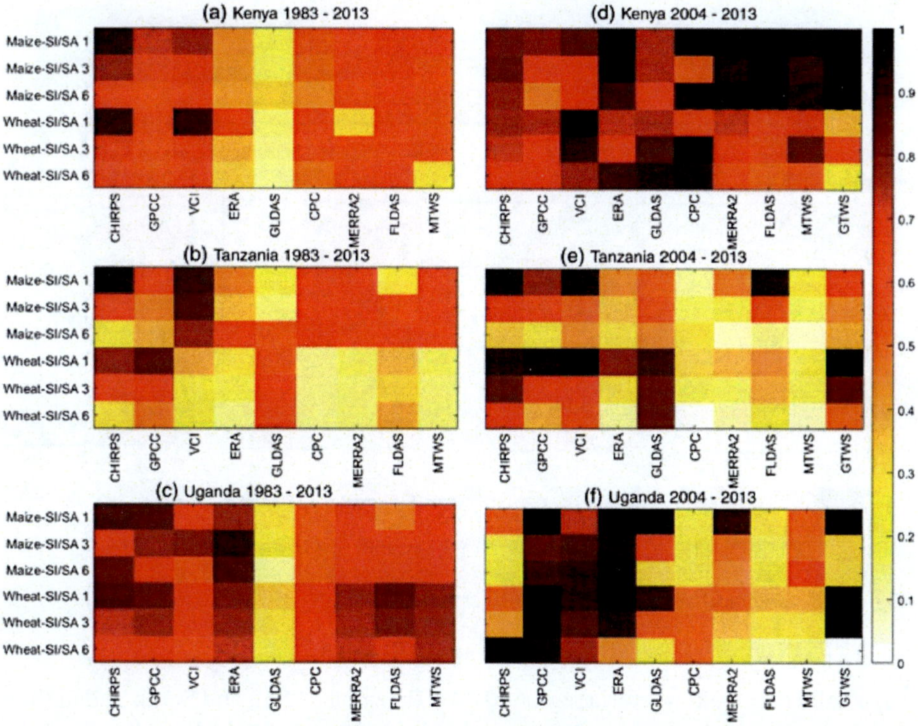

Fig. 10.15 Proportions of variability in national annual crop (maize and wheat) production (R^2) explained by various drought indices for Kenya, Tanzania, and Uganda. CHIRPS, GPCC, and VCI consistently explains relatively higher variability in crop production, while the soil moisture products have inconsistent performance (a, b, and c). The figures d, e, and f should be interpreted with care as the datasets used to fit the models are only 10 years long. Both SI and SA are computed at annual scales. (The y axis indicates the crop (maize/wheat), Standardized index SI (for a, b, and c), and Standardized Anomaly SA (for d, e and f) while 1,3 and 6 indicate the standardization time scales for the indicators on the x axis). *Source* Agutu et al. [1]

GRACE-derived TWS changes showed high positive (negative) trends over central GHA (South Sudan and Tanzania) with a maximum increase of about 20 mm/yr over the Lake Victoria Basin and central Ethiopia. High correlations (up to 0.9) were noted between monthly rainfall and GRACE TWS changes over South Sudan, western Ethiopia, and Tanzania, with a lag of 1–2 months while rainfall over arid and semi-arid regions such as northern Sudan and Somalia were least correlated, indicating that large-scale variations of TWS are mainly influenced by seasonal rainfall variations.

SPI and TSDI indices estimated over the four sICA-derived spatial regions (Lake Victoria Basin, Ethiopia–Sudanese border, South Sudan, and Tanzania) indicated several instances of severe to extreme meteorological droughts (SPI $<$1.5) resulting in extreme hydrological droughts that were observed in MERRA and GRACE TWS changes (indicated by TSDIs). Correlations between meteorological droughts (SPI) and hydrological droughts (GRACE-derived TSDI) were found to be significant

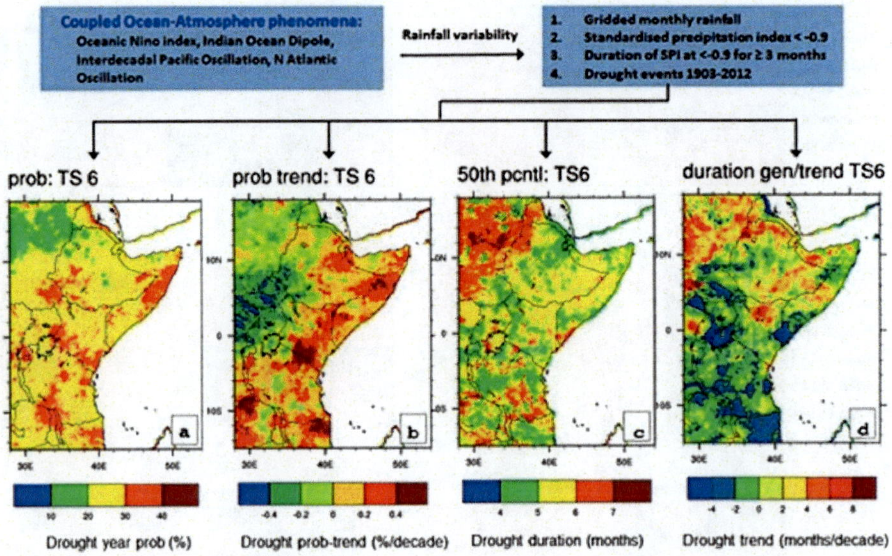

Fig. 10.16 Graphical abstract of Mpelesoka et al. [9]

especially over Lake Victoria region (0.44), Tanzania (0.52), and South Sudan (0.59), signifying that precipitation plays an important role in the hydrological budget of the regions.

While the Lake Victoria Basin experienced six major meteorological droughts between 1979 and 2014, South Sudan and Ethiopia suffered seven meteorological droughts lasting between one year to 3 and half years. The impact of prolonged meteorological droughts is clearly evident in the TWS changes over all the four regions, which resulted in multiple years of hydrological droughts.

Consistent with the previous studies e.g., [50], the Indian Ocean SST variations were found to play a larger influence on the regional TWS changes ompared to the ENSO mode. It should be noted here that correlations between MERRA and ENSO (IOD) indices were considerably lower (and often opposite). Further verifications are required to assess the quality of reanalysis products to study the long-term changes and hydrological extremes over the GHA region.

References

1. Agutu NO, Awange JL, Zerihun A, Ndehedehe CE, Kuhn M, Fukuda Y (2017) Assessing multi-satellite remote sensing, reanalysis, and land surface models' products in characterizing agricultural drought in East Africa. Remote Sens Environ 194:287–302. https://doi.org/10.1016/j.rse.2017.03.041
2. Awange JL (2021) Lake Victoria monitored from space. Springer Nature International, London

3. Awange JL, Hu KX, Khaki M (2019) The newly merged satellite remotely sensed, gauge and reanalysis-based multi-source weighted-ensemble precipitation: evaluation over Australia and Africa (1981–2016). Sci Total Environ 670:448–465. https://doi.org/10.1016/j.scitotenv.2019.03.148
4. Awange JL, Khandu Forootan E, Schumacher M, Heck B (2016) Exploring hydro-meteorological drought patterns over the Greater Horn of Africa (1979–2014) using remote sensing and reanalysis products. Adv Water Resour 94:45–59. https://doi.org/10.1016/j.advwatres.2016.04.005
5. Awange JL (2012) Environmental monitoring using GNSS: global navigation satellite system. Springer, Berlin, New York
6. Awange JL (2018) GNSS environmental sensing: revolutionizing environmental monitoring. Springer, Berlin, New York
7. Awange JL, Kiema JBK (2013) Environmental geoinformatics: monitoring and management. Springer, Heidelberg
8. Awange JL, Kiema JBK (2019) Environmental geoinformatics. Extreme hydro-climatic and food security challenges: exploiting the big data. Springer Nature, Heidelberg
9. Mpelasoka F, Awange JL, Zerihun A (2018) Influence of coupled ocean-atmosphere phenomena on the Greater Horn of Africa droughts and their implications. Sci Total Environ 610–611:691–702. https://doi.org/10.1016/j.scitotenv.2017.08.109
10. Agboma CO, Yirdaw SZ, Snelgrove KR (2009) Intercomparison of the total storage deficit index (TSDI) over two canadian prairie catchments. J Hydrol 374(3–4):351–359. https://doi.org/10.1016/j.jhydrol.2009.06.034
11. Andersen OB, Seneviratne SI, Hinderer J, Viterbo P (2005) GRACE derived terrestrial water storage depletion associated with the 2003 European heat wave. Geophy Res Lett 32(18). https://doi.org/10.1029/2005GL023574
12. Awange JL, Anyah R, Agola N, Forootan E, Omondi P (2013) Potential impacts of climate and environmental change on the stored water of lake victoria basin and economic implications. Water Resour Res 49(12):8160–8173. https://doi.org/10.1002/2013WR014350
13. Awange JL, Ferreira VG, Khandu Andam-Akorful, SA, Forootan E, Agutu NO, He XF (2015) Uncertainties in remotely-sensed precipitation data over Africa. J Climatol Int. https://doi.org/10.1002/joc.4346
14. Awange JL, Forootan E, Kuhn M, Kusche J, Heck B (2014a) Water storage changes and climate variability within the Nile basin between 2002 and 2011. Adv Water Resour 73:1–15. https://doi.org/10.1016/j.advwatres.2014.06.010
15. Awange JL, Gebremichael M, Forootan E, Wakbulcho G, Anyah R, Ferreira VG, Alemayehu T (2014b) Characterization of Ethiopian mega hydrogeological regimes using GRACE, TRMM and GLDAS datasets. Adv Water Resour 74:64–78. https://doi.org/10.1016/j.advwatres.2014.07.012
16. Awange JL, Sharifi MA, Ogonda G, Wickert J, Grafarend E, Omullo M (2008) The falling lake victoria water level: GRACE, TRIMM and CHAMP satellite analysis of the lake basin. Water Resour Manage 22:775–796. https://doi.org/10.1007/s11269-007-9191-y
17. Chen JL, Wilson CR, Tapley BD, Longuevergne L, Yang ZL, Scanlon BR (2010) Recent la plata basin drought conditions observed by satellite gravimetry. J Geophys Res Atmosph 115:D22108. https://doi.org/10.1029/2010JD014689
18. Chen JL, Wilson CR, Tapley BD, Yang ZL, Niu GY (2009) 2005 drought event in the Amazon River basin as measured by GRACE and estimated by climate models. J Geophys Res: Solid Earth 114(B5):B05404. https://doi.org/10.1029/2008JB006056
19. Cheng MK, Ries JC, Tapley BD (2013) Geocenter variations from analysis of SLR data. In: Reference Frames for Applications in Geosciences. International Association of Geodesy Symposia, vol. 138, pp. 19–25. Springer, Berlin, Heidelberg
20. Cheng MK, Tapley BD (2004) Variations in the earth's oblateness during the past 28 years. Geophys Res Lett 109(B9). https://doi.org/10.1029/2004JB003028
21. Dahle C, Flechtner F, Gruber C, König D, König R, Michalak G, Neumayer K-H (2013) GFZ GRACE Level–2 Processing Standards Document for Level–2 Product Release 0 0 05:

revised edition, January 2013. Scientific Technical Report STR12/02–Data, rev. ed. Deutsches GeoForschungsZentrum GFZ. Potsdam. https://doi.org/10.2312/GFZ.b103-1202-25
22. Forootan E (2014) Statistical signal decomposition techniques for analyzing time- variable satellite gravimetry data. PhD thesis, University of Bonn, Germany. http://hss.ulb.uni-bonn.de/2014/3766/3766.htm
23. Forootan E, Awange J, Kusche J, Heck B, Eicker A (2012) Independent patterns of water mass anomalies over Australia from satellite data and models. Remote Sens Environ 124(1):427–443. https://doi.org/10.1016/j.rse.2012.05.023
24. Forootan E, Kusche J (2012) Separation of global time-variable gravity signals into maximally independent components. J Geodesy 86(7):477–497. https://doi.org/10.1007/s00190-011-0532-5
25. Forootan E, Kusche J (2013) Separation of deterministic signals using independent component analysis (ICA). Stud Geophys Geod 57(1):17–26. https://doi.org/10.1007/s11200-012-0718-1
26. Funk C, Hoell A, Shukla S, Bladé I, Liebmann B, Roberts JB, Robertson FR, Husak G (2014) Predicting east African spring droughts using Pacific and Indian ocean sea surface temperature indices. Hydrol Earth Syst Sci Discuss 11:3111–3136. https://doi.org/10.5194/hessd-11-3111-2014
27. Houborg R, Rodell M, Li B, Reichle R, Zaitchik BF (2012) Drought indicators based on model-assimilated gravity recovery and climate experiment (GRACE) terrestrial water storage observations. Water Resour Res 48(7). https://doi.org/10.1029/2011WR011291
28. Huffman GJ, Bolvin DT, Nelkin EJ, Wolff DB, Adler RF, Gu G, Hong Y, Bowman KP, Stocker EF (2007) The TRMM multisatellite precipitation analysis (TMPA): quasi-global, multiyear, combined-sensor precipitation estimates at fine scales. J Hydrometeorol 8(1):38–55. https://doi.org/10.1175/JHM560.1
29. Kurnik B, Barbosa P, Vogt J (2011) Testing two different precipitation datasets to compute the standardized precipitation index over the Horn of Africa. Int J Remote Sens 32(21):5947–5964. https://doi.org/10.1080/01431161.2010.499380
30. Kusche J, Schmidt R, Petrovic S, Rietbroek R (2009) Decorrelated GRACE time- variable gravity solutions by GFZ, and their validation using a hydrological model. J Geod 83(10):903–913. https://doi.org/10.1007/s00190-009-0308-3
31. Landerer FW, Swenson SC (2012) Accuracy of scaled GRACE terrestrial water storage estimates. Water Resour Res 48(W04531). https://doi.org/10.1029/2011WR011453
32. Leblanc MJ, Tregoning P, Ramillien G, Tweed SO, Fakes A (2009) Basin- scale, integrated observations of the early 21st century multiyear drought in Southeast Australia. Water Resour Res 45(4):1–10. https://doi.org/10.1029/2008WR007333
33. Li B, Rodell M (2014) Evaluation of a model-based groundwater drought indicator in the conterminous U.S. J Hydrol 526:78–88. https://doi.org/10.1016/j.jhydrol.2014.09.027
34. Long D, Scanlon BR, Longuevergne L, Sun AY, Fernando DN, Save H (2013) GRACE satellite monitoring of large depletion in water storage in response to the 2011 drought in Texas. Geophys Res Lett 40(13):3395–3401. https://doi.org/10.1002/grl.50655
35. Lyon B (2014) Seasonal drought in the Greater Horn of Africa and its recent increase during the March–May long rains. J Climate 27(21):7953–7975. https://doi.org/10.1175/JCLI-D-13-00459.1
36. Marthews TR, Otto FEL, Mitchell D, Dadson SJ, Jones RG (2015) The 2014 drought in the horn of Africa: attribution of meteorological drivers. Bull Am Meteor Soc 96:83–88. https://doi.org/10.1175/BAMS-EEE_2014_ch17.1
37. McKee TB, Doesken NJ, Kliest J (1993) The relationship of drought frequency and duration to time scales. In: Proceedings of the 8th conference of applied climatology, 17–22 January, Anaheim, CA. American Meterological Society, Boston, MA, pp 184–194
38. McKee TB, Doesken NJ, Kliest J (1995) Drought monitoring with multiple time scales. 9th AMS conference on applied climatology. American Meterological Society, Boston, MA, pp 233–236

39. Narasimhan B, Srinivasan R (2005) Development and evaluation of soil moisture deficit index (SMDI) and evapotranspiration deficit index (ETDI) for agricultural drought monitoring. Agri Forest Meteorol 133(1–4):69–88. https://doi.org/10.1016/j.agrformet.2005.07.012
40. Naumann G, Dutra E, Barbosa P, Pappenberger F, Wetterhall F, Vogt VJ (2014) Comparison of drought indicators derived from multiple data sets over Africa. Hydrol Earth Syst Sci 18:1625–1640. https://doi.org/10.5194/hess-18-1625-2014
41. Nicholson SE (1997) An analysis of the enso signal in the tropical Atlantic and Western INDIAN oceans. Int J Clim 17(4):345–375. https://doi.org/10.1002/(SICI)1097-0088(19970330)17:4(345::AID-JOC127)3.0.CO;2-3
42. Omondi PA, Awange JL, Forootan E, Ogallo LA, Barakiza R, Girmaw GB, Fesseha I, Kululetera V, Kilembe C, Mbati MM, Kilavi M, King'uyu SM, Omeny PA, Njogu A, Badr EM, Musa TA, Muchiri P, Bamanya D, Komutunga E (2014) Changes in temperature and precipitation extremes over the greater horn of Africa region from 1961 to 2010. Int J Clim 34(4):1262–1277. https://doi.org/10.1002/joc.3763
43. Palmer WC (1965) Meteorological drought. Technical report research paper 45. US department of commerce, weather Bureau, Washington, DC. https://www.ncdc.noaa.gov/temp-and-precip/drought/docs/palmer.pdf
44. Rienecker MM, Suarez MJ, Gelaro R, Todling R, Bacmeister J, Liu E, Bosilovich MG, Schubert SD, Takacs L, Kim GK, Bloom S, Chen J, Collins D, Conaty A, da Silva A, Gu W, Joiner J, Koster RD, Lucchesi R, Molod A, Owens T, Pawson S, Pegion P, Redder CR, Reichle R, Robertson FR, Ruddick AG, Sienkiewicz M, Woollen J (2011) Merra: Nasas modern-era retrospective analysis for research and applications. J Clim 24:3624–3648. https://doi.org/10.1175/jcli-d-11-00015.1
45. Schneider U, Becker A, Finger P, Meyer-Christoffer A, Ziese M, Rudolf B (2014) GPCC's new land surface precipitation climatology based on quality-controlled in situ data and its role in quantifying the global water cycle. Theor Appl Clim 115(1–2):15–40. https://doi.org/10.1007/s00704-013-0860-x
46. Sheffield J, Wood EF, Chaney N, Guan K, Sadri S, Yuan X, Olang L, Amani A, Ali A, Demuth S, Ogallo L (2014) A drought monitoring and forecasting system for Sub-Sahara African water resources and food security. Bull Am Meteor Soc 95:861–882. https://doi.org/10.1175/BAMS-D-12-00124.1
47. Swenson S, Wahr J (2009) Monitoring the water balance of Lake Victoria, East Africa, from space. J Hydrol 370(1–4):163–176. https://doi.org/10.1016/j.jhydrol.2009.03.008
48. Tapley BD, Bettadpur S, Watkins M, Reigber C (2004) The gravity recovery and climate experiment: mission overview and early results. Geophys Res Lett 33:4 https://doi.org/10.1029/2004GL019779
49. Thomas AC, Reager JT, Famiglietti JS, Rodell M (2014) A GRACE-based water storage deficit approach for hydrological drought characterization. Geophys Res Lett 41(5):1537–1545. https://doi.org/10.1002/2014GL059323
50. Tierney JE, Smerdon JE, Anchukaitis KJ, Seager R (2013) Multidecadal variability in East African hydroclimate controlled by the Indian Ocean. Nature 493:389–392. https://doi.org/10.1038/nature11785
51. Viste E, Korecha D, Sorteberg A (2013) Recent drought and precipitation tendencies in Ethiopia. Theor Appl Clim 112(3):535–551. https://doi.org/10.1007/s00704-012-0746-3
52. Wahr J, Molenaar M, Bryan F (1998) Time variability of the Earth's gravity field: hydrological and oceanic effects and their possible detection using GRACE. J Geophys.Res 103(B12):30205–30229. https://doi.org/10.1029/98JB02844
53. Williams AP, Funk C (2011) A westward extension of the warm pool leads to a westward extension of the walker circulation, drying eastern Africa. Clim Dyn 37:2417–2435. https://doi.org/10.1007/s00382-010-0984-y
54. Williams AP, Funk C, Michaelsen J, Rauscher SA, Robertson I, Wils THG, Eshetu Z, Loader NJ (2012) Recent summer precipitation trends in the Greater Horn of Africa and the emerging role of Indian Oceansea surface temperature. Clim Dyn 39:2307–2328. https://doi.org/10.1007/s00382-011-1222-y

55. WMO (2012) WMO statement on the status of the global climate in 2011. Technical Report WMO-No. 1085. World Meteorological Organization. Geneva, Switzerland
56. Yirdaw SZ, Snelgrove KR, Agboma CO (2008) GRACE satellite observations of terrestrial moisture changes for drought characterization in the canadian prairie. J Hydrol 356(1–2):84–92. https://doi.org/10.1016/j.jhydrol.2008.04.004
57. Awange JL, Palancz B, Völgyesi L (2020) Hybrid imaging and visualization. Employing machine learning with mathematica - python. Springer Nature International, Berlin. ISBN 978-3-030-26152-8. https://doi.org/10.1007/978-3-030-26153-5

Index

J. Awange, *The Nile Waters*,
https://doi.org/10.1007/978-3-030-64756-8